AF412662

Computer Process Control: Modeling and Optimization

Computer Process Control: Modeling and Optimization

T. H. LEE
General Electric Company
Phoenix, Arizona

G. E. ADAMS
University of Missouri
Columbia, Missouri

W. M. GAINES
General Electric Company
Phoenix, Arizona

JOHN WILEY & SONS, INC., New York · London · Sydney

Library of Congress Catalog Card Number: 67–30914
GB 471 52207X
Printed in the United States of America

Dedicated to our wives
Nai-Hsin
Lila
Sylvia

Preface

Computer process control is one of the hybrid technologies that developed in 1958 and reached some degree of stability in the mid-sixties. The various disciplines that make up its base are *instrumentation, control theory, computer design* and *programming, statistics, operations research,* and *mathematical optimization theory,* but sufficient time has passed to allow it to develop its own character. The two areas that stand out as unique to its theory are *modeling the automated process* and *optimal control to attain economic objectives.*

This book is devoted to the theory and application of modeling and optimization, for these two areas represent the cornerstone of computer process control. Other pertinent subjects (e.g., instrumentation, programming, and computer logic) have been well covered in textbooks in their respective fields. Finally, we have concentrated on modeling and optimization because they are the least familiar to engineers and because their proper understanding can potentially produce the largest payoff.

Computer Process Control: Modeling and *Optimization* is designed primarily for three types of readers:

1. Senior and graduate engineering students.
2. Practicing engineers not familiar with computer process control but wishing to become acquainted with the field.
3. Engineers in the field who are not familiar with modeling and optimization.

This book can be used as a one-semester course by senior and graduate engineering students. We used the manuscript as lecture notes for a three-credit course, EE523, Control Computers, at Arizona State University, Tempe, Arizona, from 1962 to 1967. The examples and problems make it readily adaptable as a textbook. Practicing engineers not familiar with computer process control may use the book to get acquainted with the major topics in this field, and engineers already working in computer process control can use it as a self-teaching text to broaden their knowledge. The reader is required only to be familiar with calculus, simple statistics, and basic matrix algebra.

In preparing this material, we attempted to lay a solid foundation on the fundamental theories and practical concepts of modeling and optimization. We have tried to avoid purely mathematical theories and advanced topics still in research. The emphasis throughout is on practical applications rather than on mathematical dexterity, and most of the examples are derived from our own experience. We have adhered to the following principles in developing the mathematical derivations:

The symbols and notations are defined before every major derivation. With a few exceptions, the symbols and notations are consistent throughout the book.

The mathematical derivations are broken into small logical steps.

Each derivation depends only on what has already been developed.

All the equations are numbered and references to equations are explicit.

With the exception of Chapter 1, which is the overall introduction, the book is divided into two topics—modeling and optimization. Chapter 2 is the introduction to modeling. In chapter 3, which discusses functional models, the computer itself is the vehicle that illustrates this concept. Chapter 4 discusses analytical and experimental techniques for developing physical process models. The physical laws and regression-analysis techniques of developing these models are presented along with Kalman's technique of developing dynamic physical process models by experimentation. Chapter 5 discusses planning and operating economic models. Planning economic models used in the evaluation and selection of investment opportunities include computer process control applications. Operating economic models develop the objective functions to be optimized.

Chapter 6 introduces and summarizes the optimization problem, particularly the functions and the relationships of the physical process model, the objective function, and the constraints. Chapter 7 begins with simple calculus to develop feedforward optimal control techniques for one-variable processes and extends the treatment to include other techniques for optimizing well-behaved steady-state processes, e.g., Lagrange multiplier and linear programming. Chapter 8 presents *optimal control of moderately well-defined steady-state processes* in which the physical process, constraints, and environment are all subject to change. The feedback optimal control approach based on an incremental model is described in this chapter. Chapter 9 deals with the evolutionary optimization approach for poorly defined and well-defined steady-state processes. The various optimization strategies, the design of experiments to derive the gradient, and the effects of constraints and noise are some of the subjects discussed. A combined feedforward and feedback optimal control scheme for well-defined processes that use the EVOP and linear programming approaches is also discussed. Chapter 10, which presents techniques for optimal control of dynamic physical processes, includes the classical variational and dynamic programming approaches.

The reader who wishes to obtain an overall view of the two main subjects of modeling and optimization may read only Chapters 1, 2, and 6. These chapters present the basic concepts without going into detail. For those using the text for a one-semester course Chapters 1, 2, and 6 may be skipped as lecture material, for they contain redundant summary information and may be assigned as reading material to the students. For the reader who wants a short but intensive course in the subject it is necessary to study only Chapters 4, 5, 7, and 8. The two most important chapters in the area of modeling are Chapters 4 and 5. The significance of the material in Chapter 4—the treatment of physical models—is usually recognized. The material in Chapter 5, however—the treatment of economic models— is often neglected by engineers and particularly by engineering students. The two most important chapters in the area of optimization are Chapters 7 and 8. Chapter 7 lays the foundation for steady-state optimization techniques, and Chapter 8 makes these adaptable to computer process control applications.

We acknowledge the initial encouragement given us to undertake this task in 1961 by R. C. Berendsen, General Manager, General Electric Process Computer Business. At that time all of us were working in the General Electric Process Computer Business. Since that time other people in General Electric management have given us support and assistance: C. E. Thompson, W. D. Zarecor, V. S. Cooper, R. L. Overholtzer, B. Adler, P. A. Quantz, and L. E. Wengert. We are grateful also to our General Electric associates who contributed to this book by their critical reviews, comments, and suggestions: Harold Chestnut, C. F. Raine, T. J. Glass, W. D. Trudgen, H. J. Cadell, J. Lane, E. J. Burdick, W. M. Stanger, and R. A. Boennighausen. Above all, we are deeply grateful to Mrs. Bea Sleigher for her typing and editorial efforts during the last four years, which were truly above and beyond the call of duty.

T. H. Lee

G. E. Adams

W. M. Gaines

Phoenix, Arizona
September 1967

Contents

1

The Role of Computer Process Control Systems

1.0 INTRODUCTION

Computer process control may be defined as the technique of utilizing digital computers to control the overall process, associated equipment, and operating procedures in order to attain performance and economic objectives. In contrast, *conventional control,* utilizing regulators and instrument controllers, merely attempts to regulate individual process variables with the objective of attaining accurate and stable regulation of individual variables over a wide range of operating conditions. Thus computer process control is a much broader form of control than conventional control. The functions and complexities of computer process control are orders of magnitude greater than conventional control; for example, the functions of computer process control may include monitoring and data processing, start-up and shut-down procedures, and optimal control for maximum profit.

This chapter will present a brief historical perspective on the evolution of process control, then set forth the broad objectives of computer process control and the place of the computer process control system by implication, and close with some of the practical considerations in applying computer process control.

1.1 EVOLUTION OF PROCESS CONTROL

This section will present the historical perspective of the evolution of process control prior to 1958, at which time computer process control systems made their appearance. The various stages of process control development will be

1

discussed. *Manual methods* were used in the earliest forms of control. Then *indicating and recording instruments* were added to assist in manual control. Finally, *local automatic control* was developed which in turn evolved into *centralized automatic control.*

A. Manual Control

The earliest control was by manual means. The operator would use his senses to measure the process variables, and would intervene to control the process directly. This " by-guess-and-by-golly "* approach worked satisfac-

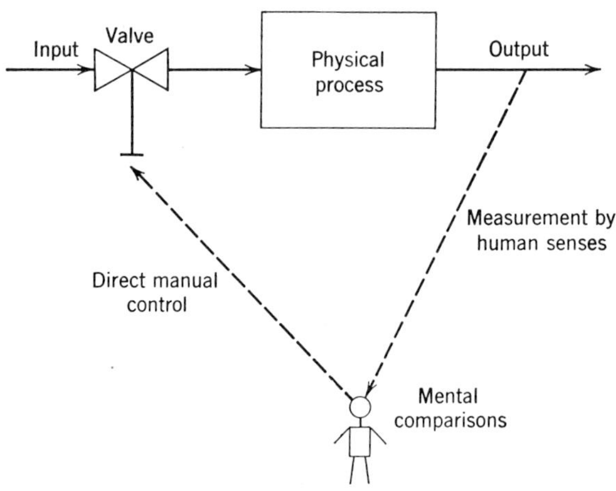

Fig. 1.1-1   Manual process control.

torily in many cases. Figure 1.1-1 shows that the operator was the feedback to control the process.

The *modus operandi* was simply to have the operator correct any deviation from the desired operating condition by changing the value of the input variable, and then observe if the correction was adequate. He would continue to make corrections in this manner until the process was brought back to proper balance.

* The story is told of how fine china was made in the old days: an operator would spit into a kiln periodically to test the temperature. If the spit disappeared quickly, the temperature was sufficient; otherwise more fuel would be added.

B. Indicating and Recording Instruments

The first link in the feedback chain to be mechanized was the sensing function. Indicating instruments replaced human senses with *fast, consistent,* and *accurate measurements.* Automatic recording instruments were added to provide the operator with a means of trend analysis and a historical record. These recording instruments eliminated the dependence on the operator to estimate trends and to log data tediously on clipboards. Figure 1.1-2 shows that the functions performed by the operator were reduced to error detection and making the corrections to the process.

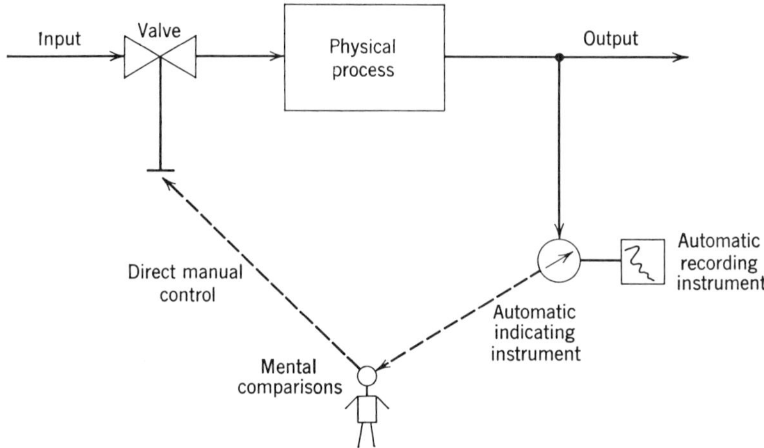

Fig. 1.1-2   Indicating and recording instruments added to process control.

C. Local Automatic Control

The introduction of the automatic controller removed the operator from the feedback loop. The continuous functions of monitoring and correcting deviations were performed by the local automatic controller, i.e., local in the sense of being located at each control valve and pump. The operator's only function was to establish the set points for the controllers. The set points, or references, were quantitative inputs to automatic controllers which in turn regulated the process variables around the set point values. Figure 1.1-3 shows the relationship between the operator and the automatic controller. Along with the development of automatic control devices to regulate material flow in the process, the mechanization of material handling was an essential developmental step toward complete automation.

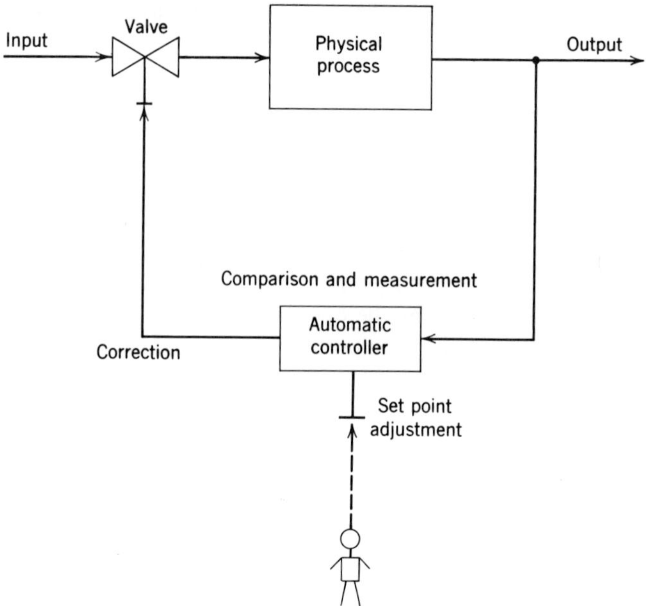

Fig. 1.1-3 Process control by local automatic controllers.

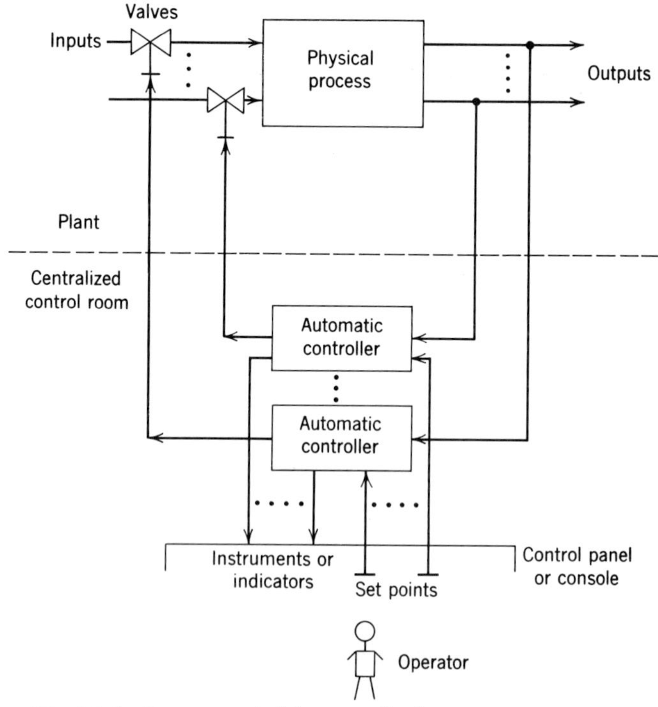

Fig. 1.1-4 Process control by centralized automatic controllers.

D. Centralized Automatic Control

In the early stages of development, automatic controllers were located physically near the process variables that were being controlled. As the number of controlled process variables increased, the monitoring and adjusting of the remotely located automatic controllers became more difficult. The solution was the development of the *centralized control room* where most of the automatic controllers were brought together and mounted on control panels, as shown in Figure 1.1-4. Even a centralized control room became unwieldy in some instances, when hundreds of process variables had to be watched by half a dozen or more operators. To alleviate the situation, *graphic panels* were used to display the *key* process variables on a schematic or process flow diagram of the plant. This allowed the operators to quickly grasp the significance of the process variables and related controls. All other instrumentation was placed behind panels or in some other inconspicuous place.

1.2 OBJECTIVES OF COMPUTER PROCESS CONTROL

As indicated in Section 1.1, when the digital computer became available for plant automation in 1958, centralized automatic control was the most advanced form of process automation available. There were many shortcomings, however, even among the best of the centralized automatic control installations. Some of these shortcomings were the following:

Lack of integration, or controlling the process as a whole.
Lack of consistency in results due to dependence on operators.
Awkward and costly data recording and processing.
Lack of capability to implement optimal control to attain economic objectives.
Increased risks of upsets or hazardous operating conditions due to slow reaction time of operators.

This section will show how the digital computer can overcome most of these deficiencies and will discuss some of the objectives of computer process control. The selection of the optimal level of automation will serve as an introduction.

A. The Optimal Level of Automation

In a free competitive society, the purpose of automation is to place a manufacturer in a more advantageous cost and profit position than his competitors. The competitive position of a plant or process can be measured in

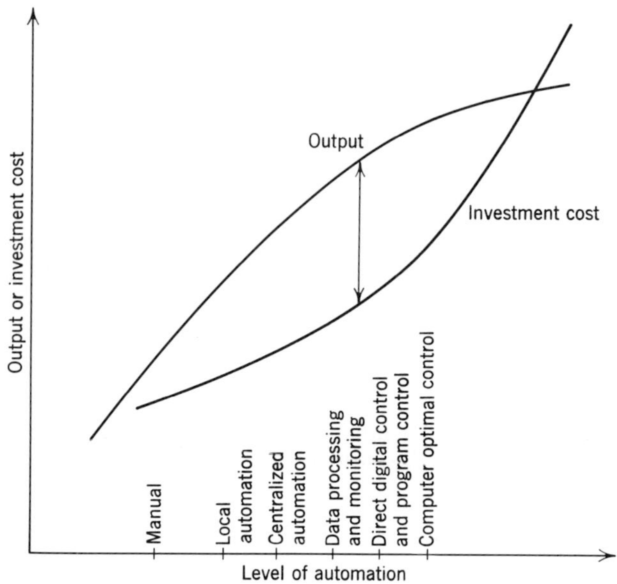

Fig. 1.2-1   Output and investment cost versus levels of automation.

terms of *productivity*. *Productivity* can be variously defined as *output per dollar of capital investment, output per man-hour of labor, output per square foot of floor space*, and *output per dollar of raw materials*. Thus the broad objective of automation is to increase productivity, as defined above.

Increasing the level of automation will tend to improve productivity, but the cost of capital investment will also be increased. Figure 1.2-1 shows a typical relationship between *output and level of automation*, and between *investment cost and level of automation*. The various levels of automation indicated include the following:

Local automatic control.
Centralized automatic control.
Data processing and monitoring.
Direct digital control and program (supervisory) control.
Computer optimal control.

Experience has indicated that typical output curves tend to saturate beyond a certain point, whereas the investment cost curves rise slowly at first and then exponentially.

The optimal level of automation is attained where the difference between the *output* and *investment cost* curves is at a maximum.* Thus it is important

* This assumes that the market for the products is not saturated, i.e., all the products can be sold.

to evaluate the output and investment cost for the alternate degrees of automation that are possible. As market conditions, raw material costs, and technologies change, these curves will also tend to change; for example, increased product prices may tend to raise the output curve, decreased costs of automation equipment may tend to lower the investment cost curve, and sharp technological breakthroughs may tend to move the sharp part of the investment cost curve further to the right. In 1955, before the availability of digital computers for process automation, the investment cost curve would have had a sharp increase before *data processing and monitoring*. This would rule out the use of computers for plant automation because of prohibitive investment cost. Up-to-date investment cost, market, and technological factors must be used to prepare these curves.

B. Computer Process Control Fulfilling a Need in Process Automation

The primary attributes of computer process control are *computational speed, storage capability*, and *decision making ability*. These characteristics of computer process control fulfilled a definite need that was not being met by centralized automatic control. The deficiencies of centralized automatic control could largely be overcome by the computer. The *computational speed* of the process control system made it possible to perform rapid calculations, so that optimal control of the overall process could be accomplished. The *storage capability, combined with the computational speed* of the process control system, allowed the data recording and processing functions to be done economically and efficiently. Perhaps the most important contribution by the computer process control system was the *ability to make decisions*.

Preprogrammed logic stored in the computer allowed the computer to relieve the operator from routine monitoring of out-of-limit variables, and to implement operating and diagnostic procedures that are consistent, fast, and reliable. Thus three major functions were made available by the computer process control system:

1. Data processing and monitoring.
2. Direct digital control and program control.
3. Optimal control.

Data processing, monitoring, and direct digital control will be discussed in Chapter 3. Program control is outside the scope of this book. Optimal control will be discussed in Chapters 6 through 10.

C. The Objectives of Computer Process Control

Section 1.2A showed that the objective of automation and computer process control is to improve productivity, which can be measured in the following ways:

Output per dollar of capital investment.
Output per man-hour of labor.
Output per square foot of floor space.
Output per dollar of raw materials.

Output per dollar of capital investment is an objective that can be met by either increasing the output or reducing the capital investment. Output can be increased by minimizing the time required for batch processes, and by using better control to increase the rate of output. Capital investment can be decreased by providing closer and faster control with an accompanying reduction in the need for large storage facilities, tanks, and additional mixing and blending stages. Capital investment can also be decreased by reducing the need for redundant or backup equipment through improved safety and diagnostic monitoring by the computer. *Output per square foot of floor space* is increased by the same factors that affect output per dollar of capital investment.

Output per dollar of labor can be increased by mechanizing the routine tasks performed by the operator, e.g., data recording and monitoring. Furthermore, because of computer process control, a given crew of operators may now monitor an integrated plant or several processes, thereby increasing the output per dollar of labor. The true significance in this area, however, is that human labor is being upgraded to do more important and meaningful tasks.

Output per dollar of raw materials can be increased by many of the optimal control techniques discussed in Chapters 6 through 10. Some examples include: minimizing the fuel cost of an electrical power system to meet a given load requirement, minimizing the cost of blending several materials to meet output specifications, and controlling the quality of the products so that deviations from the desired specifications are minimized. Optimal quality control may yield products that can sell at higher prices or produce a higher yield of acceptable products from the same amount of raw materials.

1.3 THE PLACE OF COMPUTER PROCESS CONTROL

The availability of the digital computer has added three new capabilities to process control: *rapid computations, data storage,* and *decision making capabilities.* These stimulated significant advances in process control. The result was the development of three additional stages of process control beyond centralized automatic control. The following three stages will be discussed in this section:

1. Data processing and monitoring.
2. Direct digital control and program control.
3. Optimal control.

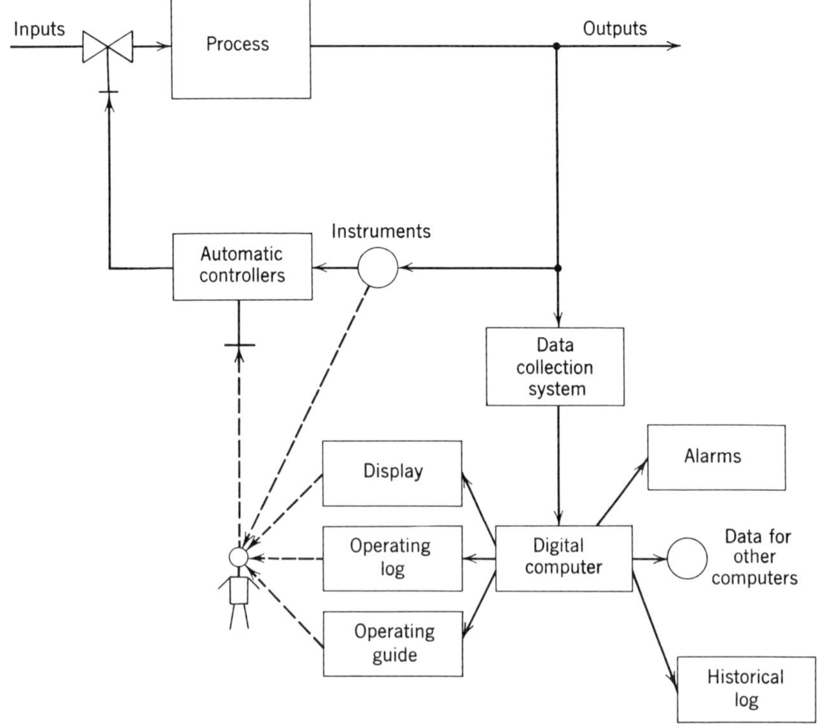

Fig. 1.3-1   Data processing and monitoring.

A. Data Processing and Monitoring

Figure 1.3-1 shows the computer used for data processing and monitoring in a process. The data collection system prepares digital process measurements for the digital computer. The digital computer provides the operator with a display of key variables and prepares the operating log that frees him from housekeeping chores. This allows the operator to concentrate on operating the process. Furthermore, the operator is given operating guides, which may be set-up or start-up instructions, upset recovery procedures, or set points for automatic controllers to improve process operation. The automatic controllers have the role of subloop control, and the primary control loop is the operator who is being assisted by the digital computer. The digital computer can also prepare a historical log, which is the summary of process operating statistics, and prepare data in a format compatible with other computers.

The data processed by the digital computer can give the operator many useful indications of the process operation. Safety monitoring may be provided by scanning process variables against high-low limits and by sounding alarms if these are exceeded. Process efficiency and other key state variables may be computed and displayed to the operator. Trend analysis may be used to alert the operator of some anticipated hazards.

B. Direct Digital Control and Program Control

Direct digital control by the computer process control system uses the computer instead of the automatic controllers to regulate the individual process variables. A single computer system can typically replace up to several hundred automatic controllers. From the standpoint of the computer, the function of the individual controllers is served by the computer on a *time-shared* basis. From the standpoint of a particular process variable being controlled, the function is served by the computer on a *sampled-data* basis. Figure 1.3-2 shows the computer process control system doing direct digital control. The reference signals may be generated by the computer process control system or they may be entered externally.

Program control is the generation of a sequence of procedural and control actions by the computer process control system. It may work through automatic controllers by sending reference signals to the set points, or it may initiate control actions through contact closures, as shown in Figure 1.3-3. Examples of program control are start-up and shut-down of processes.

C. Optimal Control

Optimal control goes beyond direct digital control and program control. It controls the process so that the overall operation is optimum. In other words, the process is controlled to maximize profit, minimize cost, minimize the elapsed time, or to control the quality of the products being produced.

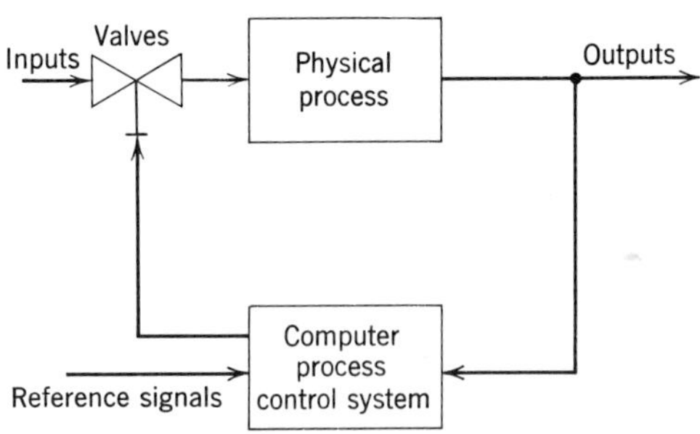

Fig. 1.3-2   Direct digital control.

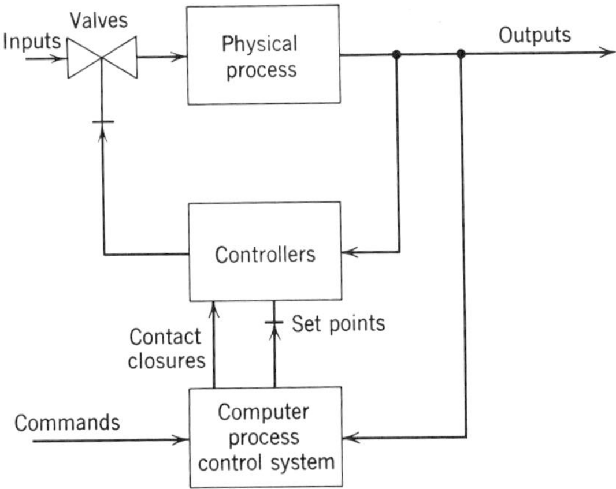

Fig. 1.3-3 Program control.

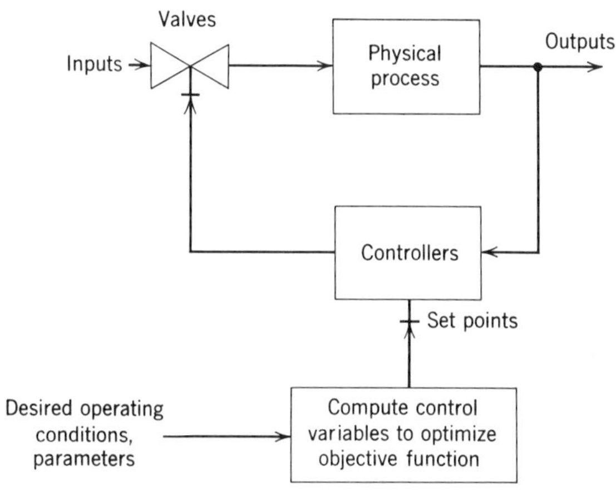

Fig. 1.3-4 Feedforward optimal control.

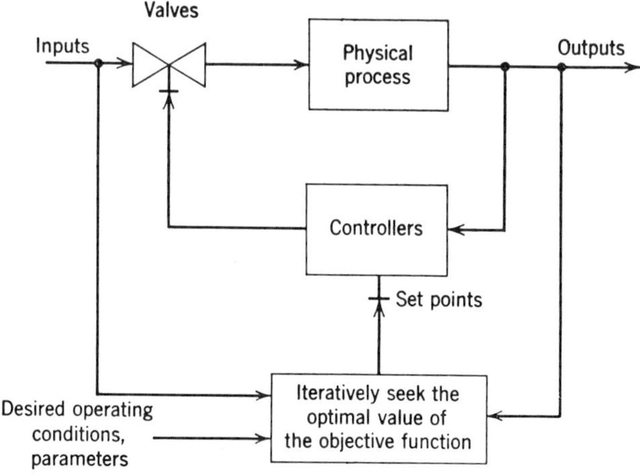

Fig. 1.3-5 Feedback optimal control.

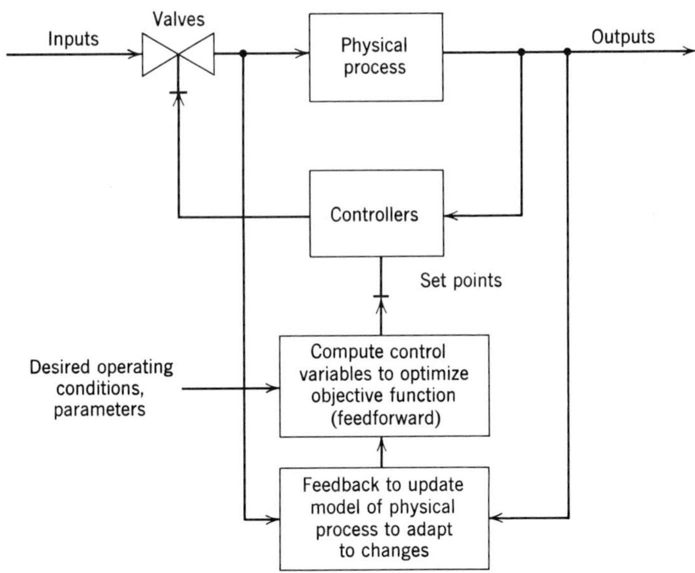

Fig. 1.3-6   Combined feedforward and feedback optimal control.

Optimal control may be implemented by either *feedforward* or *feedback* approaches. The feedforward approach requires a mathematical model of the physical process to derive the optimal control equations, as shown in Figure 1.3-4. The optimal control equations solved by the computer process control system will yield the values of the control variables required to optimize the objective function. The feedback approach may not require a model of the physical process, as shown in Figure 1.3-5. Optimal control is accomplished by direct on-line experimentation seeking to improve the objective function iteratively. Most practical applications include a combination of feedforward and feedback approaches in order to adapt to changes in the process characteristics, as shown in Figure 1.3-6.

1.4   SOME PRACTICAL CONSIDERATIONS IN APPLYING COMPUTER PROCESS CONTROL

Automation can be applied to new or existing plants. However, total automation tends to be less rewarding when it is applied to out-of-date plants. In such plants it is often necessary to undertake an extensive modernization program in order to be prepared for the addition of higher level automation systems. If such modernization is planned as part of an overall program aimed at plant automation, many of the elements of modernization can be designed for the automated process. In considering automation for either an existing or new plant, management should make sure that the equipment and facility in the plant are capable of being automated. Some of the practical considerations in the application of computer process control are explained below.

Mechanization of Material Handling Equipment

The gates, valves, dampers, feeders, and other devices controlling the flow of material must be mechanized by appropriate remote indicating and control devices.

Sequence Interlocking

Remote control of material handling equipment should have adequate sequencing and motion locking devices to prevent pileups of material caused by unexpected shut-down of a line in a group of several transport conveyers and process work machines.

Centralized Control

Immediate savings with more efficient use of manpower often result from centralized control of various departmental operations. Proper advanced planning of the progressive steps of automation can avoid repeating work done in previous steps, and insures compatibility in each step.

Size of Process Work Machines

Small numbers of large size process work machines, e.g., distillation columns, grinding mills, filters, together with their auxiliaries, are less expensive to automate than large numbers of small work machines. The cost to automate is relatively unaffected by the size of the process. Therefore savings in the instrumentation and automation of equipment are possible by using larger process operations.

Ratings of Process Work Machines

One of the benefits of automation is increased production capacity. It is important in considering the future goal of automation to recognize that all of the work machines must be of adequate size not only for the present but also for the desired future production levels under a more fully automated operation. Bottlenecks can be avoided by careful appraisals from this viewpoint.

Sensors

Progress in analytical, X-ray monitoring, and other measuring equipment has been particularly rapid. These up-to-date devices are particularly important to optimal control.

Regulators

If the process is not closely regulated by the automatic controllers, it becomes extremely difficult to superimpose a high level of control and supervision on the process. Therefore an important factor in automation is a careful consideration of the use of process regulators on all key process variables.

Overall Power System

With the higher level of automation, the dependence of the instrumentation, control equipment, and the computer system upon the power system becomes increasingly critical. The reliability of the power system and its adequacy under plant-operating conditions are important considerations.

Maintenance Capabilities

As automatic control of the process becomes more of a reality, many of the devices become more complex. The need for reliable performance may well require improved methods for quick, easy maintenance. Since a control computer is used in the plant, it is necessary that maintenance personnel be familiar with programming and digital computers as well as conventional analog-controller equipment.

1.5   SUMMARY

This chapter has provided an introduction to the subject of computer process control. It began with a historical perspective of the development of process control, and showed the benefits, functions, and some of the considerations of applying computer process control systems.

The rest of the book will concern itself mainly with the two subjects of *modeling the automated process* and *optimization techniques*. Chapter 2 will begin the subject of modeling, and Chapter 6 will begin the subject of optimization.

2

Introduction to Modeling the Automated Process

2.0 INTRODUCTION

This chapter introduces the first topic of this book—modeling, specifically, *how to model the automated process*. As shown in Chapter 1, the automated process can produce benefits ranging from quality control to profit optimization. In order to automate successfully, however, it is necessary to define the following:

The description* of the process functions.
The mathematical relationships among the physical process variables.
The economic relationships and objectives of process operation.
How the process, materials, personnel, and information should work together.

The first three areas correspond to the three types of model that will be discussed in Chapters 3 through 5. *Functional models* describe the process and computer functions in general terms and will be discussed in Chapter 3. *Physical process models* define the mathematical relationships among the physical process variables and will be discussed in Chapter 4. *Economic models* define the relationships among economic variables and specify the objectives of process operation and will be discussed in Chapter 5. *Procedural models* describe the operating procedures for the physical process, materials, personnel, and information and will not be discussed in this book beyond the material in this chapter.

* The words "description," "definition," and "model" may be used interchangeably.

15

Table 2.1-1 Characteristics of Language of Models

Language of Models	Characteristics				
	Descriptive Capability	Ambiguity	Manipulation Capability	Implementation Capability	Primary Function
English text	Good	Very ambiguous	None	Limited	Descriptive explanations and directions
Drawings and block diagrams	Good	Not ambiguous	None	Good	Design, assembly, and construction
Logical flow charts and decision tables	Fair	Not ambiguous	None	Good	Computer programming
Curves, tables, nomographs	Fair	Not ambiguous	Good	None	Expressing simple relationships between a few variables
Mathematics	Poor	Not ambiguous	Excellent	Good	Problem solution and optimization

2.1 LANGUAGE OF MODELS

The languages used for models are as varied as the backgrounds of the engineers in the field of process control and automation. For example, control engineers use *block diagrams, signal flow diagrams,* and *transfer functions*; electronics engineers use *circuit* and *logic diagrams* (*or Boolean algebra*); programmers use *logic flow charts* and *computer coding*. All of these forms of expressions are merely vehicles for stating and solving problems.

The various languages used for models may be classified into the following five types:

English text.
Drawings and block diagrams.
Logical flow charts and decision tables.
Curves, tables, and nomographs.
Mathematics.

Each of the language types has certain characteristics that make it more useful in some areas than others. Table 2.1-1 is a summary of the characteristics of the five language types. As may be deduced from the table, none of the languages are equally suitable for every function. While *mathematics* is precise, unambiguous, and useful for problem solving, it is not descriptive enough for the layman. At the other end of the spectrum, *English text* is descriptive and simple for humans to follow, but it is ambiguous and difficult to manipulate. Manipulation in this context is the ability to state the relationships among several variables so that each may be expressed in terms of the other. *Curves, tables,* and *nomographs* are expressions of frequently used mathematical relationships. They may also be experimental data presented in this form for ease of human understanding and perception. In the implementation of control equipment and preparation of computer programs, specialized languages such as *drawings, block diagrams, logical flow charts,* and *decision tables* are required.

In any definition or description of process models used for computer process control a combination of the various language types may be required to satisfy a particular requirement. When more than one language is suitable, the choice should be based on *simplicity* and *accepted usage.*

2.2   FUNCTIONAL MODELS

Functional models describe the functions performed by the major portions of a plant and the computer process control system. The first step toward automation should be the development of functional models of the plant and the setting of the boundaries of the plant so that the scope can be defined. Functional models of the plant are generally process-oriented and are usually written from the viewpoint of process engineers and designers. Functional models of the computer process control system describe the functions performed and are usually written in flow chart or equation form. Since Chapter 3 is devoted primarily to the functional models of the computer process control system, this section will briefly discuss the functional model of the plant by means of an example.

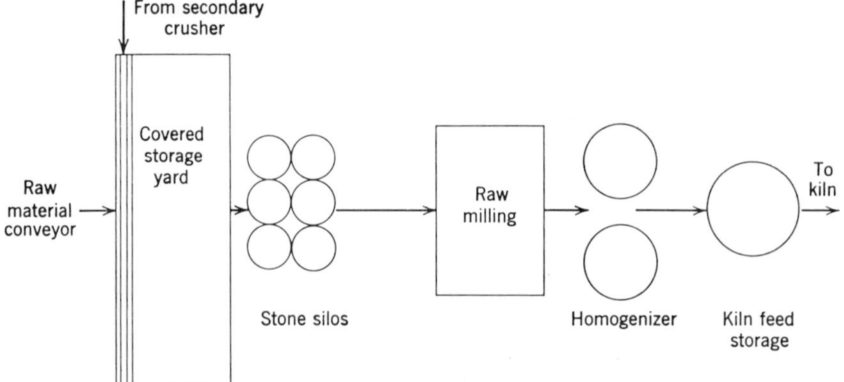

Fig. 2.2-1   Overall layout—portion of a dry cement plant.

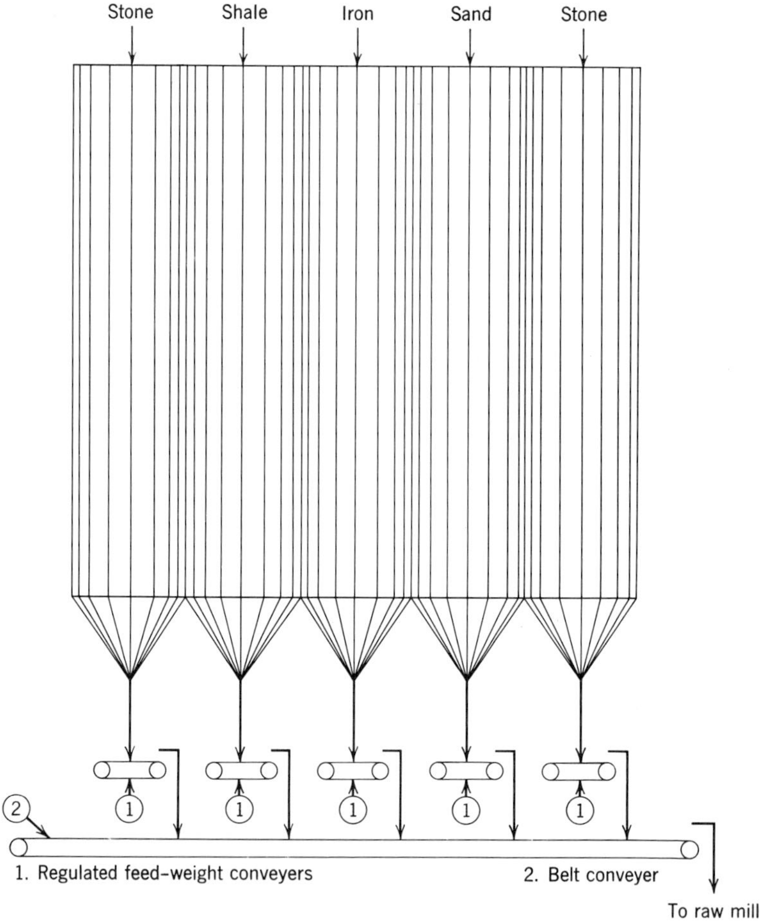

Fig. 2.2-2   Flow diagram of raw material section of a dry cement plant.

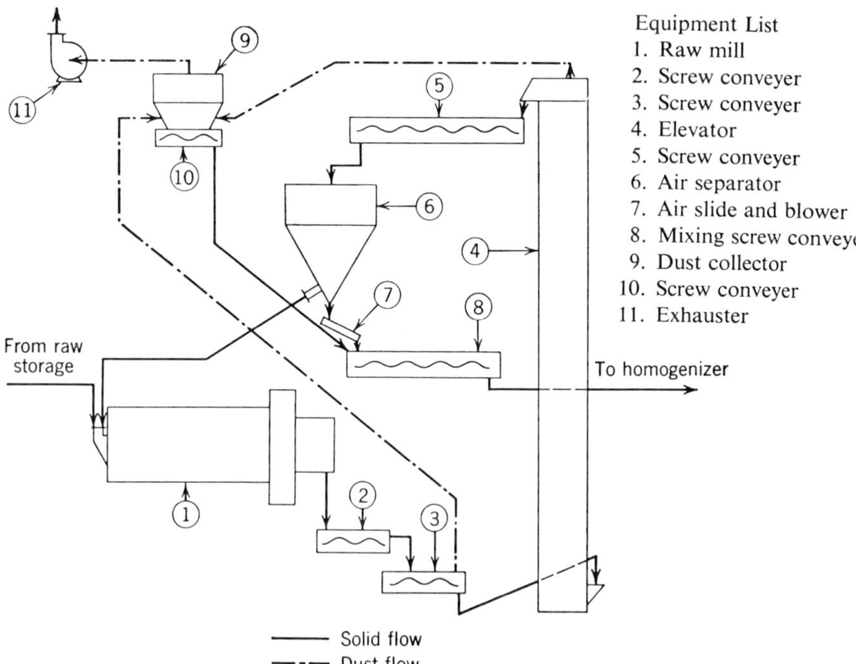

Fig. 2.2-3 Flow diagram of raw grinding section for a dry cement plant.

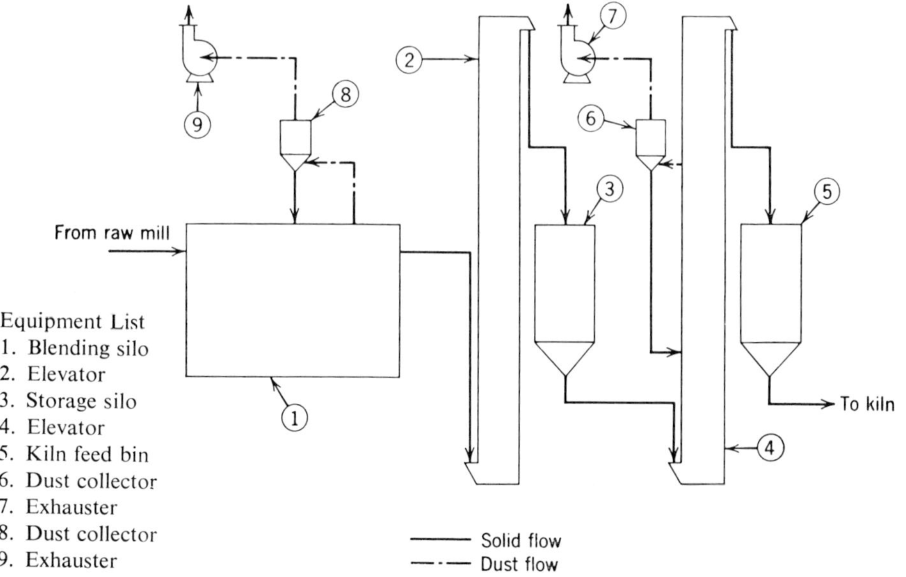

Fig. 2.2-4 Flow diagram of homogenizer and kiln feed storage section of a dry cement plant.

Figure 2.2-1 shows the overall layout of a portion of a typical cement plant. Figures 2.2-2 through 2.2-4 detail the major equipment of the plant. The drawings in Figures 2.2-1 through 2.2-4 and the English text in the next paragraph make up the functional model in an abbreviated form.

The *overall* function of this portion of the cement plant is to produce uniform kiln feed material having the proper maximum size, proper particle size gradation, and proper chemical composition for the kiln. The overall layout, Figure 2.2-1, shows the raw materials being taken from the storage bins, then being passed through the raw mill and, finally, being blended in the homogenizer. Figure 2.2-2 shows the raw material section whose function is to transfer the raw materials from the storage areas to the raw mill. The materials from the individual storage bins are withdrawn by the feeders, which are proportionally controlled so that the weight withdrawn from each bin is a fixed percentage of the total weight of material withdrawn from all the bins. Figure 2.2-3 shows the raw grinding section whose function is to establish the proper maximum particle size and proper size gradation. This is accomplished by the raw mill, air classifier, and the associated conveying and control equipment. Figure 2.2-4 shows the homogenizer and kiln feed storage section whose function is to produce a composite material of uniform physical and chemical characteristics. The blending vessel acts as a tank where jets of compressed air continually agitate the material stored within.

2.3 PHYSICAL MODELS

Physical models define the mathematical relationships among the physical variables under study. *Physical process models* define the mathematical relationships among all the variables of the physical process under study. In this book the term "physical model" is used in reference to local relationships between one physical variable and another, while "physical process model" relates *all* the physical process variables to each other.

Physical process models can be developed by analytical approaches. Sections 4.1 and 4.2 will present the laws of systems analysis and the physical characteristics of *pneumatic, hydraulic, thermal, mechanical,* and *electrical systems.* Section 4.7 will show an example of the application of these analytical techniques in developing the dynamic physical process model of a completely mixed stirred tank chemical reactor.

Physical process models can also be developed by experimental approaches. Sections 4.3, 4.4, and 4.5 will present regression techniques used for developing steady-state physical process models. Section 4.6 will present an experimental technique for developing dynamic physical process models, patterned after the technique suggested by Kalman. Section 6.1 will show how the physical process model fits into the optimization problem and the various types of physical process models.

This section will present two examples of physical process models. The notation used emphasizes that computer process control in this book is oriented toward the solution of specific problems in real life.

A. Continuous-Time Physical Process Model

As an example, consider the homogenizer or blending silo shown in Figure 2.3-1. This is the same homogenizer shown in Figure 2.2-1. The homogenizer for a dry cement process is supposed to contain a constant volume of material. This is accomplished by making the input and output rates of material flow equal. The problem is to derive the relationship between *the composition of output material flow* and *the composition of input material flow*. The materials which are critical to cement-making include calcium oxide CaO, silicon oxide SiO_2, aluminum oxide Al_2O_3, ferrous oxide Fe_2O_3, and other compounds. The relationship between *the composition of calcium oxide in the output material flow* and *the composition of calcium oxide in the input material flow* will be derived. Similar relationships can be developed for all the other materials. Let the following symbols be assigned:

C_{out} = amount of CaO taken out of the homogenizer as a result of throughput of $(R\,\Delta t)$, in tons.

C_{in} = amount of CaO taken into the homogenizer as a result of throughput of $(R\,\Delta t)$, in tons.

ΔC_h = change in the amount of CaO in the homogenizer as a result of throughput of $(R\,\Delta t)$, in tons.

T = capacity of the homogenizer, in tons.

$\% \, C_h$ = % of CaO in the homogenizer;
 = % of CaO in the output stream, if mixing is assumed to be perfect.

$\% \, C_{rm}$ = % of CaO in the output stream of the raw mill.
 = % of CaO in the input stream of the homogenizer.

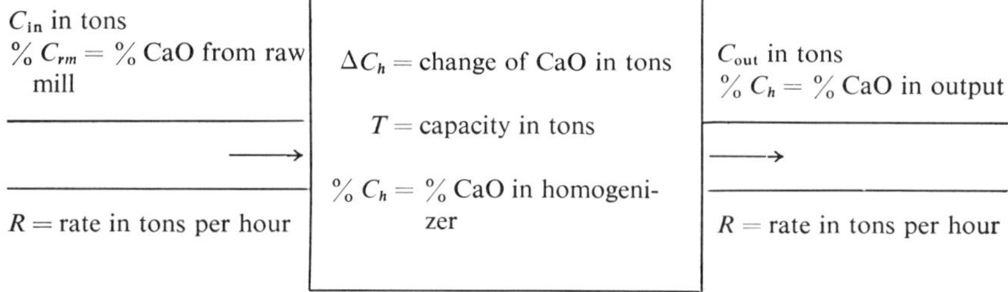

Fig. 2.3-1   Variables under study for the physical process model of homogenizer.

R = rate of material transferred into the homogenizer, in tons per hour.

         = rate of material transferred out of the homogenizer, in tons per hour.

Δt = interval of time, in hours.

Assuming that all the variables are constant over the interval of time Δt, the amount of CaO taken out of the homogenizer is

$$C_{out} = \left(\frac{\% \, C_h}{100}\right) R \, \Delta t \tag{2.3-1}$$

Similarly, the amount of CaO taken into the homogenizer is

$$C_{in} = \left(\frac{\% \, C_{rm}}{100}\right) R \, \Delta t \tag{2.3-2}$$

The increase in the amount of CaO in the homogenizer is

$$\Delta C_h = C_{in} - C_{out} \tag{2.3-3}$$

$$\Delta C_h = \left(\frac{\% \, C_{rm} - \% \, C_h}{100}\right) R \, \Delta t \tag{2.3-4}$$

Equation 2.3-4 is obtained by substituting Equations 2.3-1 and 2.3-2 into Equation 2.3-3. The increase in $\%$ CaO in the homogenizer is

$$\Delta(\% \, C_h) = \frac{\Delta C_h}{T} \, 100 \tag{2.3-5}$$

Then

$$\Delta C_h = \frac{\Delta(\% \, C_h)}{100} \, T \tag{2.3-6}$$

Substituting Equation 2.3-6 into Equation 2.3-4

$$\Delta(\% \, C_h)T = (\% \, C_{rm} - \% \, C_h)R \, \Delta t \tag{2.3-7}$$

$$\frac{\Delta(\% \, C_h)}{\Delta t} = \left(\frac{R}{T}\right)(\% \, C_{rm} - \% \, C_h) \tag{2.3-8}$$

As mentioned above, $\Delta(\% \, C_h)$, $\% \, C_{rm}$, and $\% \, C_h$ are constant in this operation. However, applying the rules of differential calculus, Equation 2.3-8 becomes 2.3-9 as Δt approaches zero, and all variables now become *continuous functions of time*.

$$\frac{d(\% \, C_h)}{dt} = \left(\frac{R}{T}\right)(\% \, C_{rm} - \% \, C_h) \tag{2.3-9}$$

The physical process model given in Equation 2.3-9 simply states that the rate of change of the percentage of CaO in the homogenizer is equal to a constant times the difference between the percent of CaO in the input, $\% \, C_{rm}$, and the percent of CaO in the homogenizer, $\% \, C_h$. The constant R/T is equal to the ratio of throughput to capacity, and has the dimension of (hours^{-1}). Similarly, let the following symbols be assigned:

$$
\begin{aligned}
\% \, S_h &= \% \text{ of } SiO_2 \text{ in the homogenizer.}\\
\% \, A_h &= \% \text{ of } Al_2O_3 \text{ in the homogenizer.}\\
\% \, F_h &= \% \text{ of } Fe_2O_3 \text{ in the homogenizer.}\\
\% \, S_{rm} &= \% \text{ of } SiO_2 \text{ in the raw mill.}\\
\% \, A_{rm} &= \% \text{ of } Al_2O_3 \text{ in the raw mill.}\\
\% \, F_{rm} &= \% \text{ of } Fe_2O_3 \text{ in the raw mill.}
\end{aligned}
$$

Using the same derivation developed in Equations 2.3-1 through 2.3-9, the following equations may be derived for SiO_2, Al_2O_3, and Fe_2O_3:

$$\frac{d(\% \, S_h)}{dt} = \left(\frac{R}{T}\right)(\% \, S_{rm} - \% \, S_h) \tag{2.3-10}$$

$$\frac{d(\% \, A_h)}{dt} = \left(\frac{R}{T}\right)(\% \, A_{rm} - \% \, A_h) \tag{2.3-11}$$

$$\frac{d(\% \, F_h)}{dt} = \left(\frac{R}{T}\right)(\% \, F_{rm} - \% \, F_h) \tag{2.3-12}$$

The physical process model of the homogenizer is the set of differential equations given by Equations 2.3-9 through 2.3-12.

B. Discrete-Time Physical Process Model

Sometimes the process is best described when the process variables are expressed as *continuous functions of time*. However, often the interest is in the *values of the variables at discrete instants of time*. Equations giving the values of the variables at discrete instants of time can be derived from the equations in continuous form.

For example, assume that in the interval

$$t_0 \leq t \leq t_0 + \Delta t \tag{2.3-13}$$

$\% \, C_{rm}$ is constant and equal to the initial value $\% \, C_{rm}(t_0)$. In this equation, t_0 is the initial value of time, Δt is the interval of time, and t is the variable. The equation relating $\% \, C_h(t_0 + \Delta t)$,* $\% \, C_h(t_0)$, and $\% \, C_{rm}(t_0)$ can be derived

---

* $\% \, C_h(t_0 + \Delta t)$ is the value of $\% \, C_h$ at time $(t_0 + \Delta t)$; $\% \, C_h(t_0)$ is the value of $\% \, C_h$ at time t_0, i.e., the initial value; and $\% \, C_{rm}(t_0)$ is the value of $\% \, C_{rm}$ at t_0.

by solving the differential equation in Equation 2.3-9:

$$\% \, C_h(t) = \left\{ 1 - \exp\left[-\frac{R}{T}(t - t_0) \right] \right\} [\% \, C_{rm}(t_0)]$$

$$+ \left\{ \exp\left[-\frac{R}{T}(t - t_0) \right] \right\} [\% \, C_h(t_0)] \qquad (2.3\text{-}14)$$

By substituting $t = t_0 + \Delta t$, Equation 2.3-14 becomes

$$\% \, C_h(t_0 + \Delta t) = \left\{ 1 - \exp\left[-\frac{R}{T}(\Delta t) \right] \right\} [\% \, C_{rm}(t_0)]$$

$$+ \left\{ \exp\left[-\frac{R}{T}(\Delta t) \right] \right\} [\% \, C_h(t_0)] \qquad (2.3\text{-}15)$$

Let P and Q be the following constants:

$$P = 1 - \exp\left[-\frac{R}{T}(\Delta t) \right] \qquad (2.3\text{-}16)$$

$$Q = \exp\left[-\frac{R}{T}(\Delta t) \right] \qquad (2.3\text{-}17)$$

Substituting P and Q in Equation 2.3-15 yields

$$\% \, C_h(t_0 + \Delta t) = P[\% \, C_{rm}(t_0)] + Q[\% C_h(t_0)] \qquad (2.3\text{-}18)$$

Equation 2.3-18 allows the calcium oxide content of the homogenizer at time $(t_0 + \Delta t)$ to be calculated as a function of the calcium oxide content at an earlier time t_0 if the calcium oxide content of the input stream is constant and known over the interval Δt. Similarly, Equations 2.3-10, 2.3-11, and 2.3-12 in continuous-time form for silicon oxide SiO_2, aluminum oxide Al_2O_3, and ferrous oxide Fe_2O_3, etc., ... can be expressed in discrete-time form as

$$\% \, S_h(t_0 + \Delta t) = P[\% \, S_{rm}(t_0)] + Q[\% \, S_h(t_0)] \qquad (2.3\text{-}19)$$

$$\% \, A_h(t_0 + \Delta t) = P[\% \, A_{rm}(t_0)] + Q[\% \, A_h(t_0)] \qquad (2.3\text{-}20)$$

$$\% \, F_h(t_0 + \Delta t) = P[\% \, F_{rm}(t_0)] + Q[\% \, F_h(t_0)] \qquad (2.3\text{-}21)$$

$$\vdots$$

2.4　ECONOMIC MODELS

Just as physical process models define the relationships among the process variables under study, *economic models define the relationships among the economic factors under study* and *specify the economic objectives.* As in the case of physical process models, economic models in equation form can be manipulated easily. Moreover, it is possible to solve for a set of conditions that will satisfy the economic objectives, e.g., maximum profit or minimum cost.

This book will deal mainly with *microeconomic models* rather than *macroeconomic models*. Microeconomic models deal with confined spheres of local phenomena, such as a plant or a segment of a particular market. Macroeconomic models deal with the overall national economy, such as gross national product, savings, and investment rate. There are two types of micro-economic model commonly used by process engineers: *planning economic models* and *operating economic models*. Chapter 5 will discuss both of these in detail. This section will introduce planning and operating economic models with examples.

A. Planning Economic Models

Planning economic models define the relationships among the economic factors under study for investment purposes. The objective is usually to attain the maximum return-on-investment, and the time scale involved may be 2 to 25 years. The economic factors pertinent to such a study may include first cost of equipment, operating and maintenance costs, interest rates and depreciation schedules, cost of raw material, and prices of products. Planning economic models provide quantitative evaluations of alternative courses of action available to management. Sections 5.2 and 5.3 will present the techniques of developing planning economic models.

As an example, consider the situation confronting management where an existing plant produces products of satisfactory quality but at higher unit cost than the competition's new plants. One alternative is to "do nothing," i.e., live with the existing plant. Another alternative is to "modernize" to reduce the unit cost of the products. A simplification is to assume that costs can be divided into two categories: (1) *fixed costs* which do not vary with volume of output and (2) *variable costs* which vary with volume of output. The fixed costs of the "modernize" alternative will be higher than those of the "do nothing" alternative because of the capital required for investment of new equipment. On the other hand, the variable costs of the modernized plant are lower than those of the existing plant because the new plant is more efficient.

Figure 2.4-1 compares the two alternatives by plotting costs against volume. The intersection of the two total cost curves is the break-even volume, i.e., the volume where the costs of the two alternatives will be equal. If the forecasted volume is less than the break-even volume, the "do nothing" alternative is more attractive. On the other hand, if the forecasted volume is

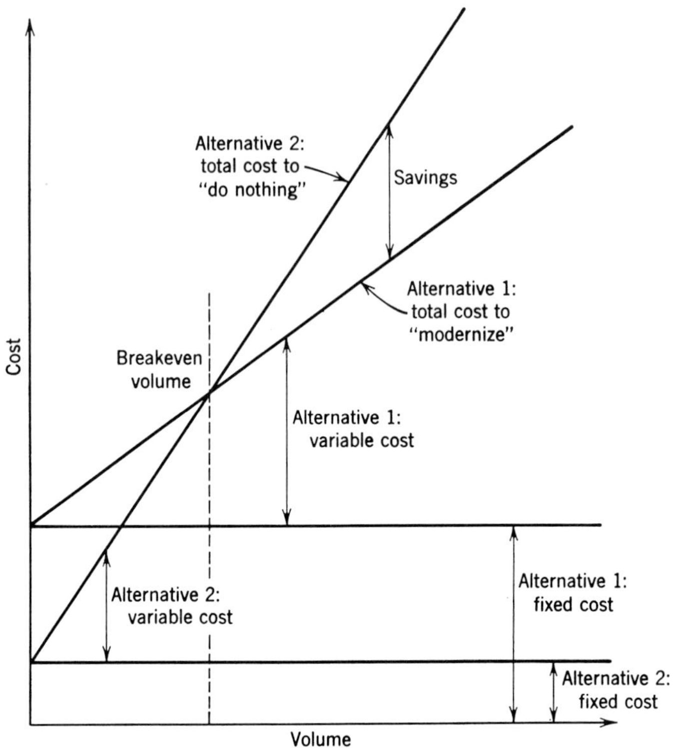

Fig. 2.4-1 Evaluation of alternatives by fixed and variable costs.

greater than the break-even volume, the "modernize" alternative is more attractive. The amount of savings due to modernization is the difference between the two total cost curves.

B. Operating Economic Models

Operating economic models define the relationships among economic factors and specify the objectives of process operation. The equipment of the plant is assumed to be fixed, and the time scale is usually in the order of a few seconds to hours. The objectives of process operation may be *minimum cost or maximum profit.* Usually, only one objective may be specified. The objective chosen stated in equation form is called the *objective function.* Sections 5.5 and 5.6 will present the techniques of developing operating economic models.

The first example of operating models is the problem of blending a number of raw materials to produce several products at minimum cost while subject to constraints. Using the example of Section 2.2, two materials—high magnesium stone and high calcium stone—are blended to form a composite material at minimum cost. A constraint on the blending is the composition of the output material: the calcium content must be between 43.0 and 43.2%, and the magnesium content must be between 0 and 2.6%. Let the following symbols be assigned:

HC = amount of high calcium stone blended, in per unit* weight.
HM = amount of high magnesium stone blended, in per unit* weight.
Q_{HC} = cost of high calcium stone, in dollars per ton.
Q_{HM} = cost of high magnesium stone, in dollars per ton.
R_{11} = ratio of calcium in high calcium stone, in per unit.
R_{12} = ratio of calcium in high magnesium stone, in per unit.
R_{21} = ratio of magnesium in high calcium stone, in per unit.
R_{22} = ratio of magnesium in high magnesium stone, in per unit.

In this specific example $R_{11} = 0.46$, $R_{12} = 0.40$, $R_{21} = 0.01$, $R_{22} = 0.04$. The problem is to solve for HC and HM, so that

$$HC \geq 0 \quad \text{and} \quad HM \geq 0 \tag{2.4.1}$$

subject to the constraints

$$HC + HM = 1 \tag{2.4-2}$$

$$0.430 \leq (0.46)(HC) + (0.40)(HM) \leq 0.432 \tag{2.4-3}$$

$$0 \leq (0.01)(HC) + (0.04)(HM) \leq 0.026 \tag{2.4-4}$$

and

$$(Q_{HC})(HC) + (Q_{HM})(HM) = \text{Minimum} \tag{2.4-5}$$

The economic model of blending at minimum cost while subject to constraints is given by Equations 2.4-1 through 2.4-5. Equation 2.4-1 states that both input material flows must be equal to or greater than zero, i.e., reverse flows are not permitted. Equation 2.4-2 is a constraint which normalizes the input material flows so their sum is equal to 1. The solution for any given volume can be found by multiplying HC and HM by a constant. Equations 2.4-3 and 2.4-4 are the blending constraints for the case where the total input material flow is equal to 1. Equation 2.4-5 is the objective function which is to be minimized. Figure 2.4-2 shows the functional model of this example.

* Per unit is percent divided by 100.

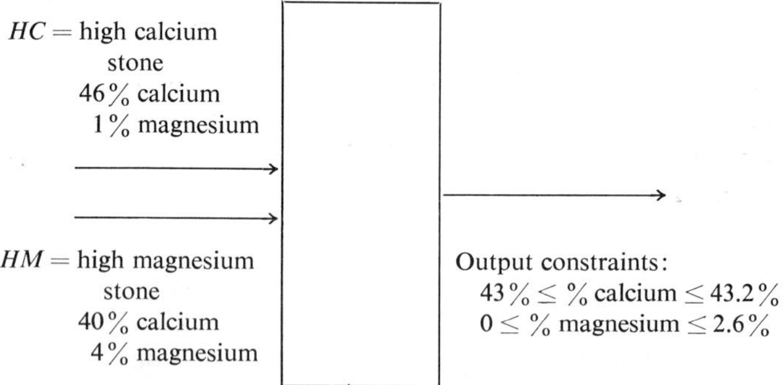

Fig. 2.4-2   Blending of high calcium stone and high magnesium stone.

C. Another Operating Economic Model Example

Another example of an operating economic model is the problem of minimizing the fuel cost required for a given load in a power generation and transmission system, as shown in Figure 2.4-3. Let the following symbols be assigned:

F_n = fuel cost of plant n, in dollars per hour.
P_n = power output of plant n, in megawatts.
P_r = power received by loads, in megawatts.
P_1 = transmission losses, in megawatts.

The problem is to find the set of P_n, so that

$$P_n \geq 0 \tag{2.4-6}$$

subject to the constraint that the total power generated is equal to the sum of the transmission losses and the load (assumed to be a constant):

$$\sum_n P_n - P_1 - P_r = 0 \tag{2.4-7}$$

with

$$\sum_n F_n = \text{Minimum} \tag{2.4-8}$$

The transmission loss P_1 can be expressed as

$$P_1 = \sum_m \sum_n P_m B_{mn} P_n \tag{2.4-9}$$

and the fuel cost for each plant F_n can be expressed as a function of the power output P_n:

$$F_n = F_n(P_n) \tag{2.4-10}$$

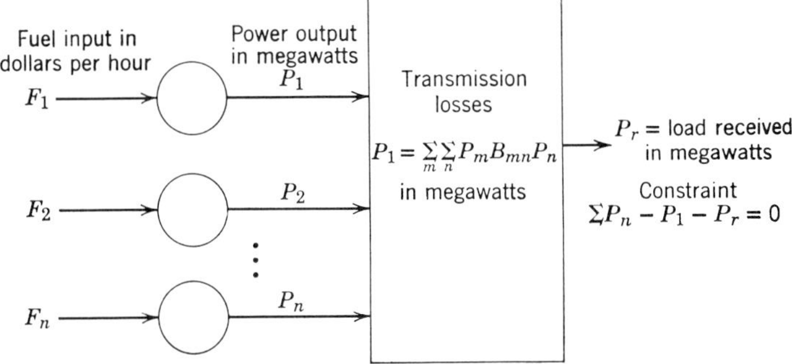

Fig. 2.4-3   Power generation and transmission network.

Substituting Equation 2.4-9 into 2.4-7 and Equation 2.4-10 into 2.4-8, the solution for minimum fuel cost operation of a power generation and transmission system is the set of P_n, so that

$$P_n \geq 0 \tag{2.4-11}$$

subject to the constraint

$$\sum_n P_n - \sum_m \sum_n P_m B_{mn} P_n - P_r = 0 \tag{2.4-12}$$

with

$$\sum_n F_n(P_n) = \text{Minimum} \tag{2.4-13}$$

Equation 2.4-13 is the objective function which is to be minimized. In contrast to the previous example, this problem is not linear. Examination of Equations 2.4-12 and 2.4-13 will show that neither the constraints nor the objective function are necessarily linear.

2.5   PROCEDURAL MODELS

In contrast to functional models, which describe the functions of the plant or computer system, *procedural models describe how a plant or computer system should be operated.* In other words, procedural models describe the operational characteristics of a system. Of particular interest to plant automation are the following two types of procedural models: (1) Information procedural models, and (2) Action and safety procedural models.

A detailed discussion of procedural models is outside the scope of this book. This section will begin with the characteristics of information, then introduce information procedural models and action and safety procedural models by means of examples.

A. Characteristics of Information

Control applications have been traditionally classified according to certain information characteristics that make the applications amenable to analysis by specific techniques. Some of these information characteristics are also applicable to information used for business data processing. These information characteristics are the following:

Steady-state or dynamic.
Linear or nonlinear.
Continuous-time or discrete-time.
Continuous-value or discrete-value.
Deterministic or stochastic.

Information is *steady-state* if the value does not depend on time or is not associated with time during the period of interest, and information is *dynamic* if the value depends on time or is associated with time. Steady-state approximations of process operations may be used for slow processes or processes that normally operate around a fixed set of conditions, e.g., turbine heat balance calculations for $\frac{1}{4}$, $\frac{1}{2}$, and full load, and linear programming solutions to blending problems. Dynamic information is required for control under transient conditions. Real-time information for process and business is generally dynamic.

Information is *linear* if its value is proportional to the value of another variable, and information is *nonlinear* if its value is not related to the value of another variable in a straight-line relationship, as shown in Figure 2.5-1.

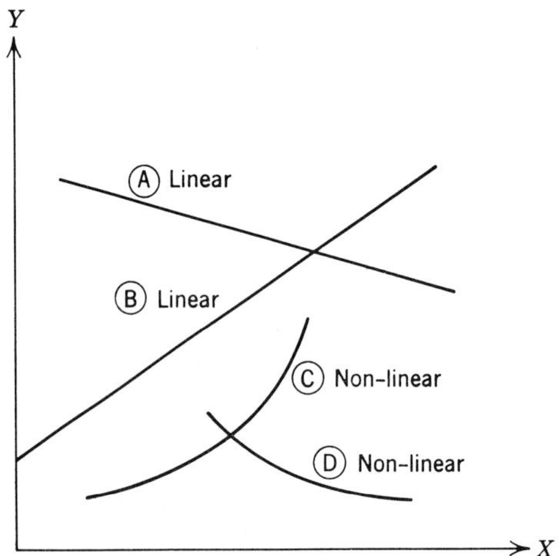

Fig. 2.5-1   Linear and nonlinear information.

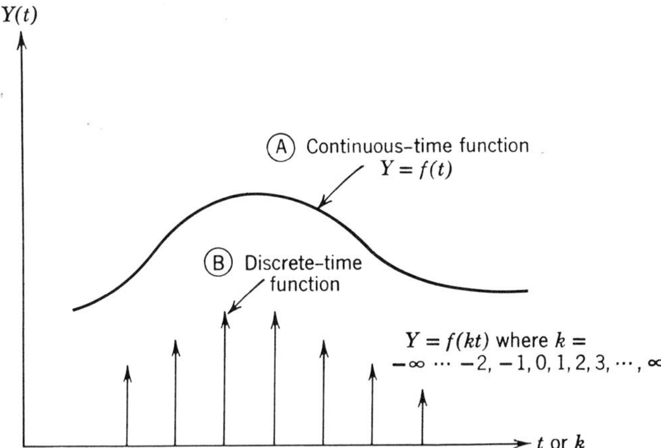

Fig. 2.5-2 Continuous-time and discrete-time information.

Examples of linear information are *resistance and capacitance of electrical networks* and *material costs*. Examples of nonlinear information are *resistance of fluid systems* and *volume-price relationship of economic models*.

Information is *continuous-time* if its value exists for every value of time, as shown by Ⓐ in Figure 2.5-2, and information is *discrete-time* if its value exists only at specific instants of time, as shown by Ⓑ in Figure 2.5-2. Continuous-time information is specified by

$$Y = f(t) \tag{2.5-1}$$

where t exists everywhere.

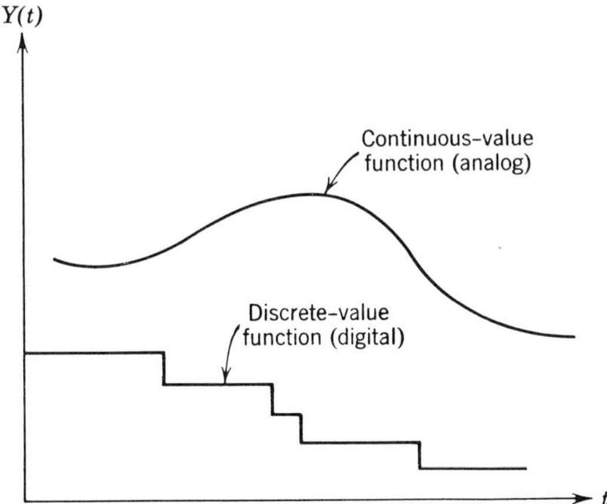

Fig. 2.5-3 Continuous-value and discrete-value information.

Discrete-time information is specified by

$$Y = f(kt) \tag{2.5-2}$$

where k is an integer ranging from $-\infty, \ldots, -3, -2, -1, 0$ to $1, 2, 3, \ldots, \infty$.

Information is *continuous-value* if a finite variable can assume an infinite number of values (analog), and information is *discrete-value* if it can assume only a finite number of values (digital). Figure 2.5-3 shows the difference between continuous-value and discrete-value information. Analog information with finite values can assume an infinite number of states, e.g., actual values of voltage, current, and pressure. Digital information can only assume only a finite number of states, e.g., 1024 states in a 10-bit register.

Information is *deterministic* if the value can always be predicted, and information is *stochastic* if the value cannot be predicted precisely. Typical

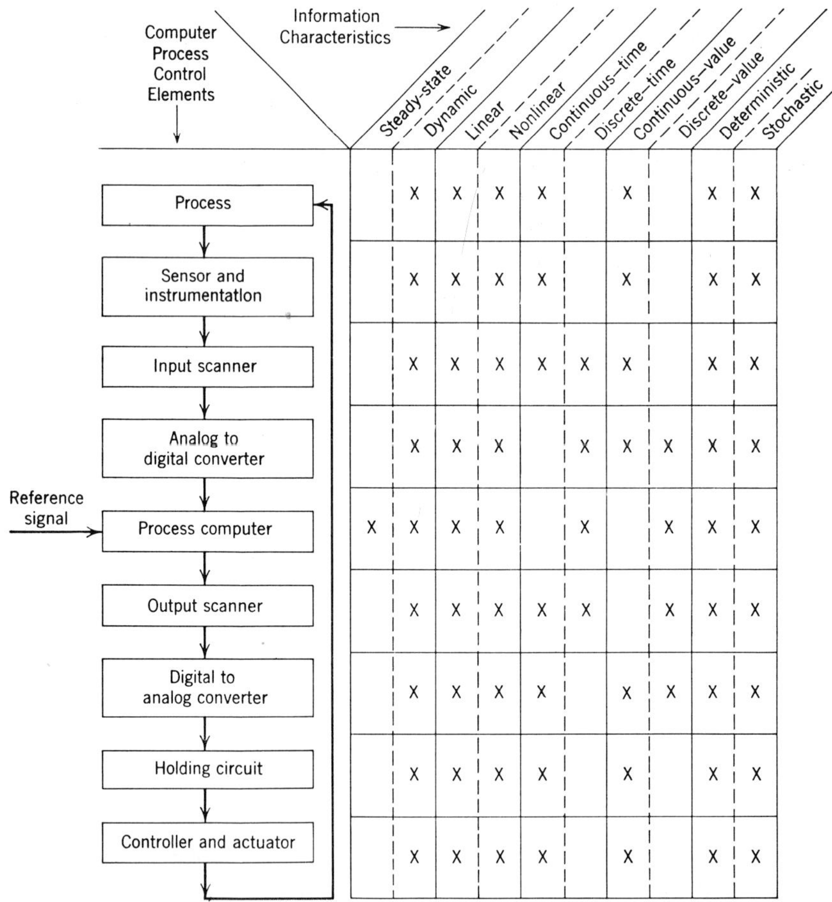

Computer Process Control Elements	Steady-state	Dynamic	Linear	Nonlinear	Continuous-time	Discrete-time	Continuous-value	Discrete-value	Deterministic	Stochastic
Process		X	X	X	X		X		X	X
Sensor and instrumentation		X	X	X	X		X		X	X
Input scanner		X	X	X	X	X	X		X	X
Analog to digital converter		X	X	X		X	X	X	X	X
Process computer	X	X	X	X		X		X	X	X
Output scanner		X	X	X	X	X		X	X	X
Digital to analog converter		X	X	X	X		X	X	X	X
Holding circuit		X	X	X	X		X		X	X
Controller and actuator		X	X	X	X		X		X	X

Fig. 2.5-4 Information characteristics for computer process control system.

deterministic information includes physical and economic parameters, e.g., diameter of a pipe, capacity of a machine, cost of raw material. Stochastic information has inherent uncertainty or randomness, which makes it impossible to predict precisely. Typical stochastic information includes process signals which contain noise, work duration times of a work station which vary according to the skill and effort of the worker, and forecasted data.

Figure 2.5-4 shows the information characteristics for a computer process control system. The block diagram as shown in Figure 2.5-4 will correspond to the functional models which will be presented in Chapter 3. Note that all the information is dynamic, although certain information in the process computer may be steady-state, e.g., information for steady-state physical process models. Information may be either linear or nonlinear because this characteristic depends on the specific element and the simplifying assumptions. Information on the input side of the input scanner and the output side of the output scanner is continuous-time, while the rest of the information is discrete-time. Similarly, information on the input side of the analog-to-digital converter and the output side of the digital-to-analog converter is continuous-value, while the rest of the information is discrete-value. Finally, information can be either deterministic or stochastic depending on the application and assumptions.

B. Information Procedural Models

Information procedural models define the content, format, and rate (or frequency) of information flow. Information models also include implications of audit and control, approvals necessary for obtaining certain types of information, precautions to guard against disastrous loss of information, and recovery procedures in case of malfunctions or breakdowns. Figure 2.5-5 shows a simplified example of *information flow* which is a part of an overall order service system to handle customer orders. An order coming into the sales office initiates three different types of document: order acknowledgement, requisition, and open order. Order acknowledgement confirms the receipt of the order with an estimated delivery date (if available). The requisition sent to the district warehouse is an order to deliver the quantity of products to the customer. An open order is a record of the order placed but not filled. Depending on whether the products on order are available in the district warehouse, the inventory system will generate shipping notices or shortage notices. Both of these notices go back to the sales office. Upon receipt of a shipping notice, the sales office updates the open order document, and prepares the following documents: invoice, accounts receivable, sales analysis. The invoice sent to the customer is the billing due for the products delivered, the accounts receivable sets up the amount that is due for collection, and the sales analysis input updates the various reports for management. There is a

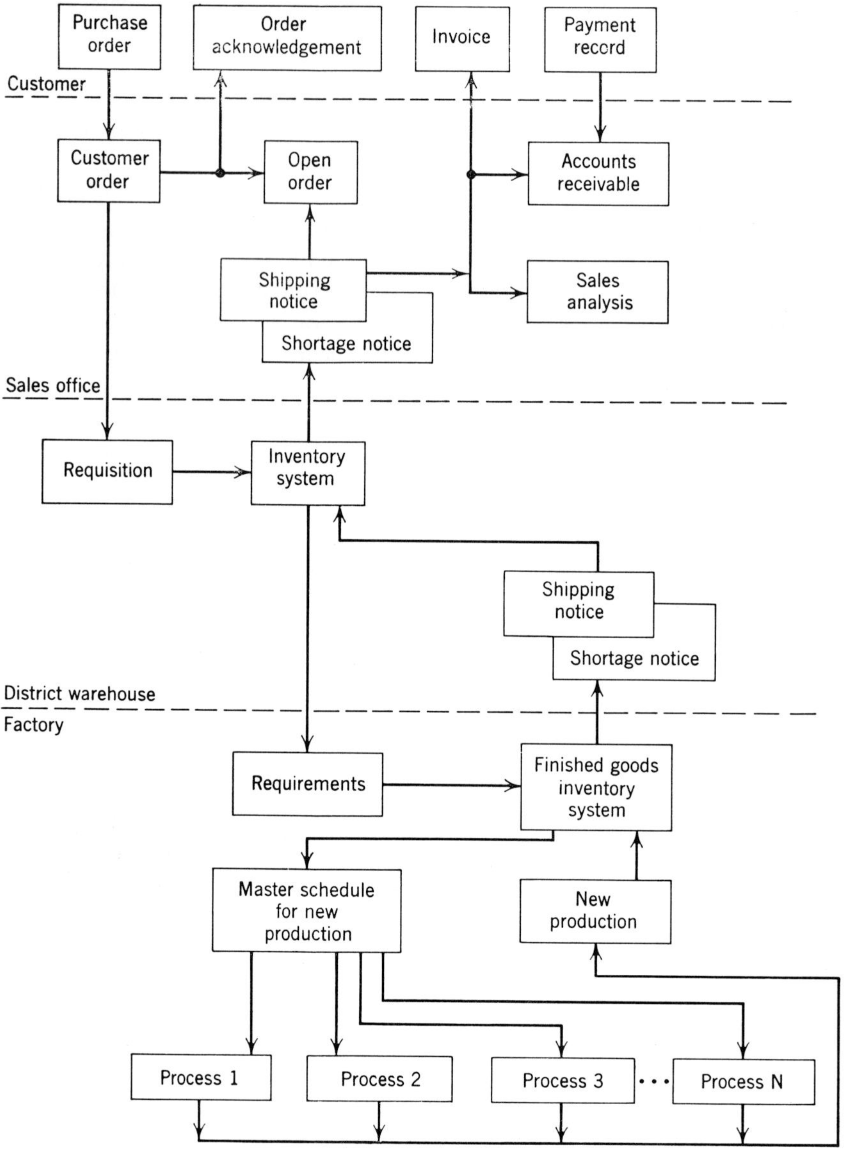

Fig. 2.5-5 Simplified information flow for handling customer orders.

similar flow of information between the district warehouse and factory. The
factory finished goods inventory system will trigger the master schedule for
new production. As the new products are produced by Process 1, 2, . . . , N,
these become additions to the factory finished goods inventory system which,
in turn, supplies the district warehouses.

C. Action and Safety Procedural Models

Action and safety procedural models refer to the actions which change the status of the plant, and the specification of operating restrictions for safety considerations. Typical action procedures include start-up, shut-down, and load changes. In the design of action and safety procedural models the plant operator is considered to be an integral part of the plant and is usually assigned the following roles:

Provide inputs to the computer system.
Monitor outputs from the instrumentation and computer systems.
Provide back-up in case of equipment failure.
Debug program and system equipment.

These roles have been assigned to the operator simply because he can perform them better and more economically than any automated system.

As an example, consider the action procedure for accelerating a steam turbine. Initially, the turbine is being rotated slowly by a small electric motor through a turning gear. The task is to get the turbine off the turning gear and to accelerate it according to a recommended acceleration curve provided by the manufacturer, as shown in Figure 2.5-6. Checks for excessive vibration and eccentricity must be made during the acceleration process. If vibration and eccentricity limits are exceeded, the turbine should be held at certain "safe speeds" which are different from the critical vibration speeds of the turbine. Acceleration may be resumed after a "hold." Manual implementation of the procedure would involve adjusting the "stop valve bypass valve" to accelerate the turbine. Semiautomatic implementation of the procedure would adjust the set point of the "full range speed governor" to accelerate the turbine. Figure 2.5-7 shows the procedural model for computer control of turbine acceleration. The control of turbine speed is by pulses sent from the control computer to a motor which operates the "stop valve bypass valve."

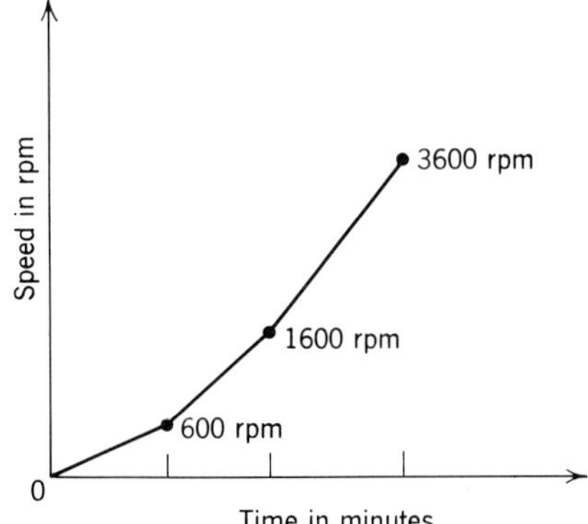

Fig. 2.5-6   Recommended turbine acceleration curve.

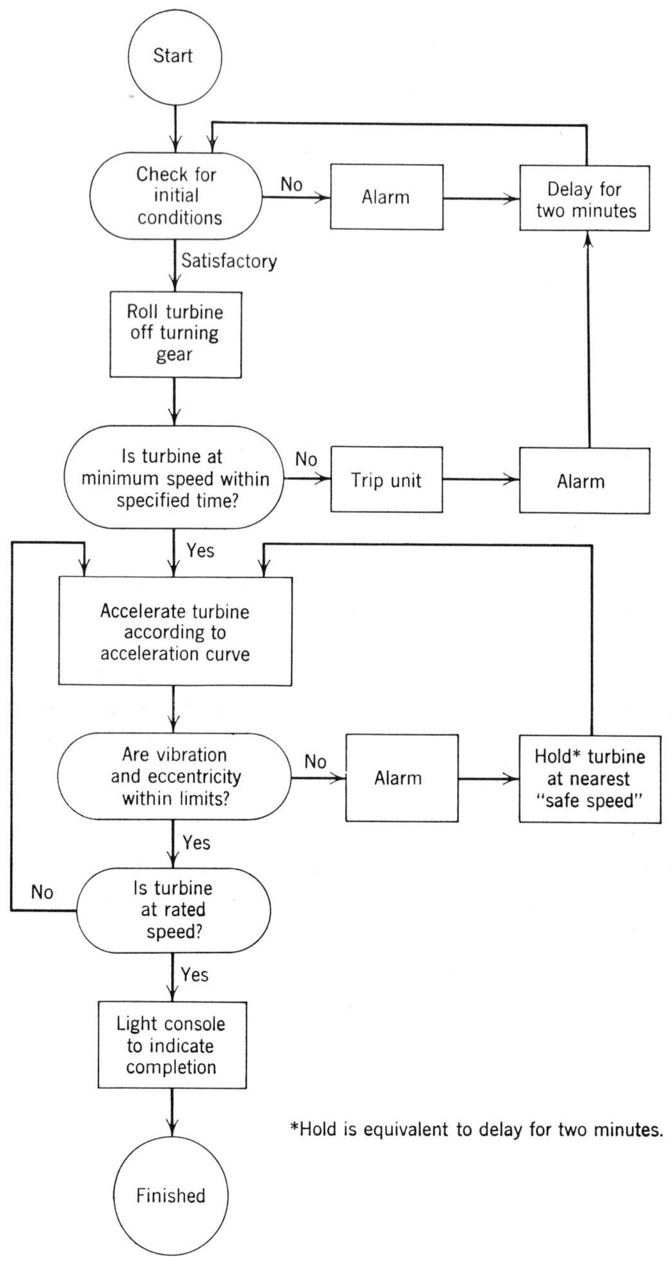

Fig. 2.5-7 Procedural model for computer control of turbine acceleration.

36

2.6   MODELS FOR AN OVERALL PLANT

Sections 2.2 through 2.5 introduced four types of models: *functional, physical, economic,* and *procedural.* This section will attempt to unify the four types of models by relating them to a plant. Examples of models for an overall plant will be used to illustrate the relationships among the four types of models. Figure 2.6-1 shows an overall plant model listing the classes of input and output variables. These classes of variables can be related to the four types of models.

A.   Classes of Input Variables

There are four classes of input variables shown in Figure 2.6-1: *procedural actions, control variables, uncontrolled variables,* and *parameters.** The *procedural actions* and *control variables* are input variables that are accessible for control purposes. Control variables are the independent variables of the process. Procedural actions can also adjust the control variables. The procedural model determines the procedural actions to be taken. Procedural actions may be connecting or disconnecting an electric circuit and opening, closing, or adjusting valves. Procedural actions are generally discrete events and most frequently take place during start-up or shut-down operations in automatic plants. The physical process model relates the control variables to the state variables. Although the control variables are adjusted by procedural

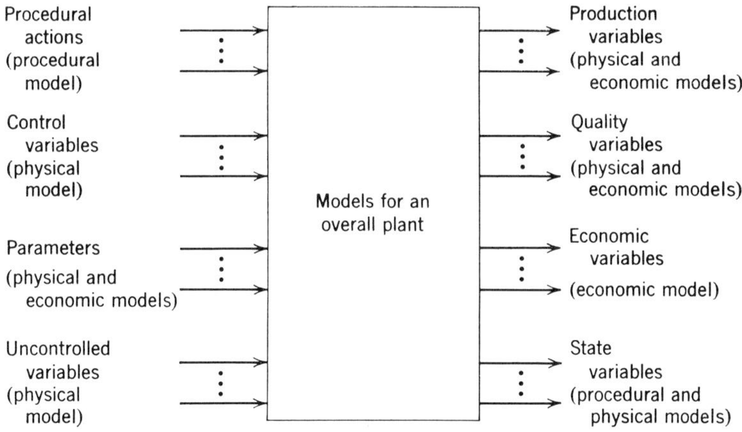

Fig. 2.6-1   Models for an overall plant.

* Parameters are really constants in a narrow sense. However, since these constants may be changed to reflect changing plant characteristics and costs, they may be classified as variables.

actions, these are typically under automatic regulation by controllers in automatic plants. Typical control variables may be steam flow rates, raw material flow rates, and power input rates.

The two other classes of input are *uncontrolled variables* and *parameters*, which may have an appreciable effect on the output variables but are not accessible for control purposes. These must be factored into the design of the physical and economic models. Uncontrolled variables may be ambient temperature, humidity, and barometric pressure. Parameters are the inherent characteristics of the plant and raw materials. Parameters generally appear as constant coefficients in physical and economic models, while uncontrolled variables generally appear as variables in physical models. Typical parameters may be: cost of raw material, chemical composition of raw materials, and physical characteristics of equipment.

B. Classes of Output Variable

There are four classes of output variable shown in Figure 2.6-1: *state variables, production variables, quality variables,* and *economic variables*. The classification is based on how the output variable is used. It is possible for a specific variable to fall into more than one class, e.g., the same variable being classified as a state variable and a quality variable. *State variables* describe the status of the plant as a function of time. State variables are generally used in the procedural model to initiate procedural action and are the dependent variables of the physical process model. Examples of state variables are reactor temperature, barrels of cement per hour, chemical composition, on-off state of equipment, and yes-no indication of whether continuous variables are within safety limits. *Production variables* indicate the rate or level of production in the plant. Production variables are included in economic models, and are usually derived from the state variables. Indications of production rate may be expressed in absolute or relative form. Absolute production variables may also include control variables, while relative production variables are usually functions of output variables only. Examples of absolute production variables are gallons of toulene per hour, barrels of cement per shift, and coils of steel per turn. An example of a relative production variable is power output in percent of rated output.

Quality variables measure the physical or chemical characteristics of the products that are important to the evaluation and control of quality and are derived from state variables. Therefore, physical process models can be used to relate quality variables to control variables, uncontrolled variables, and parameters. Quality variables may be measured directly or computed indirectly from measured variables. Directly measured quality variables are the same as state variables. Examples of directly measured quality variables are number of feet of steel coil that has pin holes, number of feet of steel coil that is overgauge or undergauge, chemical composition of output, viscosity of output, and impurities in output. Examples of indirectly computed variables are number of feet of steel coil that is defect-free* and octane number of gasoline. *Economic variables* measure the factors affecting costs and profitability of the plant and are used only in economic models. Examples of economic variables are profit, cost of raw material, value of product, fuel cost, operating and maintenance personnel cost, and depreciation cost of plant and equipment.

C. Catalytic Reforming Process Example

A small portion of a *catalytic reforming process* used in making gasoline will be used to illustrate the overall model of a plant in operation. Figure 2.6-2 shows the reformer charge tank and feed pump, which blends several streams of raw material and supplies the reformer with a feed stream of proper chemical characteristics and pressure. The classes of variables in this example are

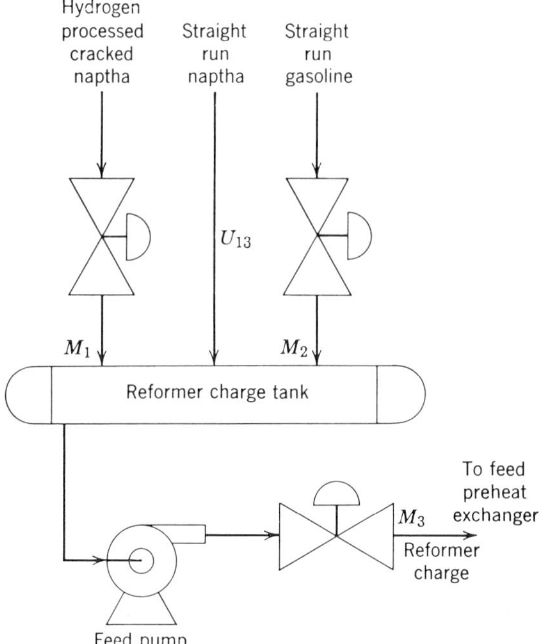

Fig. 2.6-2   Reformer charge tank and feed pump—portion of catalytic reforming process for making gasoline.

* Computed from the simultaneous absence of any defects.

Control Variables

M_1 = flow rate of hydrogen-processed cracked naphtha into reformer charge tank, in pounds per hour.

M_2 = flow rate of straight-run gasoline into reformer charge tank, in pounds per hour.

M_3 = flow rate of reformer charge from feed pump, in pounds per hour.

M_4 = power input to feed pump, in kilowatts.

Parameters

V_1 = average value of naphthene content in hydrogen-processed cracked naphtha, in percent.

V_2 = average value of aromatics content in hydrogen-processed cracked naphtha, in percent.

V_3 = average value of naphthene content in straight-run naphtha, in percent.

V_4 = average value of aromatics content in straight-run naphtha, in percent.

V_5 = average value of naphthene content in straight-run gasoline, in percent.

V_6 = average value of aromatics content in straight-run gasoline, in percent.

V_7 = average ASTM Engler 10% point of hydrogen-processed cracked naphtha.

V_8 = average ASTM Engler 10% point of straight-run naphtha.

V_9 = average ASTM Engler 10% point of straight-run gasoline.

V_{10} = average ASTM Engler 90% point of hydrogen-processed cracked naphtha.

V_{11} = average ASTM Engler 90% point of straight-run naphtha.

V_{12} = average ASTM Engler 90% point of straight-run gasoline.

V_{13} = cost of hydrogen-processed cracked naphtha, in dollars per pound.

V_{14} = cost of straight-run naphtha, in dollars per pound.

V_{15} = cost of straight-run gasoline, in dollars per pound.

V_{16} = cost of electric power, in dollars per kilowatt-hour.

Uncontrolled Variables

U_1 through U_{12} = deviations from the average values of parameters corresponding to V_1 through V_{12}.

U_{13} = flow rate of straight-run naphtha into reformer charge tank, in pounds per hour.

Procedural Actions

P_1 = adjust valve controlling M_1.

P_2 = adjust valve controlling M_2.

P_3 = adjust valve controlling M_3.

P_4 = adjust feed pump controlling S_3.

State Variables

S_1 = liquid level in reformer charge tank, in feet.
S_2 = ASTM Engler 90% point of reformer charge.
S_3 = reformer charge feed pump discharge pressure, in pounds per square inch.
S_4 = ASTM Engler 10% point of reformer charge.
S_5 = naphthene content of reformer charge, in percent.
S_6 = aromatic content of reformer charge, in percent.

Production Variables

M_3 = flow rate of reformer charge, in pounds per hour.

Quality Variables

$Q_1 = S_5$ = naphthene content in reformer charge, in percent.
$Q_2 = S_6$ = aromatic content in reformer charge, in percent.
$Q_3 = S_2$ = ASTM Engler 90% point of reformer charge.
$Q_4 = S_4$ = ASTM Engler 10% point of reformer charge.

Economic Variables

$E_1 = V_{13}$ = cost of hydrogen-processed cracked naphtha, in dollars per pound.
$E_2 = V_{14}$ = cost of straight-run naphtha, in dollars per pound.
$E_3 = V_{15}$ = cost of straight-run gasoline, in dollars per pound.
$E_4 = V_{16}$ = cost of electrical power, in dollars per kilowatt-hour.
E_5 = total cost of raw materials used, in dollars per hour.
E_6 = cost of power, in dollars per hour.
E_7 = total operating cost, in dollars per hour.

The *functional model* of the process is shown in Figure 2.6-2. The function of this process is to provide a steady but adjustable flow of feed material to the reformer having the desired chemical and physical characteristics.

The *procedural model* of the process consists of the following action statements:

1. Maintain the value of the hydrogen-processed cracked naphtha flow rate M_1 so that $M_1 = K_1 M_3$ where $K_1 < 1$.

2. Maintain the value of the straight-run gasoline flow rate M_2 so that S_1 remains within high-low limits. S_1 is the liquid level of the reformer charge tank.

3. Maintain the value of the reformer charge flow rate M_3 to obtain the desired throughput.

4. Initiate plant shut-down if the reformer charge flow rate M_3 falls below a specified limit.

5. If the liquid level of the reformer charge tank S_1 falls below a specified

limit, adjust the valves controlling M_1 and M_3 so that the new feed rate of straight-run gasoline M_2 will not cause the liquid level of the reformer charge tank to fall below the specified limit.

6. If the reformer charge feed pump discharge pressure S_3 falls below a specified limit, check to determine possible causes. If checking indicates that the feed pump has failed or a serious system leakage has occurred, initiate a plant shut-down.

The *physical process model* is given by the following set of equations:

$$\frac{dS_1}{dt} = \{M_1 + M_2 + U_{13} - M_3\}\{K_2(S_1)\} \tag{2.6-1}$$

$$S_2 = F(V_{10}, V_{11}, V_{12}, M_1, M_2, U_{13}) \tag{2.6-2}$$

$$S_3 = G(M_3, M_4) \tag{2.6-3}$$

$$S_4 = H(V_7, V_8, V_9, M_1, M_2, U_{13}) \tag{2.6-4}$$

$$S_5 = J(V_1, V_3, V_5, U_1, U_3, U_5, M_1, M_2, U_{13}) \tag{2.6-5}$$

$$S_6 = L(V_2, V_4, V_6, U_2, U_4, U_6, M_1, M_2, U_{13}) \tag{2.6-6}$$

$K_2(S_1)$ is a number relating the "net flow rate into the charge tank" to the "rate of change of the charge tank level," which is the ratio of the change in the tank level ΔS_1 to the corresponding change of the weight of the materials in the tank.

The *operating economic model* consists of the objective function to be optimized and the constraints. The objective function to be minimized is the cost of raw materials plus the cost of electric power:

$$E_7 = E_5 + E_6 \tag{2.6-7}$$

$$E_7 = (V_{14})(U_{13}) + (V_{13})(M_1) + (V_{15})(M_2) + (V_{16})(M_4) \tag{2.6-8}$$

subject to the constraints

$$0 \leq M_1 \leq K_3 \tag{2.6-9}$$

$$0 \leq M_2 \leq K_4 \tag{2.6-10}$$

$$K_5 \leq M_3 \tag{2.6-11}$$

$$R_{1L} \leq S_1 \leq R_{1U} \tag{2.6-12}$$

$$R_{3L} \leq S_3 \tag{2.6-13}$$

$$R_{2L} \leq S_2 \leq R_{2U} \tag{2.6-14}$$

$$R_{4L} \leq S_4 \leq R_{4U} \tag{2.6-15}$$

$$R_{5L} \leq S_5 \leq R_{5U} \tag{2.6-16}$$

$$R_{6L} \leq S_6 \leq R_{6U} \tag{2.6-17}$$

Equations 2.6-9 through 2.6-13 are contraints on state and control variables. Equations 2.6-14 through 2.6-17 are blending constraints that specify the quality variables of the reformer charge.

2.7   SUMMARY

At this point the reader should have a basic understanding of what models are and how they are used. By generalizing modeling to include phenomena that can be described verbally in English and by pictorial expressions it is hoped that some of the mysteries of modeling may be removed. The concepts of functional, physical, economic, and procedural models were introduced to provide the reader with a broader viewpoint of modeling. Standard engineering and operations research treatment frequently ignores functional and procedural models, with the regrettable result that a wide gulf exists between the plant operating engineers and the system design engineers.

Chapters 3 through 5 will expand the treatment of modeling by providing the basic concepts of functional, physical, and economic models, and by discussing the basic laws used for developing these models. Chapter 3 will discuss the functional models of the computer process control system, including software and hardware models. Chapter 4 will deal with the use of analytical and experimental techniques to develop physical models—steady-state and dynamic. Chapter 5 will enlarge on the basic concepts of planning and economic models and their applications to process automation.

3

Functional Models of the Computer Process Control System

3.0 INTRODUCTION

The two previous chapters have introduced the overall view of the automated process and the basic concepts of modeling the process. This chapter will turn from the process and focus attention on the computer process control system.* Basic concepts of *what the process computer* can do and how to use it* will be emphasized rather than *how it is designed and how it operates*.

The modeling concepts discussed in Chapter 2 will be used to develop *an overall mental image of the process computer*. Sufficient details will be presented to show how the basic functions are performed, and how the computer interfaces with the process itself. A simplified functional computer system will be used to present the essential functions while omitting many of the complicated design details. The proper understanding of the process computer functions is essential to the discussions of analysis and synthesis of optimization and control schemes presented in Chapters 6 through 10.

Specific models of the input-output system and the software system will be used to illustrate how the process computer is actually applied. Particular emphasis will be given to the interface of the process computer with its external environment and to the functions of a " real-time " operating system. A discussion of the commonly used process computer functions and the special function of direct digital control will conclude the chapter.

---

* The terms "computer process control system," "digital control computer," and "process computer" will be used interchangeably.

3.1   OVERALL VIEW OF THE COMPUTER PROCESS CONTROL SYSTEM

Two approaches will be used to present the overall view of the process computer. The first is the *horizontal view*, where the process computer is shown as organized into various major functions, i.e., *storage, control, arithmetic*, and *input-output*. The second is the *vertical view*, where the process computer is shown as a hierarchy of software and hardware, i.e., *compilers, assembly systems, hardware systems*, and *logic*.

Figure 3.1-1 shows the horizontal view of the process computer, and its relationship with the process, sensors, and controllers. The *storage section* of the computer is the heart of the entire system. It stores the instructions to be executed, data, tables, constants, and results. The storage section also provides working storage for input-output operations. The main functions of the storage section are the following:

1. Addressing: the ability to select one location uniquely by an address.
2. Reading: the ability to retrieve the content of any location selected.
3. Writing: the ability to store new information into any location selected. The information stored will remain in that location until new information is put in.

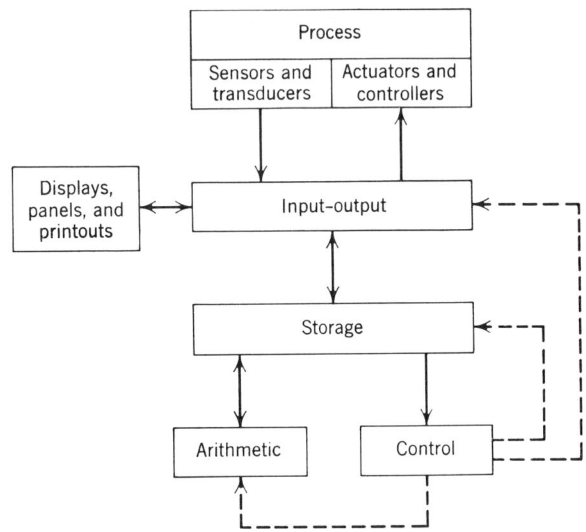

Note: Solid line is for data flow and dashed line is for control signals.

Fig. 3.1-1   Horizontal view of the process computer.

The *control section* of the computer directs the operation of the entire computer system by endlessly repeating the following steps:

1. Fetch an instruction from storage.
2. Decode the instruction fetched.
3. Execute the instruction decoded.
4. Determine the location of the next instruction to be fetched.

The *arithmetic section* performs the arithmetic and logic operations. Logic or decision operations may alter the sequence of instruction execution according to the actual values of the numbers tested or compared. The *input-output section* communicates between the computer system and the external environment by performing the following functions:

1. Select the required input-output device.
2. Select the area of storage for input-output operations.
3. Decode and encode for the particular input-output device, e.g., five-level paper tape code, punched card code, and binary coded decimal.
4. Provide buffer storage to synchronize the different data rates between storage and the input-output devices.
5. Check for errors and unusual conditions, e.g., end of a tape, double punch or blank column of a punch card, and open thermocouple.
6. Provide amplification, signal conditioning, and filtering of input signals.

Figure 3.1-2 shows the vertical view of the process computer, which is made up of a hierarchy of software and hardware. Hardware is equipment; software is programming or programming aids designed to simplify the task of programming. The hardware portion of the model is organized into four basic levels: *circuit components, logic elements, logic blocks,* and *the hardware system as seen by the user.* The lowest level of hardware is made up of tens of thousands of *circuit components,* e.g., integrated circuits, transistors, capacitors, resistors, magnetic cores, diodes, and relays. The language used in this level is wiring and circuit diagrams, and the actual computational work is done by the controlled presence or absence of electric pulses at these components with reference to time. *Logic elements,* the next level, are made up of a collection of circuit components wired together to serve specific logical functions, e.g., gates, flip-flops, and delay lines. The languages of logic elements are Boolean algebra, logic flow charts, or truth tables. The next higher level, *logic blocks,* uses the same languages as logic elements. *Logic blocks* are groupings of logic elements organized into frequently used functions, e.g., *registers, shift registers, decoding matrices,* and *adders.* The final level is the *hardware system as seen by the user,* i.e., the control, arithmetic, storage, and input-output. The hardware system is made up of logic blocks and other special devices.* The language of the computer hardware system is

---

* Typewriters, magnetic tape-drives, display panels, etc.

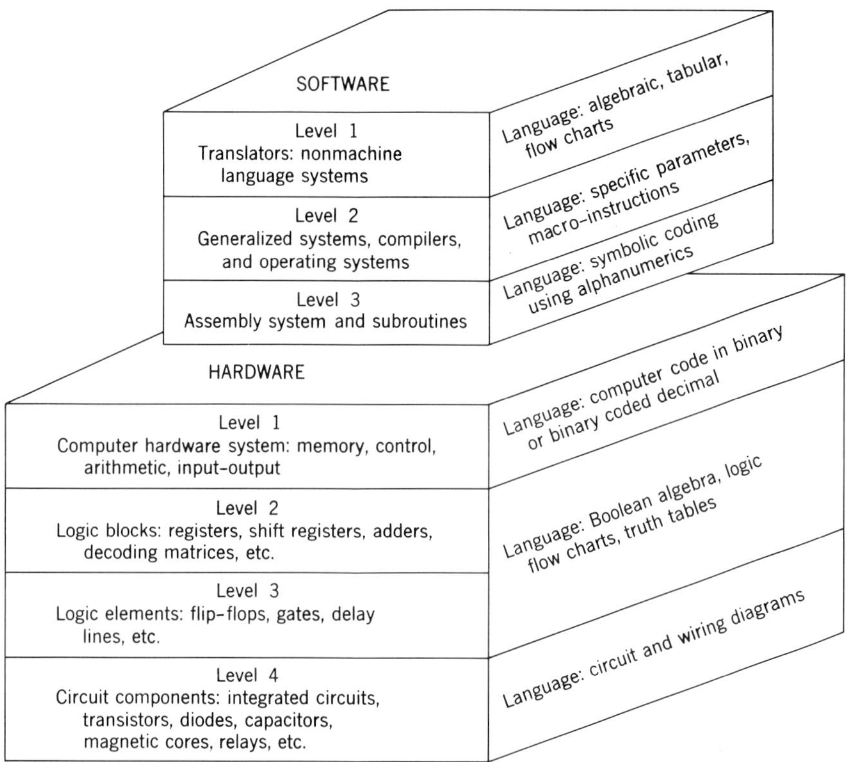

Fig. 3.1-2   Vertical view of the process computer.

the *code*, which is a language in code form used to instruct the computer system in performing its functions.

The machine language code is usually in binary or binary coded decimal and usually has the following format:

Instruction location.
Operation code.
Operand address.
Address modification (indexing, indirect and relative addressing).

Machine language coding is inefficient for direct use because of the tremendous amount of details required. The software systems are added to the hardware system to reduce the cost of programming by bringing the language closer to the natural language of the problem, i.e., mathematics and flow charts.

The software portion of the vertical view is organized into three basic levels:

Assembly system and subroutine library.
Generalized systems, compilers, and operating systems.
Translators.

The lowest level of the software system consists of the *assembly system and subroutine library*. The format of the assembly system coding is similar to the format of the machine language coding; however, it offers the following advantages:

1. It allows alphanumeric instead of binary data to be entered.
2. It allows an alphanumeric operation code to be used.
3. It allows relative and symbolic addresses to be used instead of fixed addresses and locations. This makes it possible for programs to be relocated and corrections and insertions to be made in the program with relative ease.

Subroutines are commonly used programs which are designed for easy re-use within any part of the program. Parameters entered can make general purpose subroutines specific for each use. A library of subroutines can further reduce the effort in " calling " the subroutines by standardizing the names and calling identification of the subroutines. The next level of software system consists of *generalized systems, compilers,* and *operating systems*. As in the case of hardware, this level of software has a structure of the lower level of software. *Generalized systems* are extensions of the subroutine technique to include reusable overall systems, e.g., data logging, input-output routines, matrix operations, and regression analysis. Specification of the required parameters is all that is required to make a generalized system fit a specific job. *Compilers* use pseudo- or macro-instructions to call subroutines automatically or have routines generated during program assembly. Thus the compiler technique uses software to make up for the hardware deficiency, i.e., the lack of complex operation codes built into the computer hardware. The language of compilers consists of the macro-instructions plus the assembly language. *Operating systems* are generalized programs used to control the sequence of execution of programs with provision for operator control, intervention, and override. The language of operating systems consists of the control parameters which the operator uses to set up and direct the sequence of program execution. By taking over the many housekeeping and dispatching chores from the programmer, the real-time operating system reduces the complexity of programming the process computer to the level of effort required for off-line engineering computations. The highest level of software system is the *translator level*, which frees the user from the restrictive format of the basic computer language. The language of the problem solutions and of the functions to be performed may be used at this level, e.g., algebraic expressions, flow charts, and decision tables. The software system translates this into an assembly system language or directly into machine language coding.

The implication of the vertical view of the computer is that there is a great deal of flexibility as to where hardware ends and software begins. For example, hardware macro-instructions may be considered as performing functions normally done by software; conversely, by allowing the user to have access into the subfunctions that comprise an instruction (i.e., microcoding), software may be considered as performing functions normally done by hardware. Thus, software and hardware systems must be designed as an entity to serve a given class of applications in the most effective and economical manner.

3.2   RESOURCE MODELS OF SOFTWARE

Software systems may be considered as a collection of programs, with each program performing a function and requiring certain resources. The resources required by software are *input-output devices, storage*, and *time used in execution*. The first two are analogous to capital investment, while time is expended like raw materials.* Simple resource models of programs will be introduced to acquaint the reader with some of the *cost aspects* of software, and the trade-off possibilities between the various types of resources.

A. Input-Output Devices

The first resource requirement of a program is *input-output devices*. Table 3.2-1 lists some typical input-output resource requirements. In practice, several versions of the same program may be required to serve users having different I/O resources. Translation of programs from one resource environment to another may be possible under certain circumstances.

Table 3.2-1   Input-Output Resource Requirements for
Typical Programs

Program 1	I/O Resource: None required
Program 2	I/O Resource: 100K drum storage
Program 3	I/O Resource: Paper tape reader/punch 500K disc storage
Program 4	I/O Resource: Typewriter Paper tape punch
Program 5	I/O Resource: Card reader/punch Printer 4 Magnetic tapes

* Input data is analogous to parameters rather than to raw materials in the resource models of software.

B. Storage

The second resource requirement of a program is *storage*. Storage is used for *instructions, data, constants,* and *working storage*. In most cases, programs are stored in contiguous blocks in storage. Figure 3.2-1 shows a linear array of 32,768 words of storage. The example shows the following assignment of storage:

Program	Storage Assignment	Storage Protect
a	2001–2900	Yes
b	2901–4000	No
c	4001–4896	Yes
i	$S_i - E_i$	—

Note that programs a and c are specified for *storage protect*, i.e., their contents cannot be changed by the program.

In special cases, a given program may consist of several contiguous parts separated by unrelated blocks of words. These blocks of words sandwiched between parts of a program are said to "interrupt" the normal contiguous storage allocation. Figure 3.2-2 shows the programs a, b, and c which are

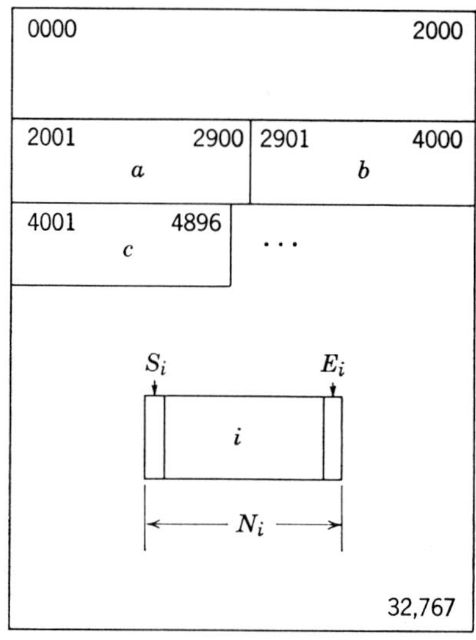

Fig. 3.2-1   Storage resource allocation using contiguous blocks for programs.

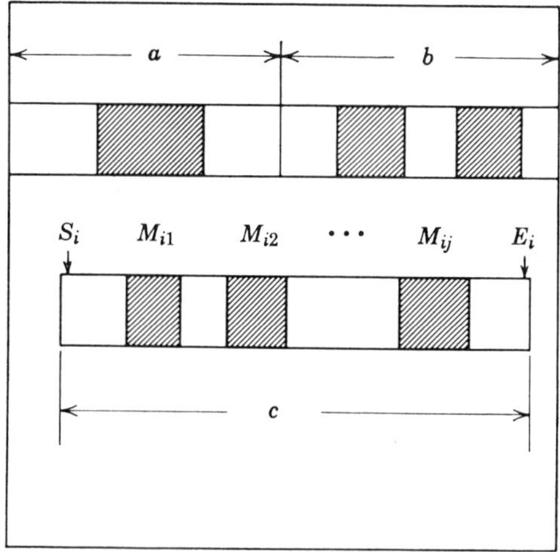

Fig. 3.2-2   Storage resource allocation using non-contiguous (interrupted) blocks for programs.

assigned blocks of storage that are interrupted by other programs, shown in the shaded areas. Thus program a is interrupted by one block, program b by two blocks, and program c by j blocks of storage.

Let the following symbols be assigned:

S_i = starting location of ith program.
N_i = number of words of storage used by ith program.
E_i = ending location of ith program.
T = total storage used by all the programs.
M_{ij} = number of words of the jth block interrupting the storage assignment of the ith program.

For the contiguous case the ending location and the total storage used are

$$E_i = S_i + (N_i - 1) \tag{3.2-1}$$

$$T = \sum_i N_i \tag{3.2-2}$$

and for the noncontiguous case the ending location is

$$E_i = S_i + (N_i - 1) + \sum_j M_{ij} \tag{3.2-3}$$

Equations 3.2-1 through 3.2-3 constitute the resource model of storage requirements.

C. Time Used in Execution

The third resource requirement of a program is the *time used in execution.*
The time used by a program is made up of fixed and variable portions. The
variable portion is dependent on the type of data and amount of data that
the program encounters each time it is executed. The data affects the time
used by a program in at least four different ways.

1. Size and values of numbers affecting arithmetic operating times.
2. Logical branching depending on the data, including the various exits
possible.
3. Latency and transfer time required for input-output equipment to
search and transfer data to and from storage, e.g., disc storage.
4. Error conditions that may cause operations to be repeated, error mes-
sages typed out, operator intervention, etc.

Again, the execution of a specific program may proceed without interruption, or it may be interrupted by another program in the midst of execution.
Figure 3.2-3 shows a sequence of programs being executed without interruption. In other words, the programs $a, b, c, \ldots, i$ are each executed to
completion before another program is started. Since no interruptions are
allowed, each program should be written so as not to exceed a certain maximum time in order to allow a reasonable schedule of processing to take place.
Figure 3.2-4 shows a sequence of programs which may be interrupted by other
programs. Program a is interrupted by program b after a_1 is completed. After
program b is completed, then a_2 is processed. In the general case, program
i is interrupted by program segments $P_{i1}(t), P_{i2}(t), \ldots, P_{ij}(t)$. The beginning
and ending times of the program, $B_i(t)$ and $F_i(t)$, are functions of time because
the times at which the program begin and end are dependent on *real-time* or
the *time of day.* Let the following symbols be assigned:

$t =$ time of day or real-time.
$B_i(t) =$ time at which the ith program begins.
$F_i(t) =$ time at which the ith program finishes.
$T_i(t) =$ time required for executing the ith program.
$G(t) =$ total time required for executing all the programs.
$D_{i1}, D_{i2}, \ldots, D_{ik} =$ data for the ith program.
$M_i =$ fixed execution time for the ith program.
$V_i =$ variable execution time for the ith program.
$P_{ij}(t) =$ the duration of the jth interrupt of the ith program.

For the uninterrupted case the time required is

$$T_i(t) = M_i + V_i(D_{i1}, D_{i2}, \ldots, D_{ik}, t) \tag{3.2-4}$$

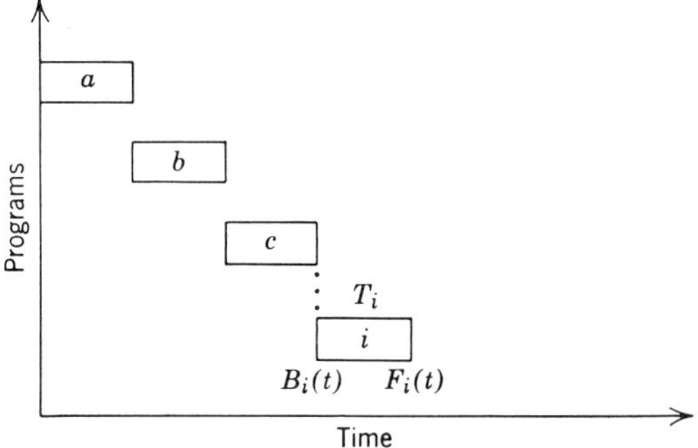

Fig. 3.2-3   Program execution without time interrupt.

V_i is a function of time t because the *time of day* or *real-time*, is also considered as the independent variable. The time at which the program finishes is

$$F_i(t) = B_i(t) + M_i + V_i(D_{i1}, D_{i2}, \ldots, D_{ik}, t) \tag{3.2-5}$$

The total time required for all the programs is

$$G(t) = \sum_i T_i(t) \tag{3.2-6}$$

For the case where interrupt is allowed, the time required is

$$T_i(t) = M_i + V_i(D_{i1}, D_{i2}, \ldots, D_{ik}, t) + \sum_j P_{ij}(t) \tag{3.2-7}$$

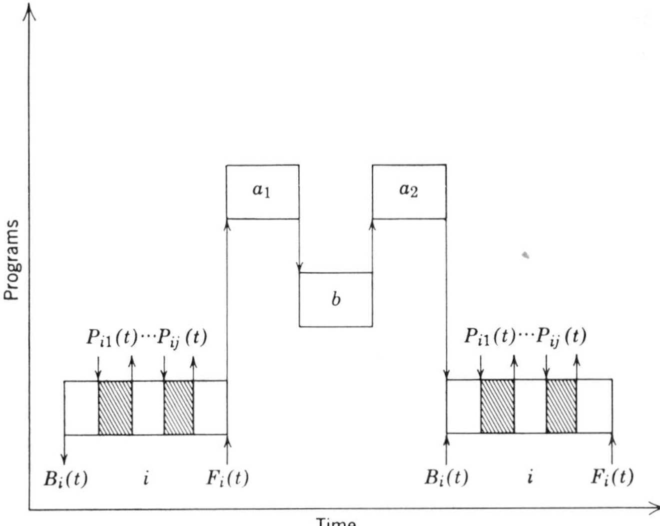

Fig. 3.2-4   Program execution with time interrupt.

The time resource model of a program is given by Equations 3.2-4 through 3.2-7.

The various resources discussed in this section may be traded-off to fit the particular requirements and constraints. Let the following symbols be assigned:

H_i = input-output resource required by the ith program.
N_i = number of words of storage used by the ith program.
T_i = time required for executing the ith program.

The trade-off relationship is expressed by the function Q

$$T_i = Q(H_i, N_i) \tag{3.2-8}$$

Typical forms of the general relationships of Equation 3.2-8 are shown in Figure 3.2-5. The trade-off relationship shows that as input-output resources H are increased, both the execution time T and memory requirements N can be decreased. For example, the availability of a drum can reduce the core storage required, increasing the number of tapes can speed up sorting time, and additional channels of input-output will allow more overlapping of input-output operations and thus a decrease in execution time. Given the same input-output resources, the speed of execution is determined by the number of words of storage allowed; for example, adding a hundred numbers by "straight line" coding will take the least amount of execution time but will require a great deal more storage. A "loop," where the same instructions are used over again after modification, however, will use less storage but take longer to execute.

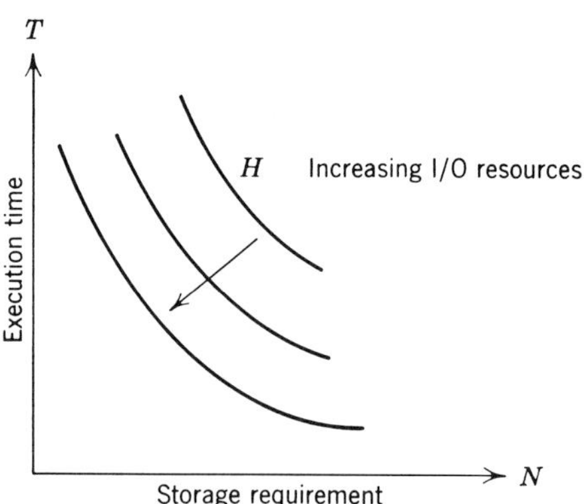

Fig. 3.2-5   Trade-off relationship between the various resources required by a program.

3.3 FUNCTIONAL MODELS OF THE REAL-TIME OPERATING SYSTEM

The function of the real-time operating system is to allocate and schedule computer resources to the programs that may be simultaneously demanding service. The computer resources for software, as discussed in Section 3.2, consist of input-output devices, storage, and time used for execution. The role of the real-time operating system is similar to that of the air-traffic controller of an airport, or the dispatcher of a manufacturing operation. Real-time operating systems are also known as executive control programs, real-time monitors, and real-time supervisors. This section will briefly describe the functional model of the real-time operating system via a discussion of the input-stimuli and the outputs, the place of the real-time operating system in the overall programming structure, and some of the subfunctions of the real-time operating system.

A. Input-Stimuli to the Real-Time Operating System

Figure 3.3-1 shows some of the major types of input-stimuli to the real-time operating system. Ⓐ shows the real-time clock and cycle timer which triggers the cyclic programs, e.g., once per 10 milliseconds and once per second. The real-time clock can also be used to measure fixed elapsed time, unless Ⓑ elapsed-time counters (which may be software) are available. Ⓒ shows digital accumulators (which may also be software) which will indicate when they are full. Ⓓ shows that alarms of analog and digital state variables and other unusual conditions may signal that diagnostic or corrective actions are required. Ⓔ shows panel switches and on-demand buttons indicating that manual intervention or keyboard inputs are required. On-demand functions include entry of process and economic parameters, customer identification, inquiries, and initiation of procedural actions. Ⓕ shows signals from input-output devices which may signal that the device is either free, busy, or has just completed the last assignment. Ⓖ shows signals from other computers indicating that the other computer is either free, busy doing data transmission, or has just completed the data transmission job. Ⓗ shows that free-time functions, such as on-line compiling, program debugging, and hardware diagnostic are required. In summary the input-stimuli may be classified into the following types:

Time-initiated:	Ⓐ and Ⓑ
Process-initiated:	Ⓒ and Ⓓ
Operator-initiated:	Ⓔ
Computer- or program-initiated:	Ⓕ, Ⓖ, and Ⓗ

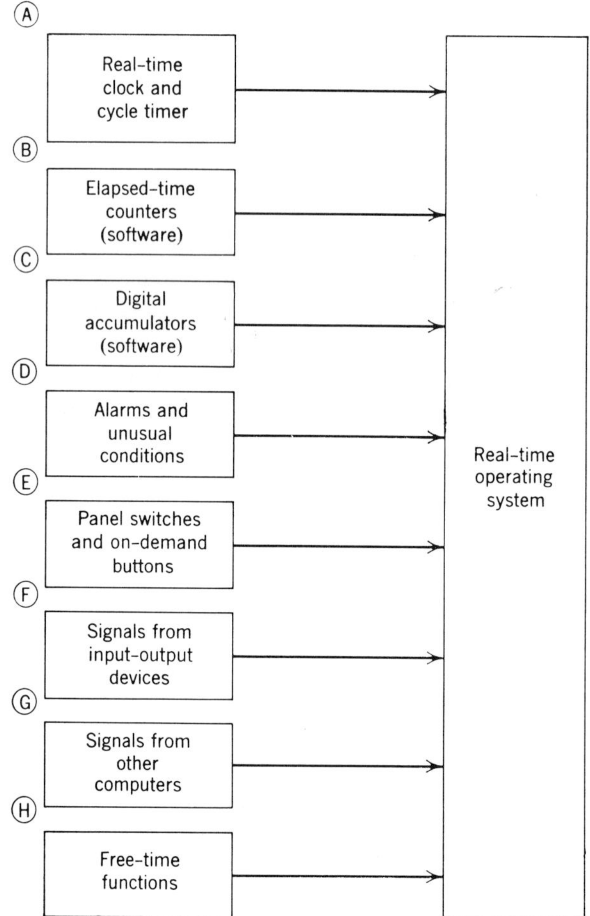

Fig. 3.3-1   Input-stimuli to the real-time operating system.

B. Outputs of the Real-Time Operating System

Figure 3.3-2 shows the outputs of the real-time operating system. Ⓐ shows the assignment of the start-times of each program requesting service. A simplified dispatching rule is to begin every cycle with the cyclic programs. After the cyclic programs are completed, the real-time operating system would assign the sequence of programs to be executed according to priority. Thus the assignment of the start-times is made indirectly, i.e., as a result of the sequence assignment. Ⓑ shows the schedule and assignment of storage required for each program to be executed. Portions of the main storage may have to be saved in the auxiliary storage to make room for the incoming programs. Also,

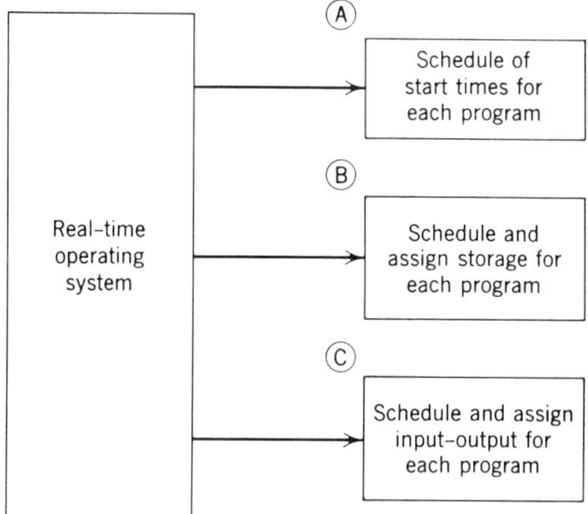

Fig. 3.3-2   Outputs of the real-time operating system.

portions of the incoming program may have to be brought in from auxiliary storage. Ⓒ shows the schedule and assignment of input-output devices required for each program to be executed. If a required input-output device is unavailable for a specific program, an alternative device may have to be made available, or the operator may have to be notified that the particular program cannot be executed because the input-output device is unavailable.

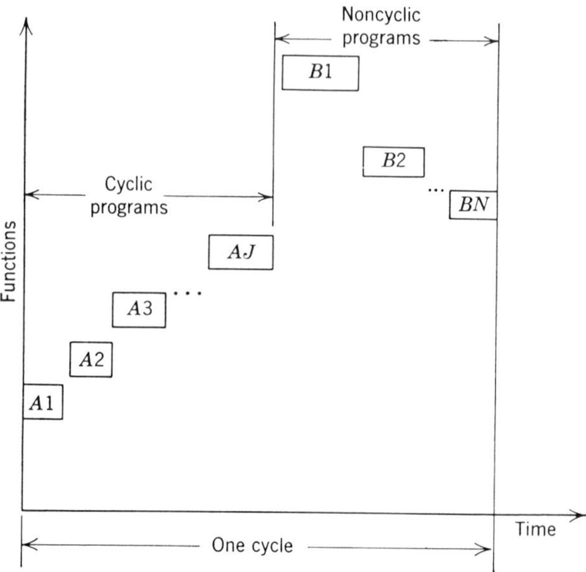

Fig. 3.3-3   Assignment of program execution sequence by the real-time operating system.

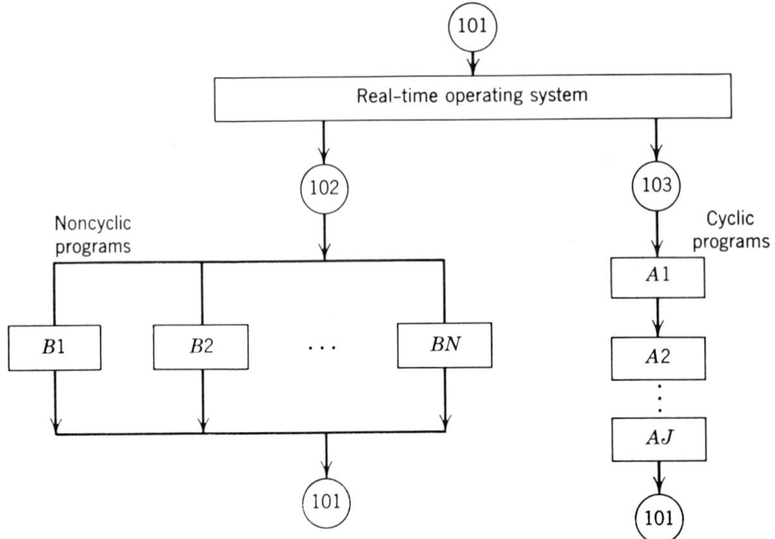

Fig. 3.3-4   Overall structure of programs for computer process control.

Figure 3.3-3 shows a typical schedule of start-time assignments. The cyclic programs are $A1$, $A2$, ..., AJ, and the noncyclic programs are $B1$, $B2$, ..., BN. The noncyclic programs are random in nature and usually activate the priority interrupt system. Figure 3.3-4 shows the overall structure of the programs for computer process control. The real-time operating system is always in control as it directs the start, interruption, resumption, and completion of the various programs.

C. Subfunctions of the Real-Time Operating System

Having established the input-stimuli and the outputs of the real-time operating system, it is merely a small step to establishing the subfunctions of the real-time operating system. These are as follows:

1. Receive an input-stimuli.
2. Remember the status that an input-stimuli has been received. The status will be removed when the corresponding program has been completed.
3. Decode the implied or built-in priority of each input-stimuli.
4. Rank the various priorities of the input-stimuli and select the highest.
5. For each program requesting service, explode into the required computer resources of storage and input-output devices.
6. Check to see that the required computer resources are available. If not, look for alternative resources. Notify the operator in case alternative resources are not available.

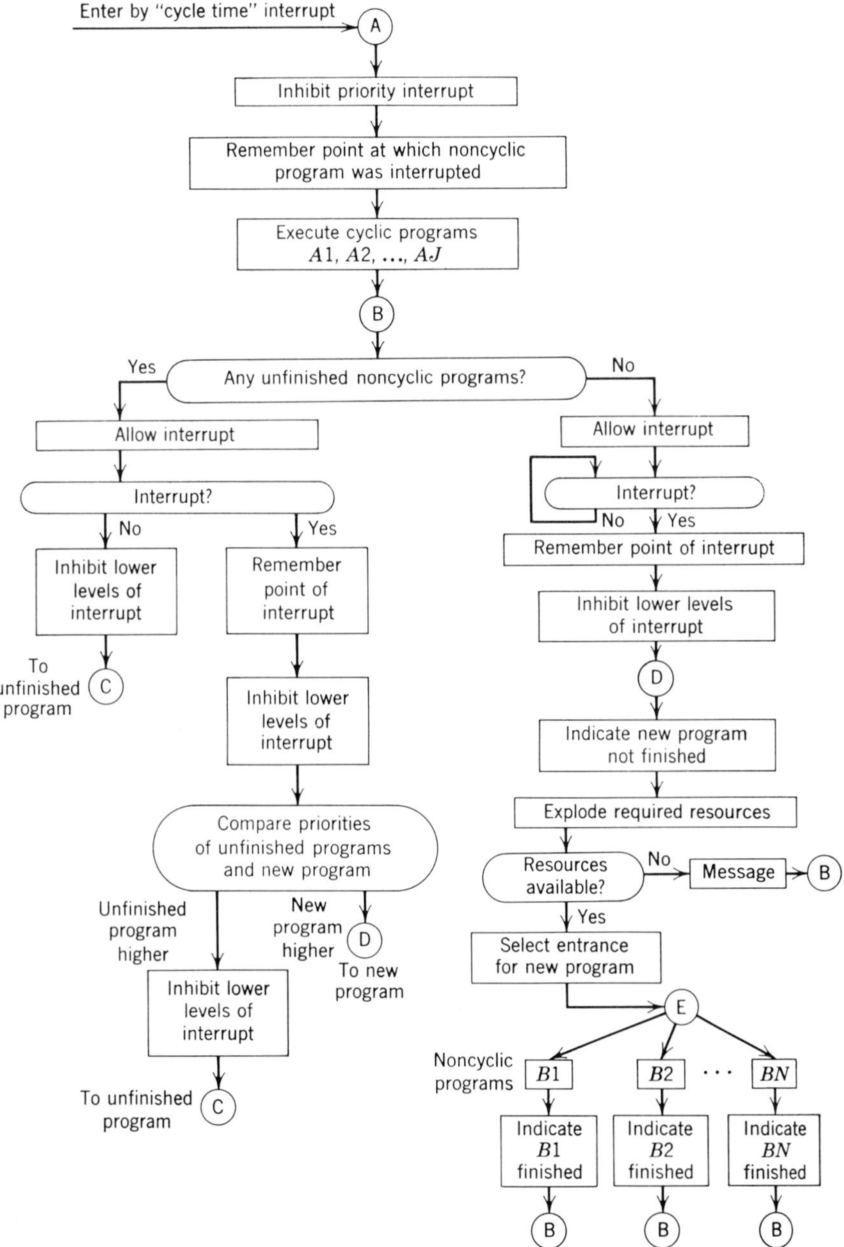

Fig. 3.3-5 Flow chart of real-time operating system.

7. Inhibit interruption for a period so that cyclic programs can be executed.
8. Allow input-stimuli to be received.
9. Remember the status of the real-time operating system.
10. Select the particular program to be executed next.

The above subfunctions do not necessarily occur in sequence.

The actual implementation of the real-time operating system can be greatly simplified through the *priority interrupt* hardware feature built into process computers. This feature allows the various input-stimuli to have preassigned priorities. When priority interrupt is not inhibited, an input-stimuli will automatically cause the sequence of program execution to be interrupted and then begin to execute the program corresponding to that particular interrupt.

Figure 3.3-5 shows a simplified flow chart of the real-time operating system which performs the subfunctions outlined in 1 to 10 above. Ⓐ is entered at the beginning of every cycle. The entrance is initiated by the real-time clock or the cycle timer. Immediately after Ⓐ, all further priority interrupts are inhibited so the cyclic functions may be assured of execution. Next, the point where a noncyclic program was interrupted by the cycle timer has to be remembered so that the program may be resumed later. Then the cyclic programs $A1$, $A2$, ..., AJ are executed. At Ⓑ the operating system is ready to execute the noncyclic programs. However, a test must be made to determine if there are any unfinished noncyclic programs. If there are unfinished programs with a higher priority than the new program, the unfinished programs must be executed first Ⓒ. If the unfinished programs have a lower priority than the new program, the new program is executed Ⓓ.

At Ⓓ, the real-time operating system remembers that the new program is to begin execution but is not finished. Then the required computer resources for the new program are exploded and a message is given if these are not available. Otherwise the selected program is entered at Ⓔ. If the program should be executed to completion, an indication is made that the program is finished. If it should be interrupted before completion, either by a higher priority noncyclic program or the cycle timer, the real-time operating system will return to the program at subsequent intervals. Once a new program is called, the required resources will not be relinquished until it is finished.

3.4   FUNCTIONAL MODELS OF THE INPUT-OUTPUT SYSTEM

This section will discuss functional models of the input-output systems that link the process sensors and controllers with the computer. Sensors and contact closures, which measure the *state variables*, *quality variables*, and *production variables,* are brought into the process computer by the *input system.* The *output system* sends signals from the computer to the controllers and actuators to initiate *procedural actions* and adjust the *control variables.*

A. Functional Model of the Input System

Figure 3.4-1 shows the functional model of the input system which brings in analog signals from the process. Practical details such as noise filtering, power supply, etc., are omitted for simplicity. Ⓐ shows electrical sensors which provide d-c voltage as output, e.g., thermocouples, tachometers, and d-c voltage measurements. The use of the transmitter is optional. It is dictated by quality and strength of the sensor signal and the distance between the sensor and the process computer. Ⓑ shows sensors which do not provide d-c voltage as output, e.g., a-c current measurements, flows, and pressures. The output from this type of sensor must be *transduced* or converted into d-c voltages and then transmitted to the process computer. Examples of transducers and transmitters of this type are thermo-converters which convert a-c current and power into d-c voltages, and electronic instrumentation systems which measure and convert pressures, flows, etc., into d-c voltages. Ⓒ shows sensors which require two transducing operations. This is typical of pneumatic instrumentation systems, which transduce and transmit measurements

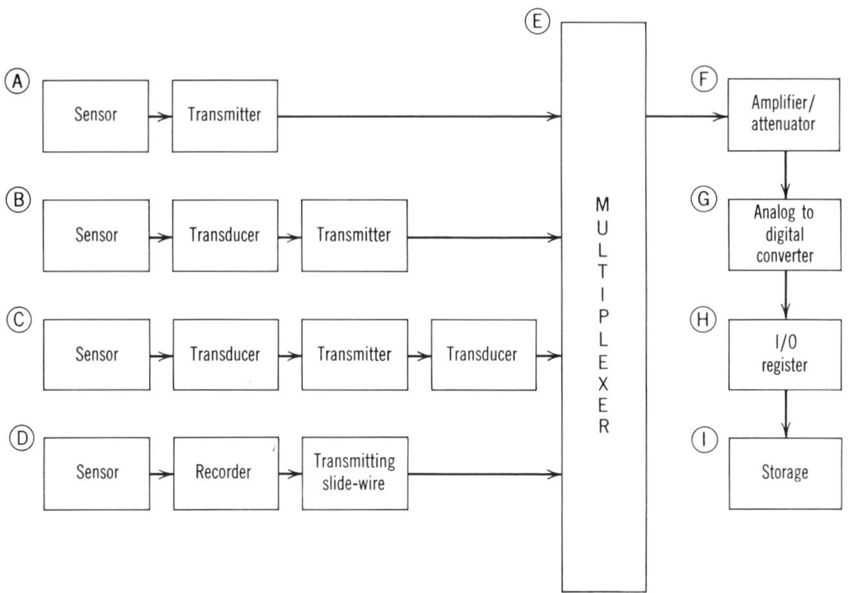

Fig. 3.4-1   Functional model of the analog input system.

of flows and pressures in the form of air pressure. A second transducer at the process computer site converts the pneumatic signal into d-c voltage for input into the computer. Ⓓ shows how existing recorders in the control room can be used for signal inputs to the process computer by using transmitting slide-wires.

Ⓔ shows the *multiplexer, scanner,* or *selector,* which connects the pair of selected signal lines to the process computer. The multiplexer allows the process computer to be "time-shared" by hundreds of pairs of signal wires. If the selection sequence is fixed, the device is sometimes referred to as a scanner or commutator. Random selection of the signal lines under program control allows much greater flexibility in planning the application. More than one multiplexer may be provided if the extra cost can be justified. Ⓕ is the *low-level amplifier* used to amplify signals having full-scale range of 10 to 50 millivolts to the one-volt level. *Attenuators* may also be included for this function to scale down higher voltages. The specific range of full-scale reading to be amplified or attenuated may be also under program control to provide more flexibility. Ⓖ is the *analog to digital converter* which converts the voltage output of the amplifier into a digital value stored in the input-output register Ⓗ. Finally, the digital number stored in the I/O register is transferred to Ⓘ, the storage location specified by the program.

Figure 3.4-2 shows the functional model of the input system which brings in digital process signals. Ⓐ and Ⓑ show digital sensors and other devices

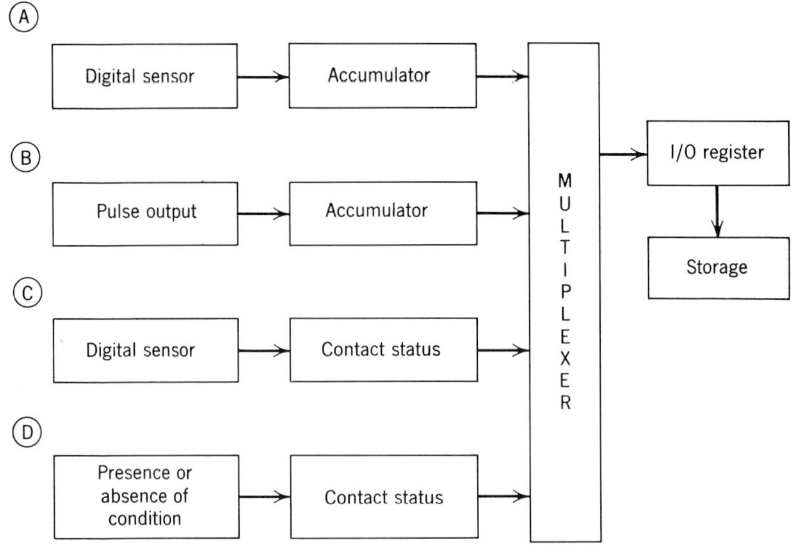

Fig. 3.4-2 Functional model of the digital input system.

which produce electrical pulses being accumulated before input to the process computer. Examples of digital sensors include turbine flow meters and kilowatt-hour meters. Examples of devices which produce electrical pulses include defect detectors for strip mill processes and signals used for tallying the movement of items in a process or a production line. Ⓒ shows the slower digital sensors which do not require external accumulators. Each time a pulse is generated, a normally opened contact will close and remain closed until interrogated by the process computer. The period between interrogation must not exceed the minimum interval between pulses. Ⓓ shows contact closures as inputs to the process computer. Actually, a voltage pulse is sent out by the process computer to interrogate the open-close status of each contact. The presence or absence of voltage at the multiplexer is read into the process computer and interpreted. Examples of contact closures include limit switches, liquid level limit indications, valves in full open or closed position, push buttons, and manual input switches.

B. Functional Model of the Output System

Figure 3.4-3 shows the functional model of an output system for indirect control of the control variables of the process, i.e., controlling the set point of the controllers which in turn adjust the actuators. The value of the set point is first sent from storage to the input-output register, and then to a selected output device through the multiplexer. Several types of output, both analog and digital, are shown in Figure 3.4-3. Ⓐ shows an analog output which uses a *servomotor* to drive the set point of the controller until the *controller reference signal* is equal to the analog voltage produced by the *digital-to-analog converter*. Ⓑ shows a digital output where the value of the number in the I/O register is converted to a corresponding number of pulses to drive a hysteresis motor which, in turn, adjusts the controller set point. The number of pulses produced by the *pulse train distributor* will match the value in the I/O register. Ⓒ shows another type of digital output where a pulse is produced whose duration corresponds to the value in the I/O register. The pulse is then used to drive a hysteresis motor. The pulse is produced by a *variable duration pulse generator*. Ⓓ is the simplest form of digital output because it consists of a computer-controlled contact closure which in turn produces a pulse of fixed duration. The fixed duration pulse is used to drive a hysteresis motor. The pulse is produced by the *fixed duration pulse generator*. Ⓔ shows *contact closure outputs* which may be used to turn equipment on or off, initiate procedural actions, and to light alarm and indicator lights.

Direct digital control, or subloop control by the process computer system, is shown in Figure 3.4-4. The procedural actions normally performed by the controller are done by the process computer system in direct digital

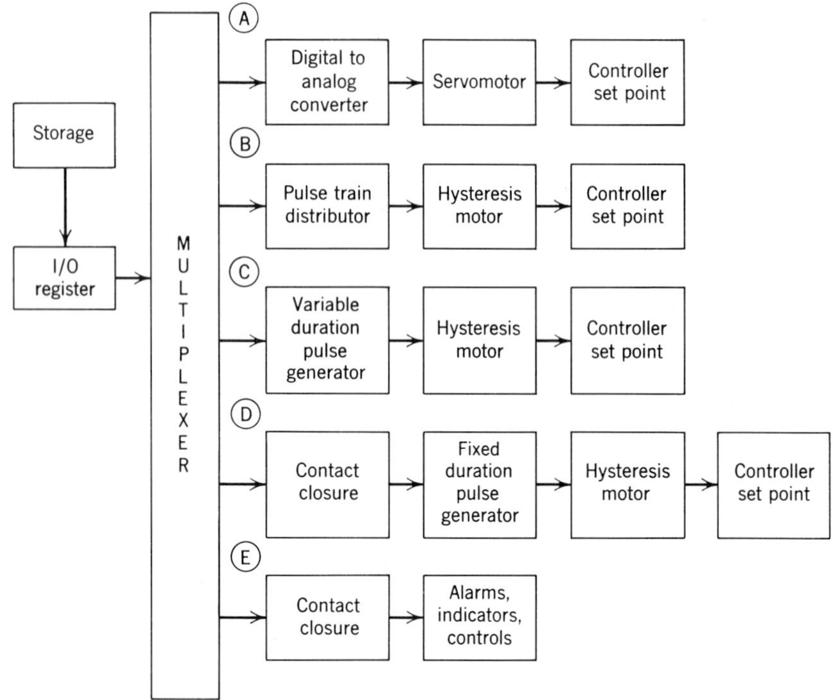

Fig. 3.4-3   Functional model of output system for indirect control.

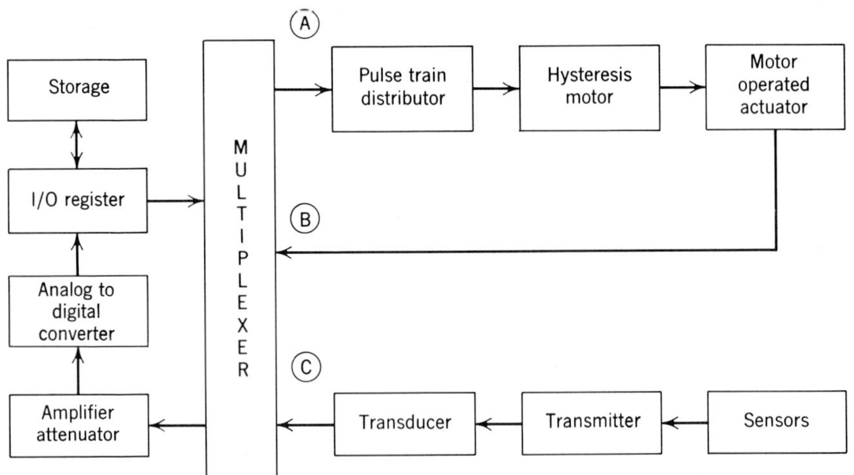

Fig. 3.4-4   Functional model of input-output system for direct digital control.

control. Ⓐ shows a hysteresis motor which directly drives a motor operated actuator. This type of output is generally preferred to the analog output because the actuator will remain in the last position in case of equipment failure, i.e., fail-safe. One of the two feedback signals required is the *actuator position* Ⓑ; the other is the *process variable* Ⓒ. Two feedback signals are provided to increase reliability and safety. It is always prudent to sense the position of the actuator Ⓑ before initiating actuator action Ⓐ.

C. Remotely Located Input-Output System

With the advance in the flexibility and reliability of communications equipment, it is possible to remotely locate the input-output system described in Sections 3.4A and 3.4B. In other words, the process computer and the input-output system may be separated in distance by several miles or several hundred miles. Figure 3.4-5 shows the functional model of a process computer Ⓐ controlling a remotely located input-output system Ⓑ via a communication network Ⓒ. The communication network may be ordinary voice-grade telephone circuits, privately leased wires, or microwave systems. The interface between the communication network and the process computer is called a modem Ⓓ. A modem is also required as an interface between the communication network and the input-output system.

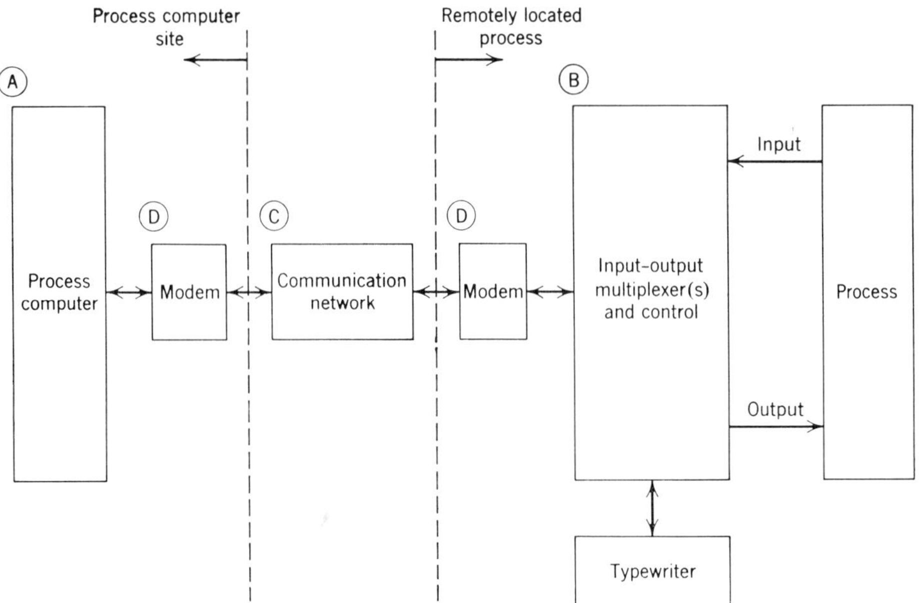

Fig. 3.4-5   Remotely located input-output systems.

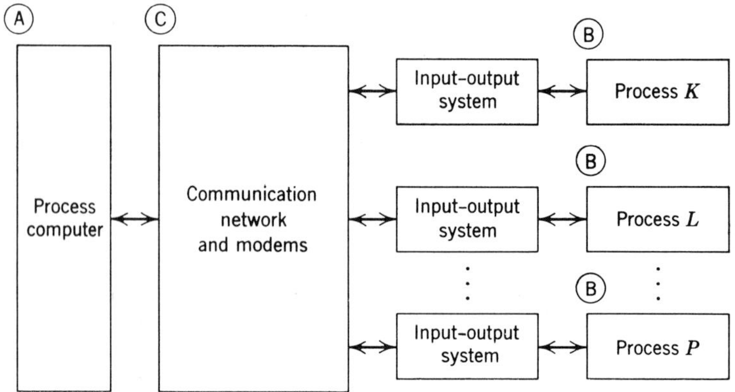

Fig. 3.4-6   One process computer controlling several remotely located processes.

Figure 3.4-6 shows a more general functional model where a single process computer system Ⓐ is controlling several remotely located processes Ⓑ through the use of data communication systems and modems Ⓒ. It is also possible to have a network of process computers, each controlling several remotely located processes with the additional capability of interchangeability so that each process computer may serve as back-up for the others in case of equipment failure. The back-up capability is especially important in direct digital control applications. Figure 3.4-7 shows that process computer Ⓐ normally controls the remotely located processes Ⓚ and Ⓛ (solid lines), and process computer Ⓑ normally controls the remotely located processes Ⓜ and Ⓝ (solid lines). However, in the event of emergency, process computer

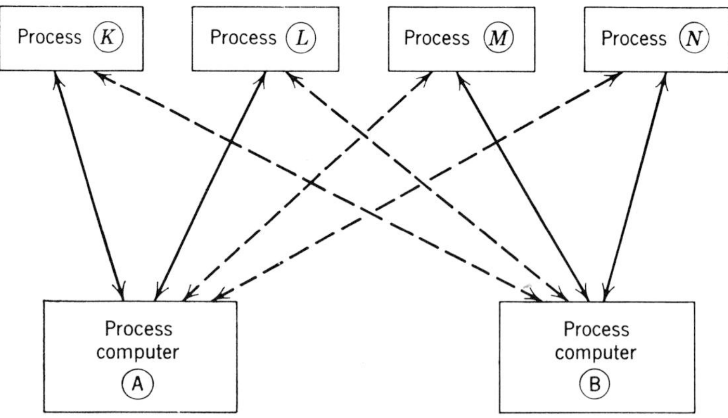

Fig. 3.4-7   Backup arrangement for process computer systems.

Ⓐ can also serve the *additional* function of controlling processes Ⓜ and Ⓝ (dashed lines), and process computer Ⓑ can be the *back-up* for processes Ⓚ and Ⓛ (dashed lines).

3.5 FUNCTIONS OF THE COMPUTER PROCESS CONTROL SYSTEM

In order to present the functions performed by the computer process control system, a set of basic functions has been selected from the infinite variety of functions that the computer process control system is capable of performing. These basic functions are as follows:

Analog and digital data acquisition.
Scaling and conversion of data.
Verification of data and reasonableness check.
Data accumulation.
Data formating and editing.
Visual displays.
Keyboard inputs.
Limit comparison and alarming.
Event and sequence recording and monitoring.
Trend recording and monitoring.
Data logging.
Computations.
Computer control actions.

In general, any application consists of a particular set of the above functions. This section will give a brief description and present the parameters for specifying some of these functions.

A. Analog and Digital Data Acquisition

The *state variables* of the process are represented by two types of data: analog and digital. Analog data generally can assume an infinite number of states, while digital data generally can have a finite number of states or two states in the binary form. Analog data refers to the numerical values of the process variables, e.g., temperature, pressure, and flow-rate. The largest class of digital data consists of yes-no indications of the conditions of specific equipment, e.g., whether the valve is closed and whether the pump is on. Other classes of digital data include the settings of the control switches, knobs and buttons of the operators panels, pulses from digital sensors, and digital data from the card reader and paper tape reader.

The analog and digital data acquisition function generally consists of the following operations:

1. Select the input point to be read.
2. Setting the scaling for the amplifier-attenuator.*
3. Selecting the storage location for input data.
4. Sampling the point and converting the reading into a digital value.*

Table 3.5-1 shows the parameters for specifying the analog data acquisition function with examples.

Table 3.5-1   Specifying the Analog Data Acquisition Function

Parameters	Example 1	Example 2	Example 3	Example 4
Point identification	241	576	825	390
Scaling	$\times 100$	$\times 1/10$	$\times 1$	$\times 10,000$
Memory location	2431	7508	1093	2047
Sampling frequency	1 sec	5 sec	1 min	Not periodic

Let the following symbols be assigned to develop the physical model relating the voltage output of the sensor and the corresponding digital value stored in memory:

$t = $ time.
$E(t) = $ voltage output of the sensor or transducer.
$V_i = $ value of the number in storage for the ith sample.
$t_i \ = $ time when the ith sample was taken.
$T \ = $ interval between the ith and the $(i + 1)$th sample.
$\alpha \ = $ time delay for the amplifier-attenuator and the analog to digital converter.
$\beta \ = $ time delay for transfer of data from the I/O register to the storage location.
$L(1) = $ time delay inherent in the process computer system.
$L(2) = $ time delay due to sampling.

Figure 3.5-1 shows the relationship between the sensor output $E(t)$ and the corresponding digital value of the number in storage, V_i, V_{i+1}, etc. $E(t)$ is a continuous function while V_i and V_{i+1}, etc., are step functions. The first time delay, $L(1)$, is inherent in the process computer system. The second time delay, $L(2)$, is determined by the sampling frequency. Thus we have the

* Some parts of these are omitted for digital data.

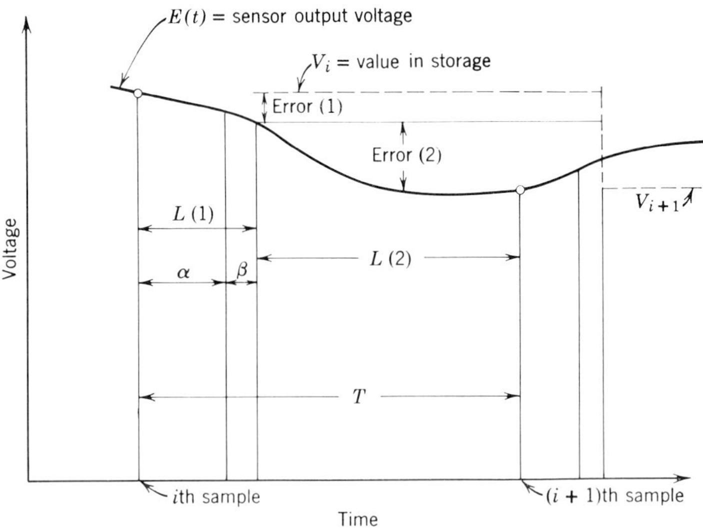

Fig. 3.5-1 Relationship between sensor output voltage and digital value in storage.

following relationships for time delays:

$$L(1) = \alpha + \beta \tag{3.5-1}$$

$$L(2) = T - (\alpha + \beta) \tag{3.5-2}$$

The error, or the difference between V_i and $E(t)$, may be analyzed at the two time delays. Thus

$$\text{Error (1)} = V_i - E(t_i + \alpha + \beta) \tag{3.5-3}$$

Error (1) is the difference between V_i and $E(t)$ at the time when the data was stored in memory. Similarly,

$$\text{Error (2)} = V_i - \text{Error (1)} - E(t) \tag{3.5-4}$$

where

$$(t_i + \alpha + \beta) < t < (t_i + T) \tag{3.5-5}$$

The maximum error is

$$\text{max Error} = \text{max }[\text{Error (2)} \pm \text{Error (1)}] \tag{3.5-6}$$

The sampling rate must be at least twice the "highest frequency component" of the signal in order to avoid excessive errors. Empirically, the "highest frequency component" can be estimated from the results of a Fourier analysis of the signal by determining the frequency above which the amplitude is less than ten percent of the amplitude at low frequencies.

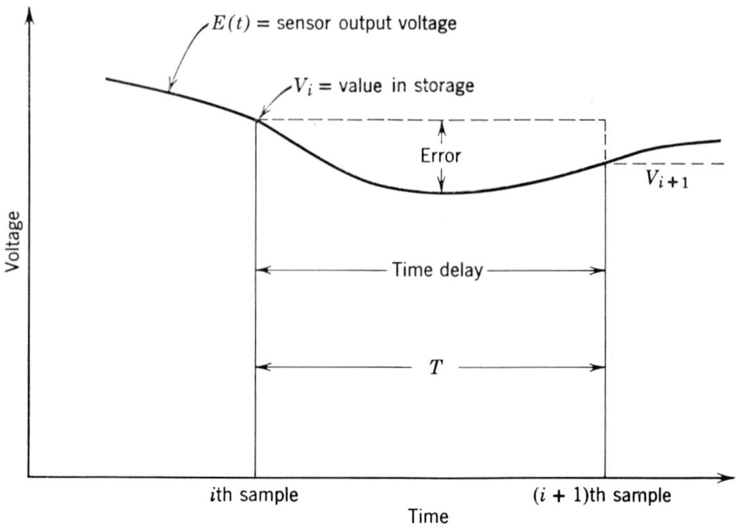

Fig. 3.5-2   Simplified relationship between sensor output voltage and value in storage.

In general the sampling period T will be much greater than $(\alpha + \beta)$, the time lag inherent in the system:

$$T >> (\alpha + \beta) \tag{3.5-7}$$

Thus the relationship of $E(t)$ and V_i can be simplified to that shown in Figure 3.5-2 when we assume $(\alpha + \beta) \to 0$. There is now only one error and one time delay.

$$\text{Error} = V_i - E(t) \tag{3.5-8}$$

where

$$t_i < t < (t_i + T)$$

and the maximum time delay is now equal to the period between samples. Thus

$$L = T \tag{3.5-9}$$

B. Scaling and Conversion of Data

The digital value of the analog signal represents the voltage of the amplified signal. This must be properly scaled to take into account the amplification or attenuation factor, and the value of the voltage should be converted into the units of the variable being measured; for example, 0.15 volt may represent

110 psi. Known repeatable calibration errors may also be factored into the conversion process to reduce the errors from the sensors and transducers.

The scaling and conversion of analog data function generally consist of the following operations:

1. Dividing by the amplifier-attenuator setting.
2. Converting to engineering units by using the appropriate formula.
3. Correcting for repeatable (calibration) errors by using the appropriate calibration formula. This may be combined into one formula used for conversion into engineering units.
4. Storing the result in a specified location.

Scaling and conversion can best be performed by a subroutine. Table 3.5-2 shows the parameters required in order to specify the data scaling and conversion subroutine.

Table 3.5-2   Specifying the Data Scaling and Conversion Subroutine

Parameters	Example 1	Example 2	Example 3	Example 4
Storage location of data	2431	7508	1093	2047
Storage location of result	3431	8508	2093	3047
Scale factor	$\times 100$	$\times 1/10$	$\times 1$	$\times 10,000$
Formula number	34	19	82	52

The flow chart of the subroutine is shown in Figure 3.5-3.

C. Data Accumulation

A common method of noise reduction is to average the values of a variable read in succession. Data accumulation may also be utilized for tallying purposes and for reporting on a per hour or per shift basis. Data accumulated and classified may be used to furnish production or quality statistics. Real-time analysis and reporting can be valuable in providing the operator with a quick profile of the process operation.

A "running average" may be used to obtain the average of the successive values of a variable, i.e., the average of the last N points being read. Accumulation of actual totals of a variable starts at the beginning of the time interval and continues to the end of the time interval. Table 3.5-3 shows the parameters used to specify the data accumulation function with examples.

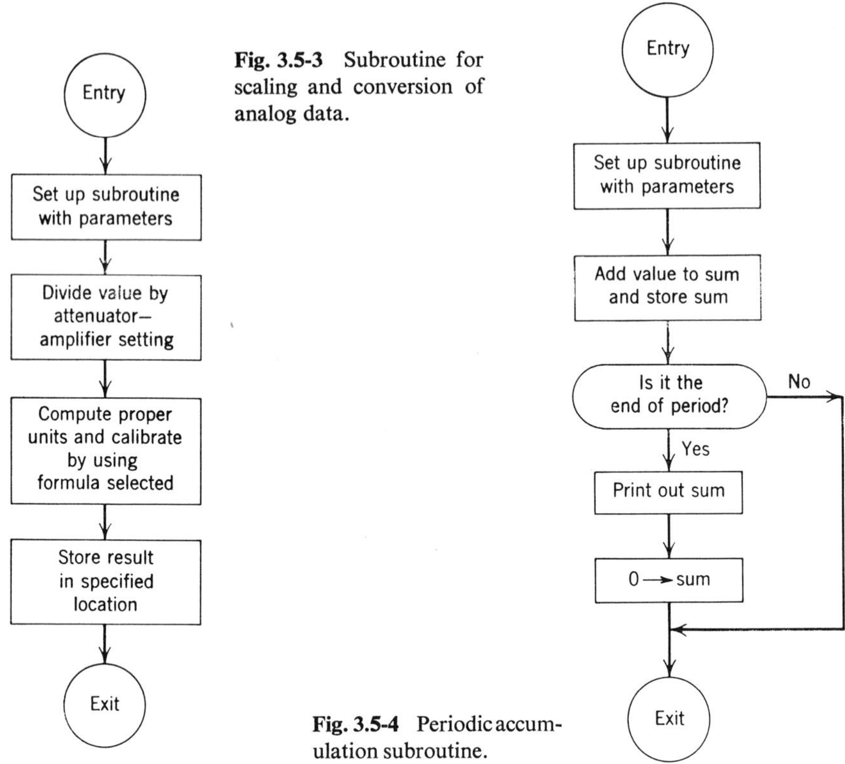

Fig. 3.5-3   Subroutine for scaling and conversion of analog data.

Fig. 3.5-4   Periodic accumulation subroutine.

Table 3.5-3   Specifying the Data Accumulation Function

Parameters	Example 1	Example 2	Example 3	Example 4
Storage location of data	3431	8508	2093	3047
Storage location of result	3432	8509	2094	3048
Period of accumulation	—	—	1 hr	8 hr
Number of points for running average	12	6	—	—

Figure 3.5-4 shows the flow chart for the periodic accumulation subroutine. Notice that the subroutine makes a test to determine whether it is at the end of the accumulation interval. If it is at the end of the interval, the sum is printed out and then cleared to zero so that a new accumulation may begin for the interval. Figure 3.5-5 shows the flow chart for computing a six-point

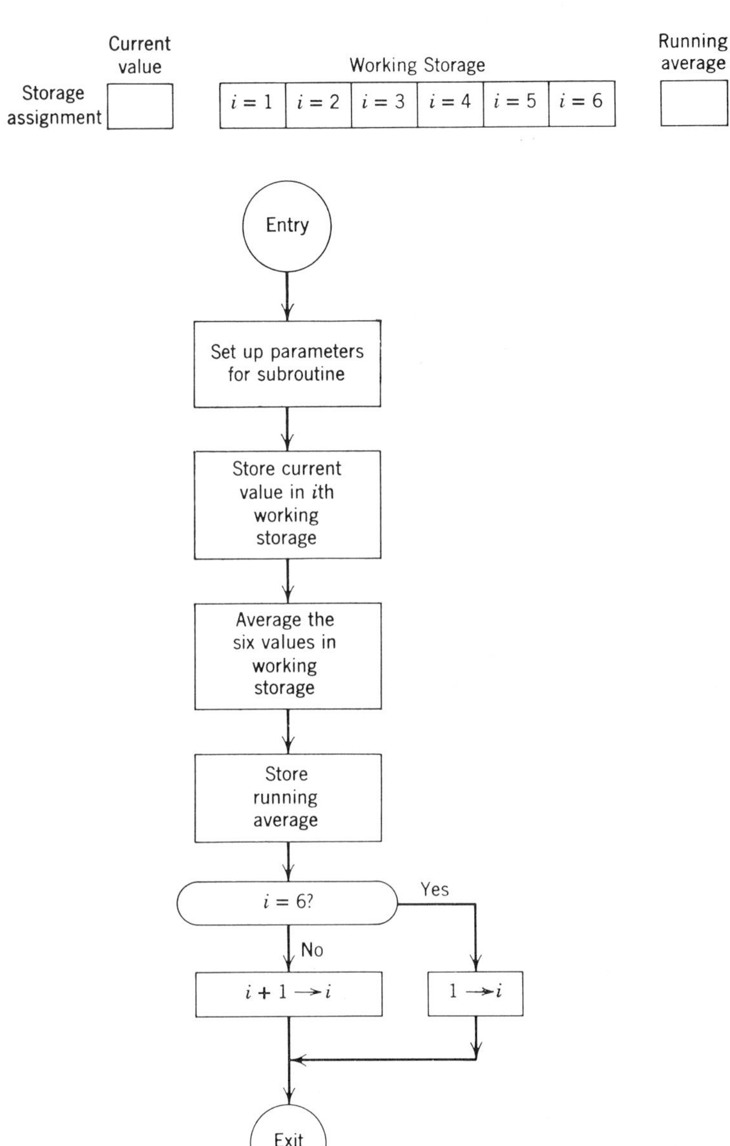

Fig. 3.5-5   Subroutine for computing a six-point running average.

running average. Notice that six locations of working storage are required. The latest current value is always stored in the location next to that of the last value stored. When the sixth storage location $i = 6$ is reached, the next current value is stored in the first location $i = 1$. Thus the six-point running average may be computed at any time by simply averaging the values of the six locations $(i = 1)$ to $(i = 6)$. A weighted average may also be used, e.g., more weight can be given to the more recent readings.

D. Data Formating and Editing

Data formating and editing is the work of arranging a series of numbers into a format that effectively conveys information when it is printed out on a logging typewriter or printer. Common functions performed include headings, spacing, insertion of commas for the thousands position, and insertion of plus and minus signs. The data format should be planned with the operator in mind to assure that it is clear and easy to read and there is no ambiguity or wrong interpretation possible.

The data formating and editing function generally consists of the following operations:

1. Converting from binary to decimal.
2. Inserting a plus or minus sign at the right-hand side of the number.
3. Inserting a decimal point.
4. Inserting a comma for the thousands and millions positions.
5. Suppressing leading zeros.
6. Storing the finished number in the proper print positions.

Table 3.5-4 shows the parameters for specifying the data formating and

Table 3.5-4 Specifying the Data Formating and Editing Function

Parameters	Example 1	Example 2	Example 3
Location of data	10,541	10,542	10,543
Location of binary point	0	4	6
Decimal point location of printout	0	2	2
Use plus sign or blank?	Blank	Blank	Blank
Use minus sign on left or right?	Right	Right	Right
Use period for decimal point?	No	Yes	Yes
Use commas for thousands position?	Yes	No	No
Suppress leading zeros from which position?	Tens	Tens	Tens
Print position of rightmost character	10	20	30
Print position of leftmost character	1	11	21

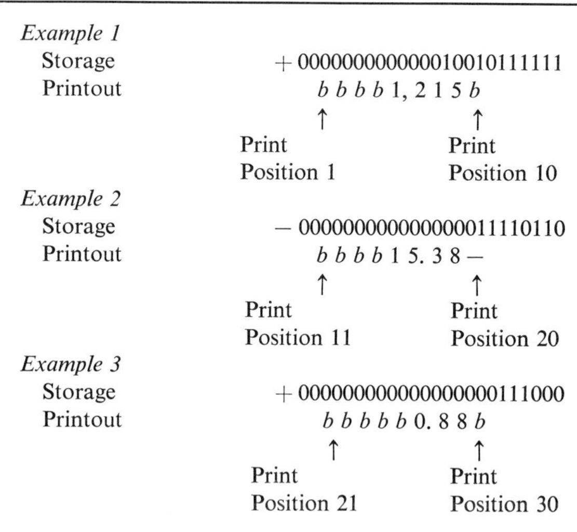

Fig. 3.5-6 Results of data formating and editing, with examples from Table 3.5-4. The character b is used to represent a blank character.

editing function with examples. Figure 3.5-6 shows the results of some of the numbers stored in location 10,541 using the parameters of Example 1, location 10,542 using Example 2, etc., ... Note that specification of the binary point in Examples 2 and 3 resulted in decimal fractions being printed out.

E. Limit Comparison and Alarming

The function of limit comparison of *state variables* serves to tell the operator whether the plant is "normal." For example, if the oxygen content of the exhaust gas from a boiler is outside the allowable range, the operator should be notified to take corrective action.

Alarming may be indicated by audio means or visually by the lighting of alarm panels or both. Also, the identification of the variable, the high-low limits, the actual value of the variable, and the time interval when the limits were exceeded should also be printed out.

Limit comparison and alarm utilizes the principle of "management by exception," i.e., only the exceptional conditions should be brought to the operator's attention. The printout for alarm should occur *once when the point goes out of limits*, and *once when the point returns to within the specified limits*. The limit comparison and alarming function consists of the following operations:

1. Comparing the value of the point against high/low limits.
2. Alarming if this is the first time that the point has exceeded limits.
3. Printing out the following when alarming:

> Time
> Point identification
> "ALARM ON"—HIGH (or LOW)
> Value of point
> Value of high/low limits

4. Shutting off the alarm if the point has exceeded the limits and has returned within limits.

5. Printing out the following when the alarm is shut off:

> Time
> Point identification
> "ALARM OFF"
> Value of point
> Value of high/low limits

Table 3.5-5 shows the parameters for specifying the limit comparison and alarm function with examples.

Table 3.5-5 Specifying the Limit Comparison and Alarm Function

Parameter	Example 1	Example 2	Example 3	Example 4
Point identification	241	576	825	390
High limit	110	2.5	—	5.8
Low limit	90	—	850	4.0
Alarm after N readings out of limits	1	2	1	1
Visual/audio alarm	Both	Visual	Visual	Both

The subroutine for the limit comparison and alarm function is shown in Figure 3.5-7. Here, an alarm occurs whenever the value of the point read is out of limit only once. As shown in Example 2 in Table 3.5-5, it is possible to specify that the value has to be out of limits two consecutive times before alarming takes place in order to reduce alarming for noisy or spurious signals. The alarm limits may also be purposely set higher to avoid excessive alarming for noisy signals. The seven possible conditions for the flow chart in Figure 3.5-7 are shown in Table 3.5-6.

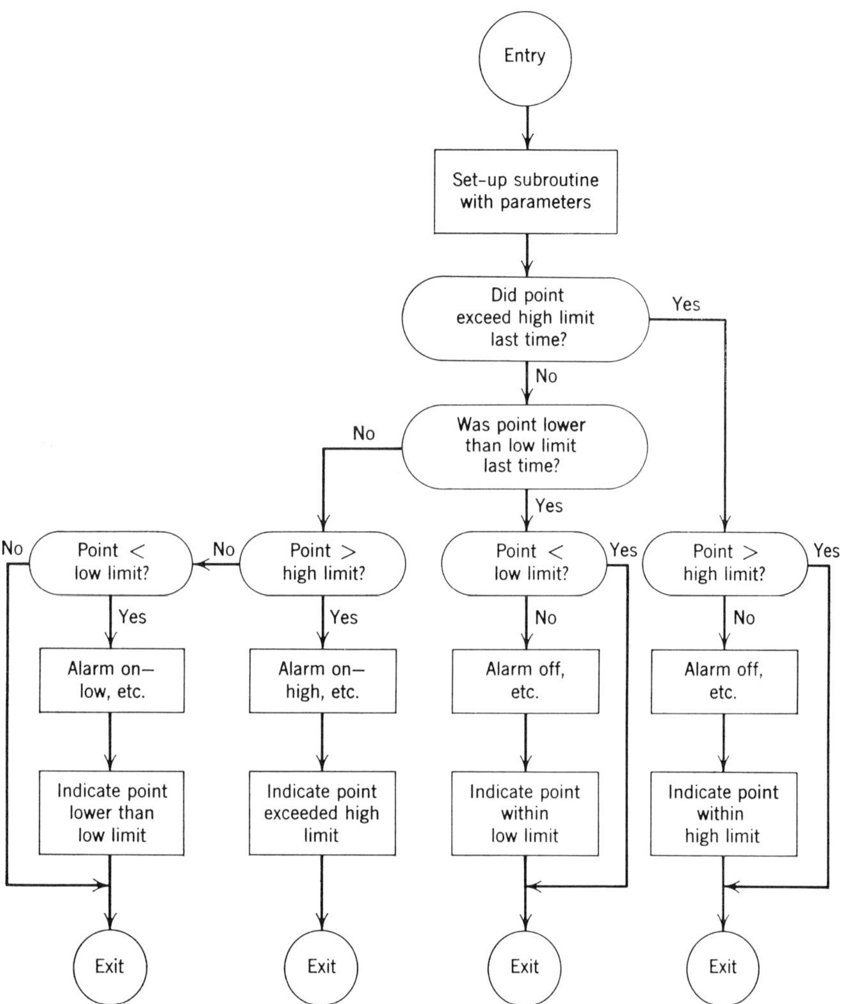

Fig. 3.5-7 Subroutine for limit comparison and alarm.

Table 3.5-6   Possible Conditions for Limit Comparison Flow Chart Shown in Fig. 3.5-7

Conditions	Case 1	Case 2	Case 3	Case 4	Case 5	Case 6	Case 7
Past Reading	Exceeded high ←— limit —→		Lower than ←— low limit—→		←——— Within limits ———→		
Current Reading	Exceeded high limit	Within limits	Lower than low limit	Within limits	Exceeded high limit	Lower than low limit	Within limits
Action	Do nothing	ALARM OFF	Do nothing	ALARM OFF	ALARM ON— HIGH	ALARM ON— LOW	Do nothing

F. Computations

Computations refer to the mathematical operations of the process computer rather than input-output and logic operations. To be more accurate, these are really arithmetical rather than mathematical functions, since the computer can only perform arithmetical operations. Thus the mathematical problem to be solved must first be transformed into arithmetical operations, i.e., by means of an algorithm. In general, the following are some of the parameters required for specifying computations:

1. Algorithm to be performed.
2. Storage location and scaling of the dependent and independent variables, constants, coefficients, tables, etc.
3. Subroutines required for the computation, such as table look-up, curve fitting, and square root.
4. Floating or fixed point operations.
5. Accuracy required for dependent variables (for iterative solutions).
6. Dimensions, such as those for matrix operations.
7. Interval of integration (for solutions of differential equations).
8. What to do in case of nonconvergence, overflow, etc.
9. Time at which computations must be completed.

As an example of a short subroutine for computation, consider the calculation of the calcium contents of the homogenizer using Equation 3.5-10:

$$C_H(t_{i+1}) = \left[1 - \Delta t\, \frac{R}{T}\right] C_H(t_i) + \Delta t\, \frac{R}{T}\, C_{RM}(t_i) \qquad (3.5\text{-}10)$$

Assume that the values Δt, R, T, $C_H(t_i)$, and $C_{RM}(t_i)$ are available in memory. The flow chart of this calculation is shown in Figure 3.5-8.

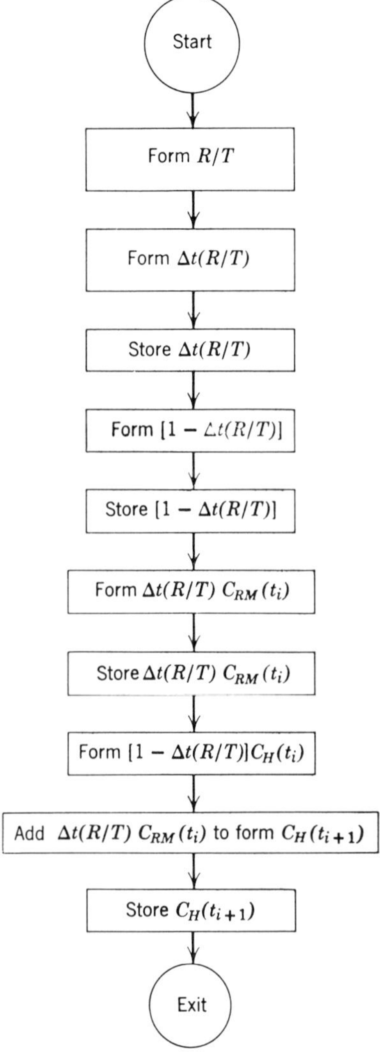

Fig. 3.5-8 Subroutine for computing Equation 3.5-10.

G. Computer Control Actions

Computer control actions taken by the computer process control system may be used for implementing:

Direct digital control.
Program control.
Optimal control.

Table 3.5-7   Specifying the Computer Control Actions

Parameters	Example 1	Example 2	Example 3
Output point identification	2	9	13
Type of control	Program	Direct digital	Optimal
Output control device	Contact closure	Pulse train	Reference signal
Storage location for output value	14,002	14,009	14,013
Safe conditions	Contacts No. 512, 513, and 514 should be closed, operator should initiate action. Contact No. 2 should not be closed longer than 30 seconds.	Position of actuator should be within range of expected value. Maximum number of pulses is ten.	Reference signal should not exceed $\pm 10\%$ change
Operator's log	No	Yes	Yes

Direct digital control will be briefly discussed in Section 3.6. Program control is outside the scope of this book. Optimal control, however, will be treated in Chapters 6 through 10.

The computer control action usually consists of the following operations:

1. Determining the required control actions by computations and logic in the process computer.
2. Checking to ensure that conditions are safe for control actions, otherwise, inhibit control action.
3. Initiating control action.
4. Typing out the control action initiated or inhibited.

The heart of the computer control function is checking to ensure that conditions are safe for control actions. Examples of safe conditions include the following:

Prior required actions have been initiated.
Critical state variables are within specified range.
Size of change in control variable will not exceed a certain percentage of the control variable, e.g., 10% change.

The operator's permission has been obtained by checking the initiate switches and manual override.

Table 3.5-7 shows the parameters required for specifying the computer control actions with examples.

3.6   DIRECT DIGITAL CONTROL

Direct digital control is the function of regulating process variables to conform to given reference signals. The process computer is replacing conventional analog controllers when it is being used for direct digital control. This section will first present the operations of direct digital control, then discuss some common algorithms for the various types of direct digital control, and finally some of the practical considerations of applying direct digital control. The detail analysis and synthesis of sampled-data systems is outside the scope of this book.

A. Operations of Direct Digital Control

Figure 3.6-1 shows a block diagram of the process computer performing direct digital control for one process variable. This is similar to the classic closed loop control, or feedback control loop. The only differences are the samplers present in the input and output. From Figure 3.6-1 the operations of direct digital control are the following ones:

1. Sampling the measured sensor output M through the input sampler.
2. Subtracting the measured sensor output M from the reference signal R to obtain the error signal E.
3. Applying the control algorithm to the error signal E to produce the output signal X.
4. Using proper safeguards before initiating output actions.
5. Using the output sampler to initiate the output signal X. The output signal X will be translated by an output controller into controls for the actuator.

By comparing these operations with the list of computer process control functions given in Section 3.5 the conclusion may be drawn that there is a great deal in common between the two lists. In other words, most of the operations required by direct digital control are already being provided by the process computer. Only the control algorithms and safeguards need to be specially programmed. Some of the safeguards were discussed in Section 3.5G. The control algorithms will be discussed in the next paragraph.

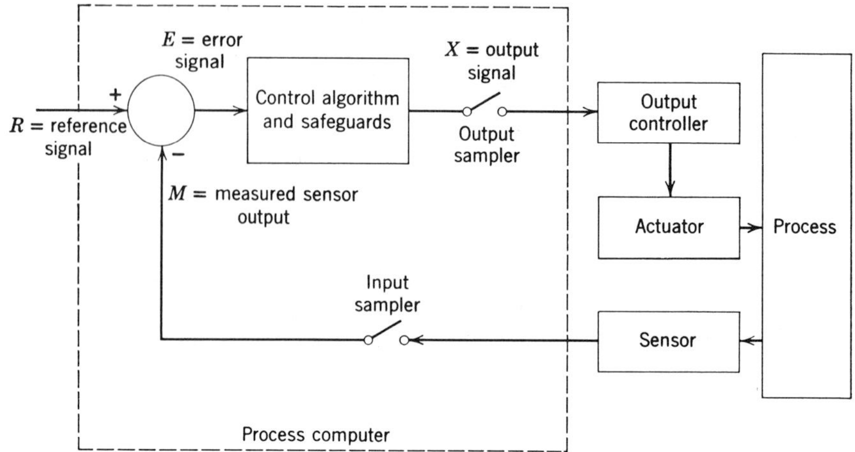

Fig. 3.6-1 Block diagram of process computer performing the direct digital control function.

B. Direct Digital Control Algorithms

The process computer can readily duplicate many of the common analog control functions. In fact, many complex analog control functions can be readily implemented, e.g., time-lags, and noninteracting multiloop controls. The following typical direct digital control algorithms are given as illustrations. Let the following symbols be assigned:

Δt $\quad$ = sampling interval.
X_n $\quad$ = output signal at time n.
X_{n-1} = output signal at time $(n-1)$.
ΔX_n = $X_n - X_{n-1}$.
E_n $\quad$ = error signal at time n.
E_{n-1} = error signal at time $(n-1)$.
ΔE_n = $E_n - E_{n-1}$.
K_p $\quad$ = proportional constant.
K_i $\quad$ = integral constant.
K_d $\quad$ = derivative constant.

For *proportional control*, the output signal is

$$X_n = K_p E_n \qquad (3.6\text{-}1)$$

For *integral control*, the output signal is

$$X_n = \sum_n K_i E_n (\Delta t) \qquad (3.6\text{-}2)$$

For *derivative control*, the output signal is

$$X_n = \frac{K_d(E_n - E_{n-1})}{(\Delta t)} \tag{3.6-3}$$

Thus the algorithm for *proportional plus integral control* is

$$X_n = K_p E_n + \sum_n K_i E_n(\Delta t) \tag{3.6-4}$$

and the algorithm for *proportional plus integral plus derivative control* is

$$X_n = K_p E_n + \sum_n K_i E_n(\Delta t) + \frac{K_d(E_n - E_{n-1})}{(\Delta t)} \tag{3.6-5}$$

For *ratio control*, where the output signal X_1 is ratioed to the value of another control signal X_2, the incremental adjustment of the output signal X_1 is

$$\Delta X_{1,n} = K_r(X_{2,n}) - X_{1,n-1} \tag{3.6-6}$$

where K_r is the ratio control constant. Since

$$\Delta X_{1,n} = X_{1,n} - X_{1,n-1} \tag{3.6-7}$$

the ratio control relationships can be shown by substituting Equation 3.6-7 into Equation 3.6-6:

$$X_{1,n} = K_r X_{2,n} \tag{3.6-8}$$

An algorithm for *nth-order single loop control* is given by

$$X_n = \sum_{i=1}^{n-1} A_i(X_{n-i}) + \sum_{i=0}^{n-2} B_i(E_{n-i}) \tag{3.6-9}$$

where A_i and B_i are constants.

C. Some Practical Considerations of Direct Digital Control

Some of the practical considerations of direct digital control include operator control and adjustment, backup in case of computer failure, and bumpless transfer from manual-to-automatic.

A special Direct Digital Control Console may be utilized to allow the operator to control and adjust the various control loops. Each loop must be individually accessible for the following adjustments:

Activate (switch to automatic).
Inhibit (switch to manual).
Set reference signal.
Adjust the control algorithm constants.

In case of computer failure a backup computer may be required to take over the function of direct digital control. The backup computer is normally devoting a small portion of its time to "following" the main direct digital control function so that the transition from one computer to the other may be made rapidly. If a backup computer is not available, means must be available for manual operation.

When transferring from manual-to-automatic operation, a special algorithm is required to reduce the large transients that may result.

3.7 SUMMARY

This chapter used the modeling technique to introduce the computer process control system. The same treatment, which introduced the physical process in Section 2.2, was used to develop the basic concepts of computer system hardware, software, and commonly used functions. At this point the reader should have a general familiarity with how to specify the commonly used functions. He should also be acquainted with the fact that the cost of each function depends on how the parameters of the function are specified. The costs are naturally related to the resources required: *storage, input-output devices*, and *time used in execution*. Common sense and economic judgment should prevail in weighing the value against cost of alternate ways of specifying a function, and in establishing priorities of execution between the different functions.

Chapter 4 will focus attention on how to develop models for the physical process, stressing both the analytical and experimental techniques.

4

Techniques for Developing Physical Process Models

4.0 INTRODUCTION

In analog subloop control the variables being regulated are generally assumed to be independent. Thus independent physical models of the specific variables being regulated have to be developed, e.g., simple input-output models of temperature, liquid level, flow rate, gas pressure, voltage, etc., as functions of the control variables. In controlling the process as a whole, however, the physical process model must be developed as an entity, since the process variables can no longer be assumed to be independent of each other.

Prior to developing the physical process model, the purpose and objectives of control should be clearly stated. Physical process models for the same process may bear little resemblance to each other if they are developed for different purposes. In general, the following design parameters should be kept in mind for the development of physical process models:

Scope or boundary of the process under study.
Depth of details.
Physical and safety constraints.
Steady-state or dynamic control (frequency components in variables).
Accuracy required.
Need and method of updating model.
State variables and available control variables.
Disturbances and other uncontrolled variables.

Development of physical process models utilizes intuition and judgment as well as the classical scientific method of *observation, classification, hypothesis,* and *test. Intuition* is required to establish the basic assumptions, major

85

relationships between key variables, and the initial approaches to establishing the physical process model. *Judgment* is required to preserve a balance between accuracy and completeness versus complexity and cost. Fortunately, well known laws of physics and chemistry are available to describe a physical process in mathematical terms.

This chapter presents *both the analytical and experimental approaches to developing physical process models.* The analytical approach is based on the basic laws of systems analysis and the characteristics of process elements. The experimental approach for steady-state models utilizes the well known technique of regression analysis, i.e., curve fitting by the least square technique. Techniques for developing linear and multiple regression models and Kalman's technique for developing dynamic models by experimentation will be presented. Finally, a dynamic physical process model for a completely mixed stirred chemical reactor will be developed as an example.

4.1   BASIC LAWS OF SYSTEMS ANALYSIS

The foundation for the development of physical process models consists of the basic laws of physics and chemistry. Developing physical process models for a plant under design or for a plant that cannot be experimented upon depends on the basic physical laws of the following types of systems: *pneumatic, hydraulic, thermal, mechanical,* and *electrical.* Even where the experimental approach is possible or necessary, using the basic laws of physics and chemistry can save much work in organizing the experiments and selecting the form of the model.

A. Unified Approach of Systems Analysis

The *unified* approach will be used in presenting the basic laws of physics and chemistry. This will enable the engineer to make the transition from his field into another field, e.g., from hydraulic to electrical and from thermal to mechanical. The following notions and definitions of systems analysis are first introduced before the physical laws of physics and chemistry are discussed: *elements, junctions, systems, quantity variable, flow variable,* and *potential variable.*

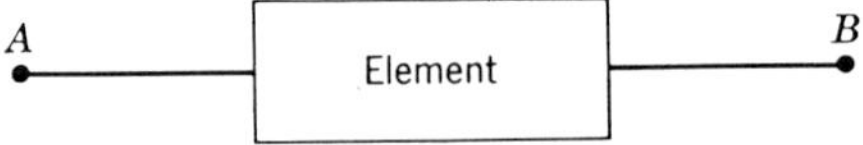

Fig. 4.1-1 Element defined by measurements performed with respect to two points in space, *A* and *B*.

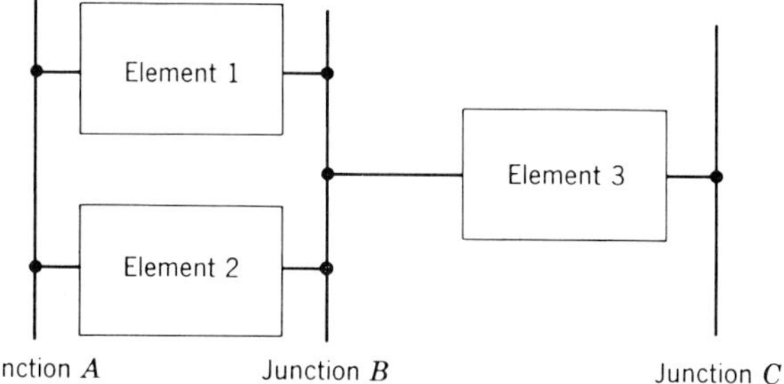

Fig. 4.1-2   Junctions defined as points which join together one or more elements.

Elements are physical objects whose behavior may be defined by measurements performed with respect to two points in space. They are the unit building blocks of a system. Elements may store, transmit, convert, and dissipate energy. Typical elements are valves, pipes, generators, motors, vessels, and heat exchangers, as shown in Figure 4.1-1.

Junctions are points which join together one or more elements. No energy storage, transmission, conversion, or dissipation takes place in junctions. In a sense, they are like the abstract points of coordinates. Typical junctions are electrical busses and connectors of pipes to vessels, as shown in Figure 4.1-2.

A *system* is a collection of elements where the net sum of the energy is zero. Or, energy generated is equal to energy converted and dissipated.

A *quantity variable* is a state variable which describes the quantity contained in the element. In certain cases, the quantity variable may also be a control variable. State and control variables were defined in Section 2.6. Junctions cannot contain quantity variables, since they only join elements together. A quantity variable is a " one-point variable," i.e., it can be defined by one point in space. Examples of quantity variables are volume, entropy, momentum, and charge.

A *flow variable* is the rate of change of the quantity variable with respect to

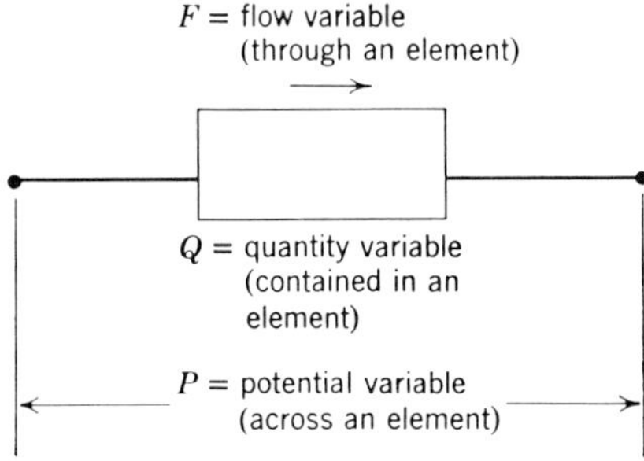

Fig. 4.1-3   Quantity, flow, and potential variables of an element.

time. It is also a "one-point variable." It may also be a state or control variable. Examples of flow variables are flow rate, entropy-flow, force, and current.

A *potential variable* is a state variable (or control variable) measured across the two ends of an element, i.e., it is a "two-point variable" in space. Any junction may have only a single value of potential relative to a reference. Examples of potential variables are pressure, temperature, velocity, and voltage. In many cases, one of the points is a reference point, e.g., melting point of ice, ground potential, and atmospheric pressure. Figure 4.1-3 shows the quantity, flow, and potential variables of an element. Table 4.1-1 summarizes the quantity variables, flow variables, and potential variables used for pneumatic, hydraulic, thermal, mechanical, and electrical systems. The units for these variables are listed below the names of the variables.

Table 4.1-1   Summary of Quantity, Flow, and Potential Variables of Physical Systems

Type of System	Quantity Variable	Flow Variable	Potential Variable
Pneumatic	Weight (lb)	Flow rate (lb/sec)	Pressure (lb/ft^2)
Hydraulic	Volume (ft^3)	Flow rate (ft^3/sec)	Head (ft)
Thermal	Entropy (Btu)	Entropy-flow (Btu/sec)	Temperature (degree)
Mechanical			
translation	Momentum (lb-sec)	Force (lb)	Velocity (ft/sec)
rotation	Angular momentum (lb-ft-sec)	Torque (lb-ft)	Angular velocity (radians/sec)
Electrical	Charge (coulomb)	Current (ampere)	Voltage (volt)

B. Two Basic Laws For Systems Analysis

There are two basic laws for systems analysis:

1. The sum of all flow variables for any junction is equal to zero, as shown in Equation 4.1-1.

2. The sum of all potential variables for any closed loop in the system is equal to zero, as shown in Equation 4.1-2.

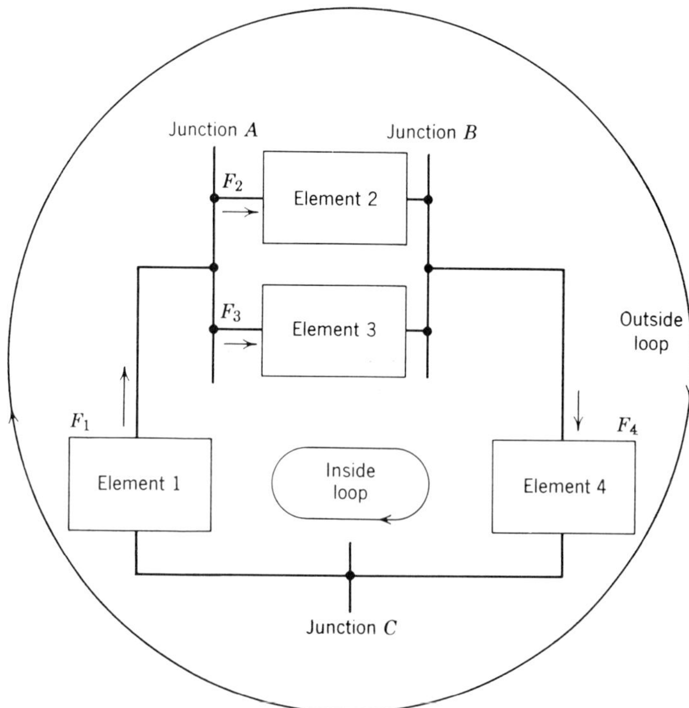

Fig. 4.1-4 Typical system used to illustrate the two laws of system analysis.

These two laws are illustrated in Figure 4.1-4.

$$\sum_i F_{ni} = 0 \qquad (4.1\text{-}1)$$

where n is the particular junction, i is the element connected to the junction, and F is the flow variable.

$$\sum_j P_{mj} = 0 \qquad (4.1\text{-}2)$$

where m is the particular loop, j is the element within the loop, and P is the potential variable.

The two laws for systems analysis are based on the *laws of conservation of energy* and *continuity of material*. An intuitive method of deriving the two laws is based on the definition of a junction. The first law is based on the definition that no junction can contain a quantity variable. Also, the second law is based on the definition that a given junction can have only *one* value of potential variable.

Using the illustration in Figure 4.1-4, we may now apply the two basic laws of systems analysis to derive the following equations:

1. For junctions A, B, and C, respectively,

$$F_1 - F_2 - F_3 = 0 \qquad (4.1\text{-}3)$$

$$F_2 + F_3 - F_4 = 0 \qquad (4.1\text{-}4)$$

$$F_4 - F_1 = 0 \qquad (4.1\text{-}5)$$

2. Around the outer and inner loop, respectively,

$$P_1 + P_2 + P_4 = 0 \qquad (4.1\text{-}6)$$

$$P_1 + P_3 + P_4 = 0 \qquad (4.1\text{-}7)$$

Note that the direction of flow should be taken into account in the flow equations. In the potential equations the potential variable is assumed to be measured in the direction of the flow variable. Thus P_1 is the potential across junctions C to A. Table 4.1-2 illustrates the two basic laws of systems analysis in the various physical systems.

Table 4.1-2 The Two Basic Laws of Systems Analysis for Physical Systems

Type of System	First Law (flow variables connected to a junction)	Second Law (potential variables around a loop)
Pneumatic	Σ (flow rate) $= 0$	Σ (pressure) $= 0$
Hydraulic	Σ (flow rate) $= 0$	Σ (head) $= 0$
Thermal	Σ (entropy flow) $= 0$	Σ (temperature) $= 0$
Mechanical		
translation	Σ (force) $= 0$	Σ (velocity) $= 0$
rotation	Σ (torque) $= 0$	Σ (angular velocity) $= 0$
Electrical	Σ (current) $= 0$	Σ (voltage) $= 0$

4.2 CHARACTERISTICS OF PHYSICAL PROCESS ELEMENTS

This section will introduce the characteristics of physical process elements, i.e., the relationship between the quantity variable and the potential variable, and the relationship between the flow variable and the potential variable. The concept of resistance and capacitance of physical process elements will be introduced. First, the general case will be developed, and then the characteristics of typical pneumatic, hydraulic, thermal, mechanical, and electrical systems will be presented.

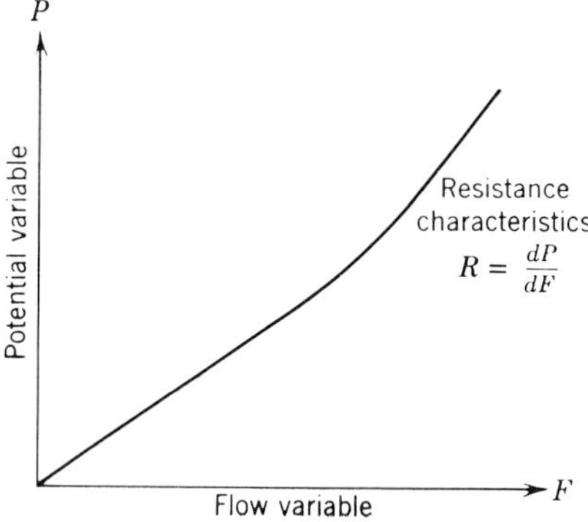

Fig. 4.2-1   Resistance characteristic for the general case.

A. General Case and Definitions

Let the following symbols be assigned:

Q = quantity variable.
F = flow variable.
P = potential variable.
R = resistance.
C = capacitance.

The *resistance* of a process element is defined as the rate of change of the potential variable with respect to the flow variable and is given by

$$R = \frac{dP}{dF} \tag{4.2-1}$$

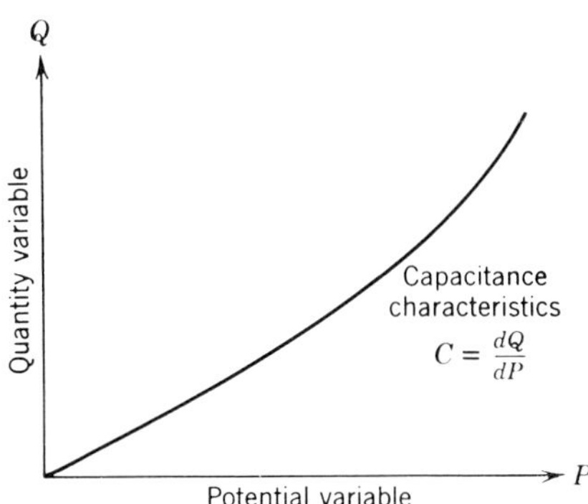

Fig. 4.2-2   Capacitance characteristic for the general case.

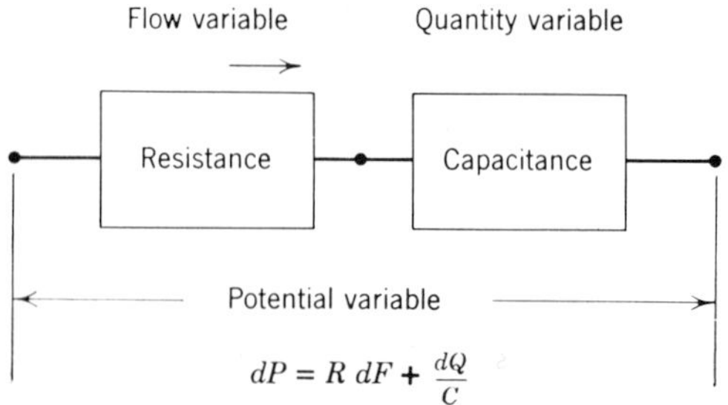

Fig. 4.2-3   Resistance and capacitance in series for the general case.

The *capacitance* of a process element is defined as the rate of change of the quantity variable with respect to the potential variable and is given by

$$C = \frac{dQ}{dP} \tag{4.2-2}$$

In special cases in which the process elements are linear

$$R = \frac{P}{F} \tag{4.2-3}$$

$$C = \frac{Q}{P} \tag{4.2-4}$$

In the general case the potential across a process element consisting of both resistance and capacitance in series is given by

$$dP = RdF + \frac{1}{C}\,dQ \tag{4.2-5}$$

Figures 4.2-1 and 4.2-2 show the general case relationship of resistance and capacitance in process elements. Figure 4.2-3 shows the process elements in series, as described in Equation 4.2-5.

B. Pneumatic Systems

Gas flow through small orifices can often be considered as *turbulent*, whereas gas flow at certain velocities through larger pipes is often considered as *laminar*. Elements which often produce turbulent flow are *orifices*, *valves,* and *small pipes.* Elements which subsequently produce laminar flow are *circular tubes* and *pipes,* as shown in Figure 4.2-4. The quantity variable for pneumatic systems is *weight*, the flow variable is *flow rate*, and the potential variable is *pressure.* Let the following symbols be assigned:

R = pneumatic resistance, in seconds per square foot.
P = pressure across the element, in pounds per square foot.
F = flow rate through the element, in pounds per second.
L = length of the pipe, in feet.
D = inside diameter of the pipe, in feet.
g = gravitational constant, in feet per second squared.
K = flow coefficient, dimensionless.
A = area of restriction, in square feet.
Y = rational expansion factor, in pounds per cubic foot.
γ = gas density, in pounds per cubic foot.
μ = absolute viscosity, in pound-seconds per square foot.

The resistance to laminar gas flow in an adiabatic condition is

$$R = \frac{dP}{dF} = \frac{128\mu L}{\pi\gamma D^4} \tag{4.2-6}$$

The resistance to turbulent gas flow in an adiabatic condition is

$$R = \frac{dP}{dF} = \frac{\sqrt{2}}{KAY}\left(\frac{\gamma V}{g}\right)^{\frac{1}{2}} \tag{4.2-7}$$

and

$$K = \sqrt{\frac{fD}{L}} \tag{4.2-8}$$

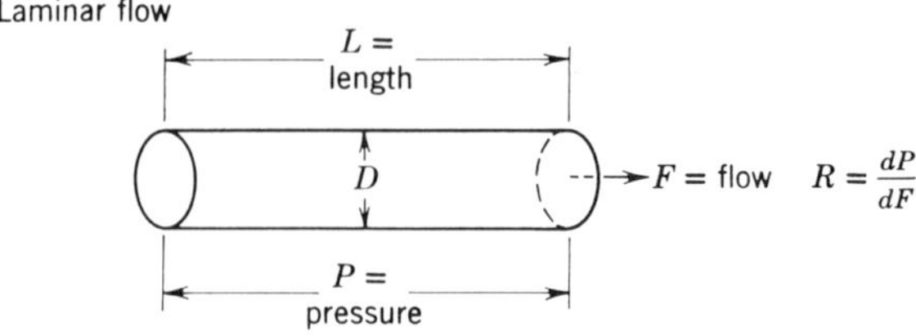

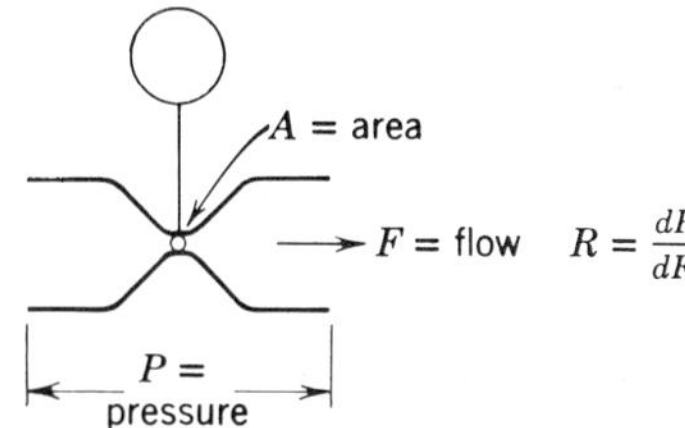

Fig. 4.2-4   Laminar and turbulent flow resistance of pneumatic and hydraulic system.

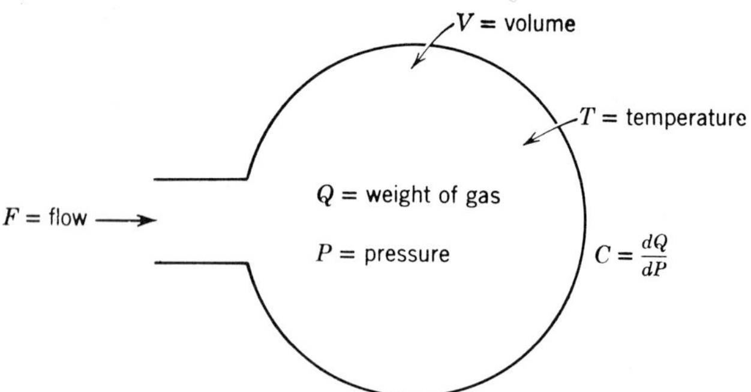

Fig. 4.2-5   Capacitance of pneumatic element.

where V is the head, f is the friction factor, D is the inside diameter of the orifice, and L is the equivalent length.

The capacitance of a pneumatic vessel will be given by using the following additional symbols:

C = pneumatic capacitance, in square feet.
Q = weight of the gas in the vessel, in pounds.
V = volume of the vessel, in cubic feet.
G = gas constant for specific gas, in feet per degree.
T = temperature, in degrees.
n = polytropic exponent.

where $n = 1.0$ for isothermal expansion, and n = ratio of specific heats for adiabatic expansion, and $n = 1.4$ for air.

The capacitance of a pneumatic element is shown in Figure 4.2-5 and is given by

$$C = \frac{dQ}{dP} = \frac{V}{nGT} \tag{4.2-9}$$

C. Hydraulic System

Hydraulic systems also have two types of flow: *turbulent flow* where the Reynolds number* is greater than 4000, and *laminar flow* where the Reynolds number is less than 2000, as shown on Figure 4.2-4. The quantity variable of hydraulic systems is *volume*, the rate variable is *flow rate*, and the potential variable is *head*. Let the following symbols be assigned:

* Reynolds number is a function of the geometry, flow rate, and viscosity.

R = hydraulic resistance, in seconds per square foot.
C = hydraulic capacitance, in square feet.
P = head, in feet.
F = liquid flow rate, in cubic feet per second.
Q = volume, in cubic feet.
L = length of the pipe, in feet.
D = inside diameter of the pipe, in feet.
g = gravitational constant, in feet per second squared.
K = flow coefficient, dimensionless.
A = area of restriction, in square feet.
γ = fluid density, in pounds per cubic foot.
μ = absolute viscosity, in pound-seconds per square foot.

The resistance of laminar hydraulic flow is

$$R = \frac{dP}{dF} = \frac{128\mu L}{\pi\gamma D^4} \tag{4.2-10}$$

The resistance of turbulent hydraulic flow is

$$R = \frac{dP}{dF} = \frac{F}{gK^2A^2} \tag{4.2-11}$$

Also

$$R = \frac{1}{KA}\sqrt{\frac{2P}{g}} \tag{4.2-12}$$

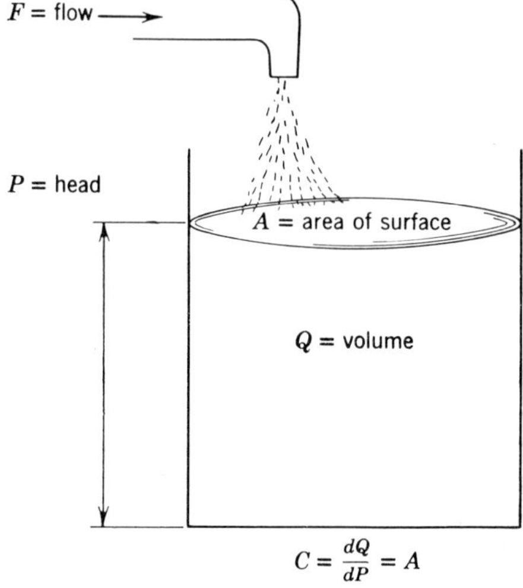

$$C = \frac{dQ}{dP} = A$$

Fig. 4.2-6   Capacitance of hydraulic element.

The hydraulic capacitance is

$$C = \frac{dQ}{dF} = A \qquad\qquad (4.2\text{-}13)$$

A is the cross section of the tank at the liquid surface. For a tank with a uniform cross section the hydraulic capacitance is constant and is equal to the cross-sectional area, as shown on Figure 4.2-6.

D. Thermal Systems

There are three types of heat transfer: *conduction, convection,* and *radiation.* Conduction takes place when heat is transferred through a solid object. Convection takes place through the movement of the liquid or gas in a container. Radiation is the transfer of heat from the surface of one object to another object at a distance. The quantity variable of thermal systems is *entropy or heat,* the flow variable is *entropy-flow or heat-flow,* and the potential variable is *temperature.* Figure 4.2-7 shows an element having resistance to heat transfer. Let the following symbols be assigned:

F = heat flow through the element, in Btu per second.
P = temperature across the element, in degrees.
R = thermal resistance of the element, in degree-seconds per Btu.
C = thermal capacitance of the element, in Btu per degree.
K = thermal conductivity of the element, in Btu per foot-second-degree.
H = convection coefficient of the element, in Btu per square foot-second-degree.
A = cross sectional area of the element, in square feet.
X = thickness of the element, in feet.

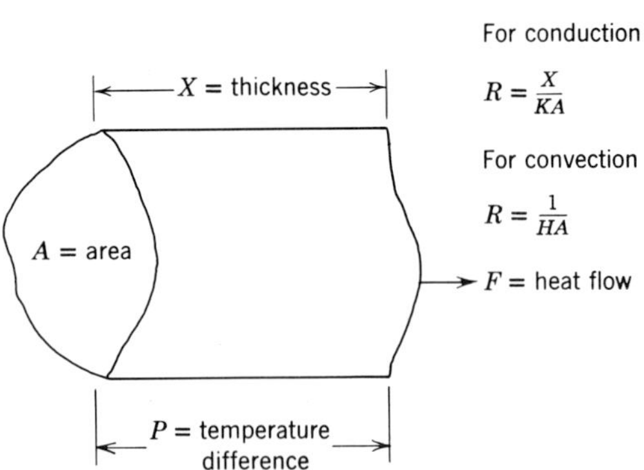

Fig. 4.2-7   Thermal resistance for heat transfer by conduction and convection.

The thermal resistance for conduction through a homogeneous material is

$$R = \frac{dP}{dF} = \frac{X}{KA} \tag{4.2-14}$$

and the thermal resistance for forced convection is

$$R = \frac{dP}{dF} = \frac{1}{HA} \tag{4.2-15}$$

Notice that the thickness of the conductor is a factor for conduction but is not a factor in convection. The distance of the container for convection is factored into the convection coefficient.

Let the following symbols be assigned:

Q = heat stored, in Btu.
W = weight of block, in pounds.
S = specific heat at constant pressure, in Btu per degree-pound.

The thermal capacitance of a block is given by

$$C = \frac{dQ}{dP} = WS \tag{4.2-16}$$

E. Mechanical Systems

There are two kinds of mechanical systems: *translational* and *rotational*. Mechanical systems have an additional characteristic that differentiate them from pneumatic, hydraulic, and thermal systems. In addition to resistance and capacitance, mechanical systems have the characteristics of *mass*, or *moment of inertia*.

Translational Mechanical Systems

Figure 4.2-8 shows a translational mechanical system which consists of three types of elements: *spring, damping,* and *mass.* Let the following symbols be assigned:

Q = momentum, in pound-seconds.
F = force, in pounds.
P = velocity, in feet per second.
t = time, in seconds.
K = spring constant, in pounds per foot.
D = viscous damping coefficient, in pound-seconds per foot.
M = mass, in pound-seconds squared per foot.

The quantity variable of a translational mechanical system is *momentum,*

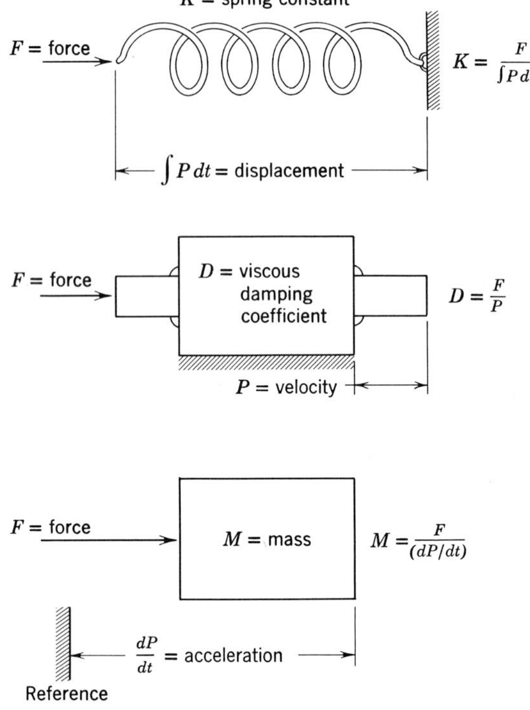

Fig. 4.2-8 Characteristics of translational mechanical system.

the flow variable is *force*, and the potential variable is *velocity*. The spring constant is the inverse of the derivative of displacement with respect to force. Thus the spring constant is analogous to capacitance. The spring constant is given by

$$K = \frac{1}{d(\int P\,dt)/dF} \qquad (4.2\text{-}17)$$

For a linear spring system, the spring constant is equal to the force applied divided by the displacement yielded due to the application of the force. Thus

$$K = \frac{F}{\int P\,dt} \qquad (4.2\text{-}18)$$

Viscous damping is the result of the force resisting the rate of change of displacement of a cylinder in a hydraulic dashpot. It is analogous to resistance and is given by

$$D = \frac{dQ/dt}{P} \qquad (4.2\text{-}19)$$

For a linear dashpot the viscous damping coefficient is the force divided by the velocity

$$D = \frac{F}{P} \tag{4.2-20}$$

Mass is defined as the derivative of the quantity variable with respect to the potential variable

$$M = \frac{dQ}{dP} \tag{4.2-21}$$

In other words, it is the derivative of momentum with respect to velocity. It may also be expressed as the force applied to an object divided by the acceleration resulting from the application of the force. Thus

$$M = \frac{(dQ/dt)}{(dP/dt)} = \frac{F}{(dP/dt)} \tag{4.2-22}$$

Rotational Mechanical Systems

Figure 4.2-9 shows a rotational mechanical system which consists of three elements: *rotational spring* (analogous to capacitance), *rotational viscous damping* (analogous to resistance), and *moment of inertia*. Let the following symbols be assigned:

$Q =$ angular momentum, in pound-feet-seconds.
$F =$ torque, in pound-feet.
$P =$ angular velocity, in radians per second.
$t =$ time, in seconds.
$K =$ rotational spring constant, in pound-feet per radian.
$D =$ rotational viscous damping coefficient, in pound-feet-seconds per radian.
$M =$ moment of inertia, in pound-feet-seconds squared per radian.

The quantity variable of a rotational mechanical system is *angular momentum*, the flow variable is *torque*, and the potential variable is *angular velocity*.

The rotational spring constant is equal to the torque divided by the angular displacement:

$$K = \frac{F}{\int P\,dt} \tag{4.2-23}$$

The rotational viscous damping coefficient is equal to the torque divided by the angular velocity, or the angular momentum divided by the angular displacement:

$$D = \frac{F}{P} = \frac{Q}{\int P\,dt} \tag{4.2-24}$$

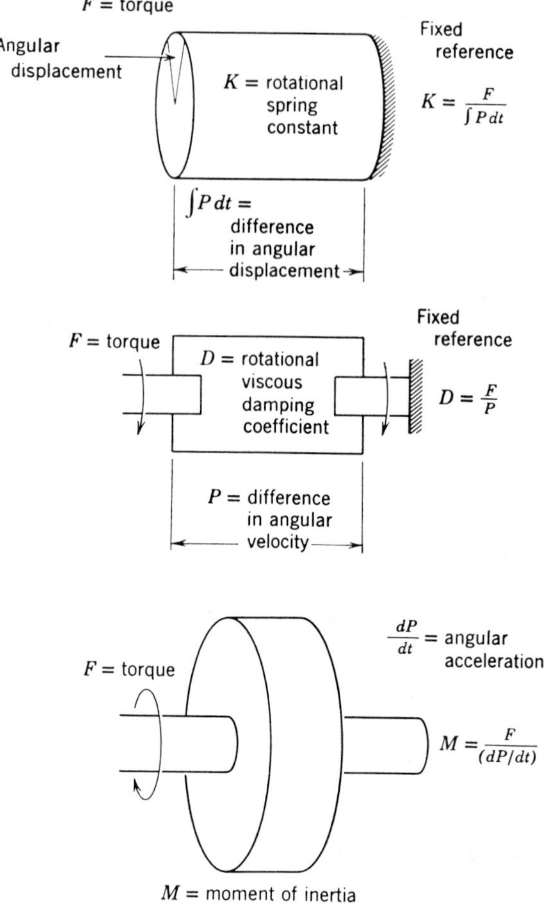

Fig. 4.2-9 Characteristics of rotational mechanical system.

The moment of inertia is equal to torque divided by the angular acceleration or angular momentum divided by the angular velocity:

$$M = \frac{Q}{P} = \frac{F}{(dP/dt)} \tag{4.2-25}$$

F. Electrical System

The quantity variable of an electrical system is *charge*, the flow variable is *current*, and the potential variable is *voltage*. The three characteristics of an electrical system are *resistance*, *capacitance*, and *inductance*, as shown in

Figure 4.2-10. Let the following symbols be assigned:

Q = charge, in coulombs.
F = current, in amperes.
P = voltage, in volts.
t = time.
R = resistance, in ohms.
C = capacitance, in farads.
L = inductance, in henrys.

The resistance is equal to the derivative of the voltage across an element with respect to the current:

$$R = \frac{dP}{dF} \qquad (4.2\text{-}26)$$

For a linear system

$$R = \frac{P}{F} \qquad (4.2\text{-}27)$$

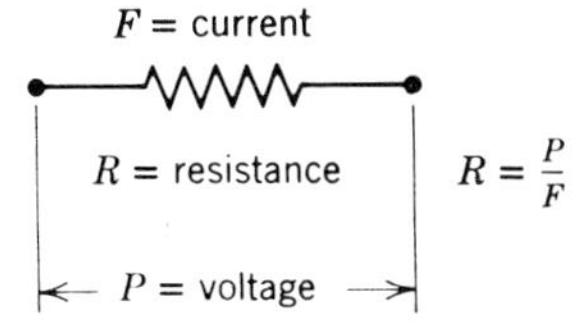

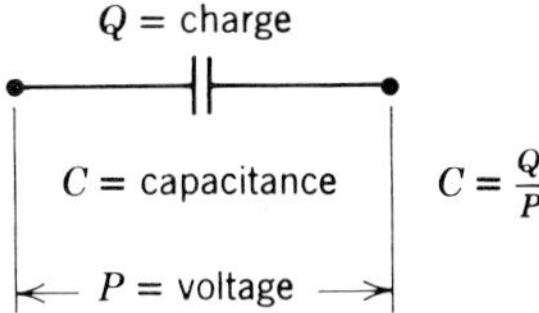

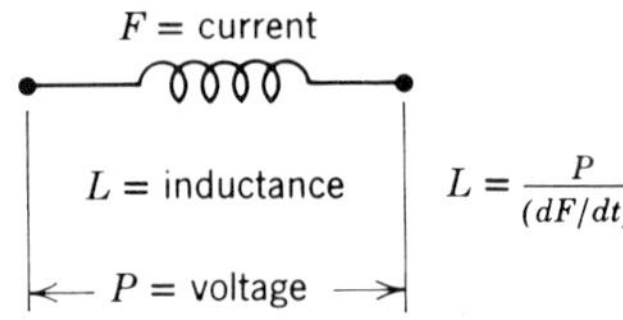

Fig. 4.2-10 Characteristics of electrical system.

The capacitance is equal to the derivative of the charge contained in an element with respect to the voltage across the element:

$$C = \frac{dQ}{dP} \qquad (4.2\text{-}28)$$

For a linear system

$$C = \frac{Q}{P} \qquad (4.2\text{-}29)$$

The inductance is equal to the derivative of the voltage across an element with respect to the rate of change of current:

$$L = \frac{dP}{d(dF/dt)} \qquad (4.2\text{-}30)$$

For a linear system

$$L = \frac{P}{(dF/dt)} \qquad (4.2\text{-}31)$$

4.3   LINEAR REGRESSION MODELS

Experimental techniques may be used to develop steady-state process models for existing plants, and to check the theoretical models developed by using the physical laws discussed in Sections 4.1 and 4.2. Also, experiments may have to be conducted periodically to update or calibrate the coefficients of the physical process models. *Regression analysis is a technique of deriving a model that "best fits" a set of experimental data.* The so-called "best fit" does not necessarily make the model fit a particular set of data perfectly, i.e., the error between the model and any set of data is not necessarily zero. The "best fit" is understood to mean that the error function, which relates the differences between the model and the data, is minimized. The error function to be minimized is the sum of the squares of the differences between each data point and the value predicted by the model. This is referred to as *curve fitting by the least squares technique.*

This section will derive a linear regression model for two variables, Section 4.4 will discuss testing for significance, and Section 4.5 will derive the multiple linear and nonlinear regression models.

A.   Derivation for the Linear Regression Model

Figure 4.3-1 shows n sets of experimental data, (X_1, Y_1), (X_2, Y_2), ..., (X_n, Y_n). The model is assumed to be a straight line:

$$\hat{Y} = \phi(X) = A_0 + A_1 X \tag{4.3-1}$$

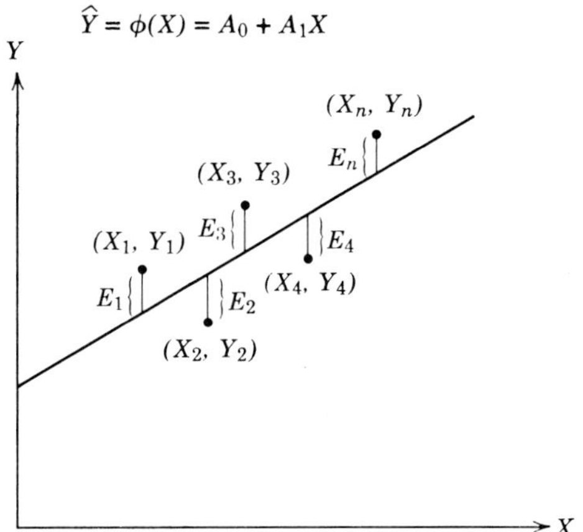

Fig. 4.3-1   Linear regression model.

where $\hat{Y}$ is the predicted value using the regression model. The coefficients A_0 and A_1 are to be derived so that the sum of the squared errors between the data predicted by the model and the experimental data is a minimum. In Figure 4.3-1 the error for each data point is defined as the distance along the vertical line from each point to the straight line of the model.

Let the values of $\hat{Y}$ on the straight line be

$$\hat{Y}_1 = A_0 + A_1 X_1 \qquad \text{at } X = X_1$$

$$\hat{Y}_2 = A_0 + A_1 X_2 \qquad \text{at } X = X_2$$

$$\vdots \qquad\qquad\qquad \vdots \qquad\qquad\qquad (4.3\text{-}2)$$

$$\hat{Y}_n = A_0 + A_1 X_n \qquad \text{at } X = X_n$$

Then the errors are

$$E_1 = (\hat{Y}_1 - Y_1) = (A_0 + A_1 X_1 - Y_1) \qquad \text{at } X = X_1$$

$$E_2 = (\hat{Y}_2 - Y_2) = (A_0 + A_1 X_2 - Y_2) \qquad \text{at } X = X_2$$

$$\vdots \qquad\qquad\qquad\qquad\qquad \vdots \qquad\qquad (4.3\text{-}3)$$

$$E_n = (\hat{Y}_n - Y_n) = (A_0 + A_1 X_n - Y_n) \qquad \text{at } X = X_n$$

The error function F is defined as

$$F = E_1{}^2 + E_2{}^2 + \cdots + E_n{}^2 \qquad (4.3\text{-}4)$$

$$\therefore \quad F = (A_0 + A_1 X_1 - Y_1)^2$$

$$+ (A_0 + A_1 X_2 - Y_2)^2$$

$$\vdots$$

$$+ (A_0 + A_1 X_n - Y_n)^2 \qquad (4.3\text{-}5)$$

Or

$$F = \sum_{i=1}^{n} (A_0 + A_1 X_i - Y_i)^2 \qquad (4.3\text{-}6)$$

The technique of calculus will be used to derive the values of A_0 and A_1 so as to make the function F a minimum. From Chapter 7 the conditions of minimum are

$$\frac{\partial F}{\partial A_0} = 0 \quad \text{and} \quad \frac{\partial F}{\partial A_1} = 0 \qquad (4.3\text{-}7)$$

$$\frac{\partial F}{\partial A_0} = \frac{\partial}{\partial A_0} \sum_{i=1}^{n} (A_0 + A_1 X_i - Y_i)^2$$

$$= \sum_{i=1}^{n} \frac{\partial}{\partial A_0} (A_0 + A_1 X_i - Y_i)^2$$

$$= \sum_{i=1}^{n} 2(A_0 + A_1 X_i - Y_i)$$

$$= 2\left(nA_0 + A_1 \sum_{i=1}^{n} X_i - \sum_{i=1}^{n} Y_i \right) = 0 \tag{4.3-8}$$

$$\therefore \quad nA_0 + \left(\sum_{i=1}^{n} X_i \right) A_1 = \sum_{i=1}^{n} Y_i \tag{4.3-9}$$

Similarly,

$$\frac{\partial F}{\partial A_1} = \sum_{i=1}^{n} \frac{\partial}{\partial A_1} (A_0 + A_1 X_i - Y_i)^2$$

$$= \sum_{i=1}^{n} 2(A_0 + A_1 X_i - Y_i)(X_i)$$

$$= 2A_0 \sum_{i=1}^{n} X_i + 2A_1 \sum_{i=1}^{n} X_i^2 - 2 \sum_{i=1}^{n} Y_i X_i = 0 \tag{4.3-10}$$

$$\therefore \quad \left(\sum_{i=1}^{n} X_1 \right) A_0 + \left(\sum_{i=1}^{n} X_i^2 \right) A_1 = \sum_{i=1}^{n} X_i Y_i \tag{4.3-11}$$

Equations 4.3-9 and 4.3-11 are the two simultaneous linear algebraic equations which can be used to solve for A_0 and A_1. In matrix notation Equations 4.3-9 and 4.3-11 become

$$\begin{pmatrix} n & \sum_{i=1}^{n} X_i \\ \sum_{i=1}^{n} X_i & \sum_{i=1}^{n} X_i^2 \end{pmatrix} \cdot \begin{pmatrix} A_0 \\ A_1 \end{pmatrix} = \begin{pmatrix} \sum_{i=1}^{n} Y_i \\ \sum_{i=1}^{n} X_i Y_i \end{pmatrix} \tag{4.3-12}$$

Solving Equation 4.3-12,

$$A_0 = \frac{\left(\sum_{i=1}^{n} Y_i \right)\left(\sum_{i=1}^{n} X_i^2 \right) - \left(\sum_{i=1}^{n} X_i \right)\left(\sum_{i=1}^{n} X_i Y_i \right)}{n\left(\sum_{i=1}^{n} X_i^2 \right) - \left(\sum_{i=1}^{n} X_i \right)^2} \tag{4.3-13}$$

$$A_1 = \frac{n\left(\sum_{i=1}^{n} X_i Y_i \right) - \left(\sum_{i=1}^{n} X_i \right)\left(\sum_{i=1}^{n} Y_i \right)}{n\left(\sum_{i=1}^{n} X_i^2 \right) - \left(\sum_{i=1}^{n} X_i \right)^2} \tag{4.3-14}$$

where n is the number of samples in the experimental data.

The *standard estimate of error* of the regression model is given by the standard deviation s

$$s = \left[\frac{\sum_{i=1}^{n} (E_i^2)}{(n-2)} \right]^{1/2} \tag{4.3-15}$$

$$s = \left[\frac{\sum_{i=1}^{n} (A_0 + A_1 X_i - Y_i)^2}{(n-2)} \right]^{1/2} \tag{4.3-16}$$

The standard deviation is thus the square root of the sum of the errors divided by the degrees of freedom $(n-2)$. For the data points with normal distribution approximately 66% of the points will fall within one standard deviation of the model, and 95% of the points will fall within two standard deviations. Thus, if the standard deviation of a particular regression model is 5, 95% of the experimental observations will be within ± 10 from the straight line, as shown in Figure 4.3-2. The standard estimate of error is an important indication of whether the regression model is meaningful or not. A large standard estimate of error may indicate that the model does not fit the process from which the set of experimental data was obtained. Another reason for a large standard estimate of error may be due to the inherent variability in the data, in which case a larger set of data may have to be taken to correct the problem.

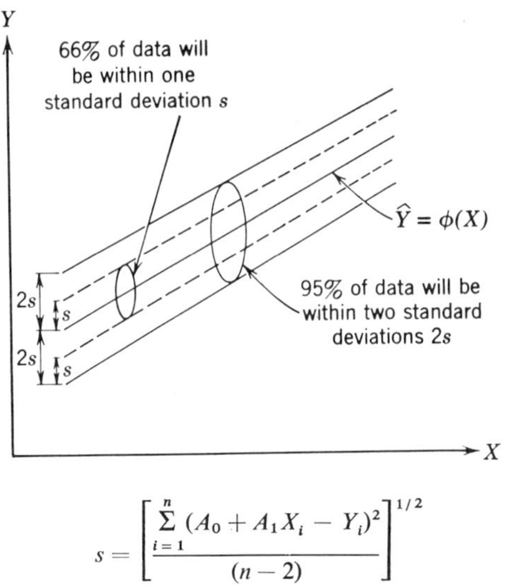

Fig. 4.3-2 Standard estimate of error for linear regression model of two variables.

X_i	Y_i	X_i^2	$(X_i \cdot Y_i)$
13.2	200	174.4	2640
15.9	250	252.8	3975
16.5	220	272.2	3630
19.6	223	384.1	4370
20.5	273	420.2	5596
23.5	242	552.2	5687
25.1	290	630.0	7279
27.5	277	756.2	7618
28.5	300	812.2	8550
29.6	333	876.2	9857
30.5	300	930.2	9150
34.0	285	1156.0	9690
$\sum_{i=1}^{n}$ 284.4	3193	7216.7	78042

$$A_0 = \frac{\left(\sum_{i=1}^{n} Y_i\right)\left(\sum_{i=1}^{n} X_i^2\right) - \left(\sum_{i=1}^{n} X_i\right)\left(\sum_{i=1}^{n} X_i Y_i\right)}{n\left(\sum_{i=1}^{n} X_i^2\right) - \left(\sum_{i=1}^{n} X_i\right)^2} = \frac{(23.0429 - 22.1951) \cdot 10^6}{(86.607 - 80.8834) \cdot 10^3}$$

$$A_0 = 148.3$$

$$A_1 = \frac{n\left(\sum_{i=1}^{n} X_i Y_i\right) - \left(\sum_{i=1}^{n} X_i\right)\left(\sum_{i=1}^{n} Y_i\right)}{n\left(\sum_{i=1}^{n} X_i^2\right) - \left(\sum_{i=1}^{n} X_i\right)^2} = \frac{(936.50 - 908.09) \cdot 10^3}{(86.607 - 80.8834) \cdot 10^3}$$

$$A_1 = 4.96$$

$$\therefore \quad \hat{Y} = 148.3 + 4.96\,(X)$$

Fig. 4.3-3 Computations for the coefficients of the linear regression model.

B. Example of Linear Regression Model for Two Variables

Figure 4.3-3 shows an example with 12 sets of experimental data. The problem is to derive a linear regression model. Figure 4.3-3 also shows some of the computations used to derive the coefficients of the straight line by using Equations 4.3-13 and 4.3-14. The equation of the regression model is

$$\hat{Y} = 148.3 + 4.96(X) \tag{4.3-17}$$

The straight line is plotted on Figure 4.3-4. The standard error of estimate given by Equation 4.3-16 is

$$s = 24.8 \tag{4.3-18}$$

This means that 66% of the data points are within $\hat{Y} \pm 24.8$, and 95% of the data points are within $\hat{Y} \pm 49.6$.

X	13.2	15.9	16.5	19.6	20.5	23.5	25.1	27.5	28.5	29.6	30.5	34.0
Y	200	250	220	223	273	242	290	277	300	333	300	285

$$\hat{Y} = \phi(X) = 148.3 + 4.96X$$

$$s = 24.8$$

$$2s = 49.6$$

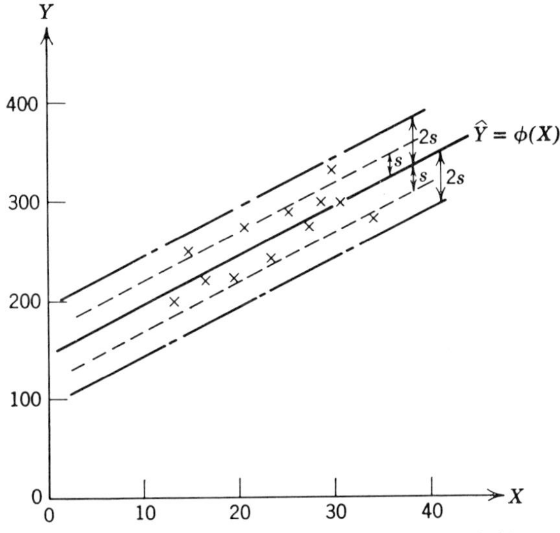

Fig. 4.3-4 Linear regression model for two variables.

C. Translating the Linear Regression Model to Go through the Mean

A technique that is frequently used to simplify computations is to translate the coordinates first so that the origin of the new coordinates *goes through the mean* of the set of experimental data and then develop the model in the new coordinates. Let the following symbols be assigned:

(X, Y): data in old coordinates.
(X', Y'): data in new coordinates.
$(\overline{X}, \overline{Y})$: the mean in old coordinates.
$\hat{Y}'$: regression model in new coordinates.

The mean is defined by

$$\overline{X} = \frac{\sum\limits_{i=1}^{n} X_i}{n} \tag{4.3-19}$$

$$\overline{Y} = \frac{\sum\limits_{i=1}^{n} Y_i}{n} \tag{4.3-20}$$

The translation of the coordinates is defined by

$$X' = X - \overline{X} \tag{4.3-21}$$

$$Y' = Y - \overline{Y} \tag{4.3-22}$$

The regression model in the new coordinates is

$$\hat{Y}' = A_1 X' \tag{4.3-23}$$

where A_1 is given by

$$A_1 = \frac{\sum_{i=1}^{n} (X_i' Y_i')}{\sum_{i=1}^{n} (X_i')^2} \tag{4.3-24}$$

Substituting Equations 4.3-21 and 4.3-22 into Equation 4.3-23, the regression model in the old coordinates is

$$(\hat{Y} - \overline{Y}) = A_1 (X - \overline{X}) \tag{4.3-25}$$

or

$$\hat{Y} = \overline{Y} + A_1 (X - \overline{X}) \tag{4.3-26}$$

A_0 can be found by comparing Equations 4.3-26 and 4.3-1:

$$A_0 = \overline{Y} - A_1 \overline{X} \tag{4.3-27}$$

Translating the model to *go through the mean* can make numerical solutions much better behaved. The denominator for A_1 in Equation 4.3-14 may be so small that division difficulties may result. The denominator for A_1 in Equation 4.3-24 poses no such problem.

4.4   SIGNIFICANCE OF REGRESSION MODELS

After a regression model is developed, the next step is to analyze the significance of the model. In other words, is the hypothesis based on the regression model sufficiently valid to make predictions? What is the precision that may be attached to the estimates made by the regression model? A detailed treatment of *tests of significance and analysis of variance* is outside the scope of this book. This section will only present the basic concepts and give some elementary tests. The main purpose is to make the reader aware that regression models are subject to errors, and objective tests are available to measure the validity of regression models.

Some of the basic concepts and definitions will be discussed first, then two tests will be presented for predicting the precision and analyzing the significance.

A. Basic Concepts and Definitions

The precision and significance of regression models may be analyzed by examining the following variances:

Variance of data about the mean, as shown in Figure 4.4-1.
Variance of data about the regression line, as shown in Figure 4.4-2.
Variance of the regression line about the mean, as shown in Figure 4.4-3.

Let the following symbols be assigned:

(X_i, Y_i) = experimental data $(i = 1)$ to $(i = n)$.
$\hat{Y} = \phi(X)$ = regression model.
$(X_i, \hat{Y}_i)$ = points on the regression line $(i = 1)$ to $(i = n)$.
$(\overline{X}, \overline{Y})$ = mean.
SS_t = variance of data about the mean; also called sum of squares—total.
SS_r = variance of data about the regression line; also called sum of squares—residual.
SS_b = variance of regression line about the mean; also called sum of squares—component.

The variance of the data about the mean is

$$SS_t = \sum_{i=1}^{n} (Y_i - \overline{Y})^2 \qquad (4.4\text{-}1)$$

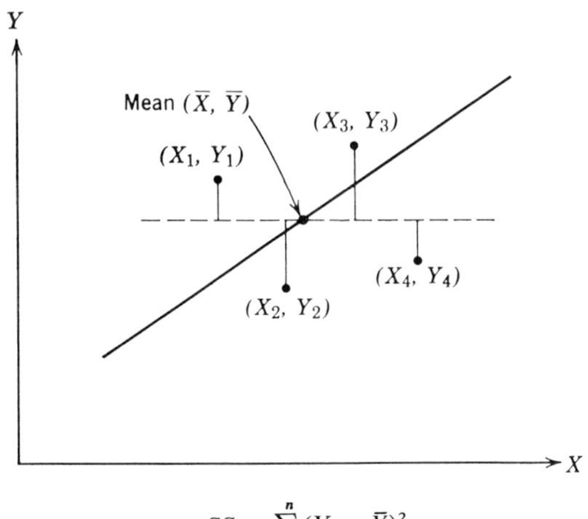

$$SS_t = \sum_{i=1}^{n} (Y_i - \overline{Y})^2$$

Fig. 4.4-1 SS_t = variance of data about the mean.

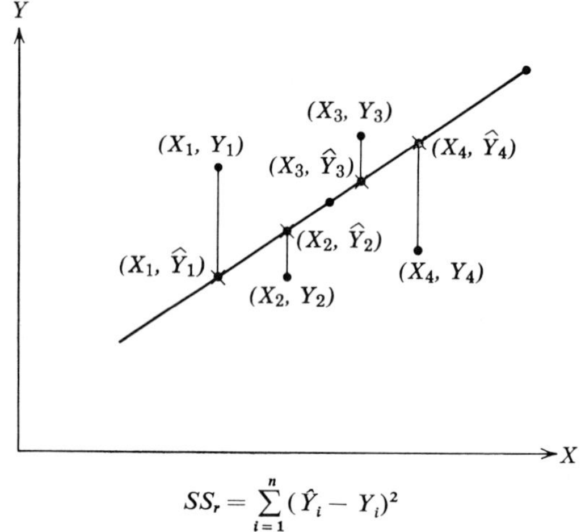

$$SS_r = \sum_{i=1}^{n} (\hat{Y}_i - Y_i)^2$$

Fig. 4.4-2 $SS_r =$ variance of data about the regression line.

The variance of the data about the regression line is

$$SS_r = \sum_{i=1}^{n} (Y_i - \hat{Y}_i)^2 \tag{4.4-2}$$

The variance of the regression line about the mean is

$$SS_b = \sum_{i=1}^{n} (\hat{Y}_i - \overline{Y})^2 \tag{4.4-3}$$

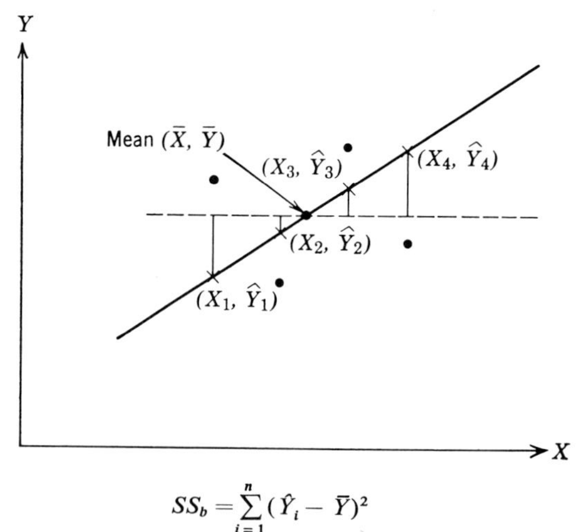

$$SS_b = \sum_{i=1}^{n} (\hat{Y}_i - \overline{Y})^2$$

Fig. 4.4-3 $SS_b =$ variance of the regression line about the mean.

A brief derivation will be given to relate the three types of variances. Equation 4.4-2 may be expanded by adding and subtracting $\overline{Y}$.

$$SS_r = \sum_{i=1}^{n} (Y_i - \hat{Y}_i)^2$$

$$= \sum_{i=1}^{n} [(Y_i - \overline{Y}) - (\hat{Y}_i - \overline{Y})]^2 \tag{4.4-4}$$

$$SS_r = \sum_{i=1}^{n} [(Y_i - \overline{Y})^2 + (\hat{Y}_i - \overline{Y})^2 - 2(Y_i - \overline{Y})(\hat{Y}_i - \overline{Y})]$$

$$= \sum_{i=1}^{n} (Y_i - \overline{Y})^2 + \left[\sum_{i=1}^{n} (\hat{Y}_i - \overline{Y})^2 - 2 \sum_{i=1}^{n} (Y_i - \overline{Y})(\hat{Y}_i - \overline{Y}) \right] \tag{4.4-5}$$

From Equation 4.3-25

$$(\hat{Y}_i - \overline{Y}) = A_1(X_i - \overline{X}) \tag{4.4-6}$$

Substituting Equation 4.4-6 into Equation 4.4-5,

$$SS_r = \sum_{i=1}^{n} (Y_i - \overline{Y})^2 + \left[\sum_{i=1}^{n} (\hat{Y}_i - \overline{Y})^2 - \sum_{i=1}^{n} 2(A_1)(Y_i - \overline{Y})(X_i - \overline{X}) \right] \tag{4.4-7}$$

From Equation 4.3-24

$$\sum_{i=1}^{n} (Y_i - \overline{Y})(X_i - \overline{X}) = (A_1) \sum_{i=1}^{n} (X_i - \overline{X})^2 \tag{4.4-8}$$

Substituting Equation 4.4-8 into Equation 4.4-7,

$$SS_r = \sum_{i=1}^{n} (Y_i - \overline{Y})^2 + \sum_{i=1}^{n} (\hat{Y}_i - \overline{Y})^2 - 2(A_1)^2 \sum_{i=1}^{n} (X_i - \overline{X})^2 \tag{4.4-9}$$

Substituting Equation 4.4-6 into Equation 4.4-9,

$$SS_r = \sum_{i=1}^{n} (Y_i - \overline{Y})^2 + \sum_{i=1}^{n} (\hat{Y}_i - \overline{Y})^2 - 2 \sum_{i=1}^{n} (Y_i - \overline{Y})^2 \tag{4.4-10}$$

$$\therefore \quad SS_r = \sum_{i=1}^{n} (Y_i - \overline{Y})^2 - \sum_{i=1}^{n} (\hat{Y}_i - \overline{Y})^2 \tag{4.4-11}$$

Substituting Equation 4.4-2 into 4.4-11,

$$\sum_{i=1}^{n} (Y_i - \hat{Y}_i)^2 = \sum_{i=1}^{n} (Y_i - \overline{Y})^2 - \sum_{i=1}^{n} (\hat{Y}_i - \overline{Y})^2 \tag{4.4-12}$$

Equation 4.4-12 may be rewritten as

$$\sum_{i=1}^{n} (Y_i - \overline{Y})^2 = \sum_{i=1}^{n} (Y_i - \hat{Y}_i)^2 + \sum_{i=1}^{n} (\hat{Y}_i - \overline{Y})^2 \tag{4.4-13}$$

Using the definitions of SS_t, SS_r, and SS_b, Equation 4.4-13 becomes

$$SS_t = SS_r + SS_b \qquad (4.4\text{-}14)$$

Equation 4.4-14, stated in words, becomes

$$\begin{pmatrix} SS \text{ of data} \\ \text{about} \\ \text{the mean} \end{pmatrix} = \begin{pmatrix} SS \text{ of data about} \\ \text{the regression} \\ \text{line} \end{pmatrix} + \begin{pmatrix} SS \text{ of the regression} \\ \text{line about the} \\ \text{mean} \end{pmatrix} \qquad (4.4\text{-}15)$$

Equations 4.4-13 through 4.4-15 provide the fundamental relationship between the three types of variances. This relationship will be used for the testing of significance and estimating the precision of regression models.

B. R^2—the Coefficient of Determination

A simple indication of the precision of regression models is the coefficient of determination R^2. By definition, R^2 is the ratio of SS_b to SS_t, i.e., the ratio of "SS of the regression line about the mean" to "SS of data about the the mean." Thus

$$R^2 = \frac{SS_b}{SS_t} \qquad (4.4\text{-}16)$$

For the regression model to be a good predictor, R^2 should be nearly equal to one:

$$R^2 \simeq 1 \qquad (4.4\text{-}17)$$

When R^2 is equal to 1, all the data will fit exactly on the regression line. Thus for $R^2 = 1$

$$SS_r = 0 \qquad (4.4\text{-}18)$$

and

$$SS_t = SS_b \qquad (4.4\text{-}19)$$

C. Testing for Significance

The procedure for testing for significance was developed by Fisher in 1956. Basically, it is an objective test to distinguish between *effects that are due to random variations (or errors)* and *effects that are due to the variations of the control variables in the experiments*. This is necessary to determine if the model developed is sufficiently valid for prediction purposes. The regression model may be regarded as a hypothesis based on the set of experimental data. The Fisher F-test provides a rule for comparing the given hypothesis against a set of predetermined observed data which already establishes an *unacceptable hypothesis of low probability*. If the given hypothesis belongs to this set, it is rejected as an unacceptable model. If the given hypothesis is outside this set, it is considered an acceptable model.

Let the following symbols be assigned:

$F_{1,(rn-K-1),\alpha}$ = F-distribution value at 1 and $(rn - K - 1)$ degrees of freedom, with certain chance of error α. These values of F are obtained from standard statistical tables.

n = number of experimental points.

r = number of times the experiment is repeated; e.g., $r = 1$ for one set of experimental data.

K = number of control variables; e.g., $K = 1$ for two variable regression models.

α = certain chance of error; thus the confidence level is $(1 - \alpha)$.

A_j = regression coefficient of jth variable; A_1 is the coefficient of two variable linear regression models.

MS = mean square, equals SS divided by the degrees of freedom.

The Fisher F-test is used to test the hypothesis that the regression coefficient A_j is equal to 0 by making the comparison:

$$R_j > F_{1,(rn-K-1),\alpha} \tag{4.4-20}$$

where R_j is defined as the ratio of MS_b over MS_r. Thus

$$R_j = \frac{MS_b}{MS_r} \tag{4.4-21}$$

If the test in Equation 4.4-20 is satisfied, the hypothesis that the regression coefficient A_j is equal to 0 can be rejected, with a confidence level of $(1 - \alpha)$. In other words, the F-test provides a comparison of the variance of the regression coefficient distribution against the variance of the total data distribution, all with a confidence level of $(1 - \alpha)$.

Table 4.4-1   Analysis of Variance for Linear Regression

Sources of Variation	Sum of Squares	Degree of Freedom	Mean Square
Due to regression	$SS_b = \sum_{i=1}^{n}(\hat{Y}_i - \bar{Y})^2$	1	$MS_b = SS_b$
About regression (residual)	$SS_r = \sum_{i=1}^{n}(Y_i - \hat{Y}_i)^2$	$(n-2)$	$MS_r = \dfrac{SS_r}{(n-2)}$
About mean	$SS_t = \sum_{i=1}^{n}(Y_i - \bar{Y})^2$	$(n-1)$	$MS_t = \dfrac{SS_t}{(n-1)}$

An analysis of variance table is shown in Table 4.4-1. The degrees of freedom for n-experimental points are (1) for SS_b, $(n-2)$ for SS_r, and $(n-1)$ for SS_t. The mean square is obtained by dividing the SS by the degrees of freedom. For example, from the F-statistic table in Tables 4.4-2 and 4.4-3,

$$F_{1,12,0.05} = 4.75. \tag{4.4-22}$$

Thus, if a linear regression model derived from 14 data points has

$$R_j = MS_b \div MS_r = 12.92, \tag{4.4-23}$$

$$\therefore \quad R_j > F_{1,12,0.05}. \tag{4.4-24}$$

The hypothesis that the regression coefficient A_j is equal to 0 can be rejected, with a confidence level of 0.95. On the other hand, if

$$R_j = 2.36, \tag{4.4-25}$$

$$\therefore \quad R_j < F_{1,12,0.05}. \tag{4.4-26}$$

Then the hypothesis that the regression coefficient A_j is equal to 0 may be accepted, with a confidence level of 0.95.

Equation 4.4-24 may be interpreted as indicating that the fit of the regression model is *adequate*, whereas Equation 4.4-26 indicates that the fit is *inadequate*.

4.5   MULTIPLE REGRESSION MODELS

In many situations a model may consist of one variable expressed as a function of several variables. Thus

$$\hat{Y} = \phi(X_1, X_2, \ldots, X_K) \tag{4.5-1}$$

where $\hat{Y}$ is the predicted state variable, and $X_1, X_2, \ldots, X_K$ are the control variables.

A. Multiple Linear Regression Models

Again the problem is to derive the set of coefficients $A_0, A_1, A_2, \ldots, A_K$ for Equation 4.5-2 so that the sum of the error squared between the regression model and the set of n experimental data is minimum.

The multiple linear regression model is given by

$$\hat{Y} = A_0 + \sum_{j=1}^{K} A_j X_j \tag{4.5-2}$$

Table 4.4-2 F-Distribution, 95% Confidence Level

Degrees of Freedom for Numerator of F-Ratio

	1	2	3	4	5	6	7	8	9	10	12	15	20	24	30	40	60	120	∞
1	161.4	199.5	215.7	224.6	230.2	234.0	236.8	238.9	240.5	241.9	243.9	245.9	248.0	249.1	250.1	251.1	252.2	253.3	254.3
2	18.51	19.00	19.16	19.25	19.30	19.33	19.35	19.37	19.38	19.40	19.41	19.43	19.45	19.45	19.46	19.47	19.48	19.49	19.50
3	10.13	9.55	9.28	9.12	9.01	8.94	8.89	8.85	8.81	8.79	8.74	8.70	8.66	8.64	8.62	8.59	8.57	8.55	8.53
4	7.71	6.94	6.59	6.39	6.26	6.16	6.09	6.04	6.00	5.96	5.91	5.86	5.80	5.77	5.75	5.72	5.69	5.66	5.63
5	6.61	5.79	5.41	5.19	5.05	4.95	4.88	4.82	4.77	4.74	4.68	4.62	4.56	4.53	4.50	4.46	4.43	4.40	4.36
6	5.99	5.14	4.76	4.53	4.39	4.28	4.21	4.15	4.10	4.06	4.00	3.94	3.87	3.84	3.81	3.77	3.74	3.70	3.67
7	5.59	4.74	4.35	4.12	3.97	3.87	3.79	3.73	3.68	3.64	3.57	3.51	3.44	3.41	3.38	3.34	3.30	3.27	3.23
8	5.32	4.46	4.07	3.84	3.69	3.58	3.50	3.44	3.39	3.35	3.28	3.22	3.15	3.12	3.08	3.04	3.01	2.97	2.93
9	5.12	4.26	3.86	3.63	3.48	3.37	3.29	3.23	3.18	3.14	3.07	3.01	2.94	2.90	2.86	2.83	2.79	2.75	2.71
10	4.96	4.10	3.71	3.48	3.33	3.22	3.14	3.07	3.02	2.98	2.91	2.85	2.77	2.74	2.70	2.66	2.62	2.58	2.54
11	4.84	3.98	3.59	3.36	3.20	3.09	3.01	2.95	2.90	2.85	2.79	2.72	2.65	2.61	2.57	2.53	2.49	2.45	2.40
12	4.75	3.89	3.49	3.26	3.11	3.00	2.91	2.85	2.80	2.75	2.69	2.62	2.54	2.51	2.47	2.43	2.38	2.34	2.30
13	4.67	3.81	3.41	3.18	3.03	2.92	2.83	2.77	2.71	2.67	2.60	2.53	2.46	2.42	2.38	2.34	2.30	2.25	2.21
14	4.60	3.74	3.34	3.11	2.96	2.85	2.76	2.70	2.65	2.60	2.53	2.46	2.39	2.35	2.31	2.27	2.22	2.18	2.13
15	4.54	3.68	3.29	3.06	2.90	2.79	2.71	2.64	2.59	2.54	2.48	2.40	2.33	2.29	2.25	2.20	2.16	2.11	2.07
16	4.49	3.63	3.24	3.01	2.85	2.74	2.66	2.59	2.54	2.49	2.42	2.35	2.28	2.24	2.19	2.15	2.11	2.06	2.01
17	4.45	3.59	3.20	2.96	2.81	2.70	2.61	2.55	2.49	2.45	2.38	2.31	2.23	2.19	2.15	2.10	2.06	2.01	1.96
18	4.41	3.55	3.16	2.93	2.77	2.66	2.58	2.51	2.46	2.41	2.34	2.27	2.19	2.15	2.11	2.06	2.02	1.97	1.92
19	4.38	3.52	3.13	2.90	2.74	2.63	2.54	2.48	2.42	2.38	2.31	2.23	2.16	2.11	2.07	2.03	1.98	1.93	1.88
20	4.35	3.49	3.10	2.87	2.71	2.60	2.51	2.45	2.39	2.35	2.28	2.20	2.12	2.08	2.04	1.99	1.95	1.90	1.84
21	4.32	3.47	3.07	2.84	2.68	2.57	2.49	2.42	2.37	2.32	2.25	2.18	2.10	2.05	2.01	1.96	1.92	1.87	1.81
22	4.30	3.44	3.05	2.82	2.66	2.55	2.46	2.40	2.34	2.30	2.23	2.15	2.07	2.03	1.98	1.94	1.89	1.84	1.78
23	4.28	3.42	3.03	2.80	2.64	2.53	2.44	2.37	2.32	2.27	2.20	2.13	2.05	2.01	1.96	1.91	1.86	1.81	1.76
24	4.26	3.40	3.01	2.78	2.62	2.51	2.42	2.36	2.30	2.25	2.18	2.11	2.03	1.98	1.94	1.89	1.84	1.79	1.73
25	4.24	3.39	2.99	2.76	2.60	2.49	2.40	2.34	2.28	2.24	2.16	2.09	2.01	1.96	1.92	1.87	1.82	1.77	1.71
26	4.23	3.37	2.98	2.74	2.59	2.47	2.39	2.32	2.27	2.22	2.15	2.07	1.99	1.95	1.90	1.85	1.80	1.75	1.69
27	4.21	3.35	2.96	2.73	2.57	2.46	2.37	2.31	2.25	2.20	2.13	2.06	1.97	1.93	1.88	1.84	1.79	1.73	1.67
28	4.20	3.34	2.95	2.71	2.56	2.45	2.36	2.29	2.24	2.19	2.12	2.04	1.96	1.91	1.87	1.82	1.77	1.71	1.65
29	4.18	3.33	2.93	2.70	2.55	2.43	2.35	2.28	2.22	2.18	2.10	2.03	1.94	1.90	1.85	1.81	1.75	1.70	1.64
30	4.17	3.32	2.92	2.69	2.53	2.42	2.33	2.27	2.21	2.16	2.09	2.01	1.93	1.89	1.84	1.79	1.74	1.68	1.62
40	4.08	3.23	2.84	2.61	2.45	2.34	2.25	2.18	2.12	2.08	2.00	1.92	1.84	1.79	1.74	1.69	1.64	1.58	1.51
60	4.00	3.15	2.76	2.53	2.37	2.25	2.17	2.10	2.04	1.99	1.92	1.84	1.75	1.70	1.65	1.59	1.53	1.47	1.39
120	3.92	3.07	2.68	2.45	2.29	2.17	2.09	2.02	1.96	1.91	1.83	1.75	1.66	1.61	1.55	1.50	1.43	1.35	1.25
∞	3.84	3.00	2.60	2.37	2.21	2.10	2.01	1.94	1.88	1.83	1.75	1.67	1.57	1.52	1.46	1.39	1.32	1.22	1.00

Degrees of Freedom for Denominator of F-Ratio

Reproduced with permission from E. S. Pearson and H. O. Hartley, *Biometrika Tables for Statisticians*, Vol. 1, Cambridge University Press, New York, 1954.

Table 4.4-3 F-Distribution, 99% Confidence Level

Degrees of Freedom for Numerator of F-Ratio

Degrees of Freedom for Denominator of F-Ratio

	1	2	3	4	5	6	7	8	9	10	12	15	20	24	30	40	60	120	∞
1	4052	4999.5	5403	5625	5764	5859	5928	5982	6022	6056	6106	6157	6209	6235	6261	6287	6313	6339	6366
2	98.50	99.00	99.17	99.25	99.30	99.33	99.36	99.37	99.39	99.40	99.42	99.43	99.45	99.46	99.47	99.47	99.48	99.49	99.50
3	34.12	30.82	29.46	28.71	28.24	27.91	27.67	27.49	27.35	27.23	27.05	26.87	26.69	26.60	26.50	26.41	26.32	26.22	26.13
4	21.20	18.00	16.69	15.98	15.52	15.21	14.98	14.80	14.66	14.55	14.37	14.20	14.02	13.93	13.84	13.75	13.65	13.56	13.46
5	16.26	13.27	12.06	11.39	10.97	10.67	10.46	10.29	10.16	10.05	9.89	9.72	9.55	9.47	9.38	9.29	9.20	9.11	9.02
6	13.75	10.92	9.78	9.15	8.75	8.47	8.26	8.10	7.98	7.87	7.72	7.56	7.40	7.31	7.23	7.14	7.06	6.97	6.88
7	12.25	9.55	8.45	7.85	7.46	7.19	6.99	6.84	6.72	6.62	6.47	6.31	6.16	6.07	5.99	5.91	5.82	5.74	5.65
8	11.26	8.65	7.59	7.01	6.63	6.37	6.18	6.03	5.91	5.81	5.67	5.52	5.36	5.28	5.20	5.12	5.03	4.95	4.86
9	10.56	8.02	6.99	6.42	6.06	5.80	5.61	5.47	5.35	5.26	5.11	4.96	4.81	4.73	4.65	4.57	4.48	4.40	4.31
10	10.04	7.56	6.55	5.99	5.64	5.39	5.20	5.06	4.94	4.85	4.71	4.56	4.41	4.33	4.25	4.17	4.08	4.00	3.91
11	9.65	7.21	6.22	5.67	5.32	5.07	4.89	4.74	4.63	4.54	4.40	4.25	4.10	4.02	3.94	3.86	3.78	3.69	3.60
12	9.33	6.93	5.95	5.41	5.06	4.82	4.64	4.50	4.39	4.30	4.16	4.01	3.86	3.78	3.70	3.62	3.54	3.45	3.36
13	9.07	6.70	5.74	5.21	4.86	4.62	4.44	4.30	4.19	4.10	3.96	3.82	3.66	3.59	3.51	3.43	3.34	3.25	3.17
14	8.86	6.51	5.56	5.04	4.69	4.46	4.28	4.14	4.03	3.94	3.80	3.66	3.51	3.43	3.35	3.27	3.18	3.09	3.00
15	8.68	6.36	5.42	4.89	4.56	4.32	4.14	4.00	3.89	3.80	3.67	3.52	3.37	3.29	3.21	3.13	3.05	2.96	2.87
16	8.53	6.23	5.29	4.77	4.44	4.20	4.03	3.89	3.78	3.69	3.55	3.41	3.26	3.18	3.10	3.02	2.93	2.84	2.75
17	8.40	6.11	5.18	4.67	4.34	4.10	3.93	3.79	3.68	3.59	3.46	3.31	3.16	3.08	3.00	2.92	2.83	2.75	2.65
18	8.29	6.01	5.09	4.58	4.25	4.01	3.84	3.71	3.60	3.51	3.37	3.23	3.08	3.00	2.92	2.84	2.75	2.66	2.57
19	8.18	5.93	5.01	4.50	4.17	3.94	3.77	3.63	3.52	3.43	3.30	3.15	3.00	2.92	2.84	2.76	2.67	2.58	2.49
20	8.10	5.85	4.94	4.43	4.10	3.87	3.70	3.56	3.46	3.37	3.23	3.09	2.94	2.86	2.78	2.69	2.61	2.52	2.42
21	8.02	5.78	4.87	4.37	4.04	3.81	3.64	3.51	3.40	3.31	3.17	3.03	2.88	2.80	2.72	2.64	2.55	2.46	2.36
22	7.95	5.72	4.82	4.31	3.99	3.76	3.59	3.45	3.35	3.26	3.12	2.98	2.83	2.75	2.67	2.58	2.50	2.40	2.31
23	7.88	5.66	4.76	4.26	3.94	3.71	3.54	3.41	3.30	3.21	3.07	2.93	2.78	2.70	2.62	2.54	2.45	2.35	2.26
24	7.82	5.61	4.72	4.22	3.90	3.67	3.50	3.36	3.26	3.17	3.03	2.89	2.74	2.66	2.58	2.49	2.40	2.31	2.21
25	7.77	5.57	4.68	4.18	3.85	3.63	3.46	3.32	3.22	3.13	2.99	2.85	2.70	2.62	2.54	2.45	2.36	2.27	2.17
26	7.72	5.53	4.64	4.14	3.82	3.59	3.42	3.29	3.18	3.09	2.96	2.81	2.66	2.58	2.50	2.42	2.33	2.23	2.13
27	7.68	5.49	4.60	4.11	3.78	3.56	3.39	3.26	3.15	3.06	2.93	2.78	2.63	2.55	2.47	2.38	2.29	2.20	2.10
28	7.64	5.45	4.57	4.07	3.75	3.53	3.36	3.23	3.12	3.03	2.90	2.75	2.60	2.52	2.44	2.35	2.26	2.17	2.06
29	7.60	5.42	4.54	4.04	3.73	3.50	3.33	3.20	3.09	3.00	2.87	2.73	2.57	2.49	2.41	2.33	2.23	2.14	2.03
30	7.56	5.39	4.51	4.02	3.70	3.47	3.30	3.17	3.07	2.98	2.84	2.70	2.55	2.47	2.39	2.30	2.21	2.11	2.01
40	7.31	5.18	4.31	3.83	3.51	3.29	3.12	2.99	2.89	2.80	2.66	2.52	2.37	2.29	2.20	2.11	2.02	1.92	1.80
60	7.08	4.98	4.13	3.65	3.34	3.12	2.95	2.82	2.72	2.63	2.50	2.35	2.20	2.12	2.03	1.94	1.84	1.73	1.60
120	6.85	4.79	3.95	3.48	3.17	2.96	2.79	2.66	2.56	2.47	2.34	2.19	2.03	1.95	1.86	1.76	1.66	1.53	1.38
∞	6.63	4.61	3.78	3.32	3.02	2.80	2.64	2.51	2.41	2.32	2.18	2.04	1.88	1.79	1.70	1.59	1.47	1.32	1.00

The error function is

$$F = (A_0 + A_1 X_{11} + A_2 X_{21} + \cdots + A_K X_{K1} - Y_1)^2$$
$$+ (A_0 + A_1 X_{12} + A_2 X_{22} + \cdots + A_2 X_{K2} - Y_2)^2$$
$$\vdots$$
$$+ (A_0 + A_1 X_{1n} + A_2 X_{2n} + \cdots + A_K X_{Kn} - Y_n)^2 \tag{4.5-3}$$

To minimize F set

$$\frac{\partial F}{\partial A_0} = \frac{\partial F}{\partial A_1} = \frac{\partial F}{\partial A_2} \cdots = \frac{\partial F}{\partial A_K} = 0 \tag{4.5-4}$$

The following set of simultaneous linear equations in matrix notation are obtained to solve for the coefficients of the multiple linear regression model, $A_0, A_1, A_2, \ldots, A_K$:

$$\begin{pmatrix} n & \sum_{i=1}^{n}(X_{1i}) & \sum_{i=1}^{n}(X_{2i}) & \cdots & \sum_{i=1}^{n}(X_{Ki}) \\ \sum_{i=1}^{n}(X_{1i}) & \sum_{i=1}^{n}(X_{1i})^2 & \sum_{i=1}^{n}(X_{1i}X_{2i}) & \cdots & \sum_{i=1}^{n}(X_{1i}X_{Ki}) \\ \sum_{i=1}^{n}(X_{2i}) & \sum_{i=1}^{n}(X_{2i}X_{1i}) & \sum_{i=1}^{n}(X_{2i})^2 & \cdots & \sum_{i=1}^{n}(X_{2i}X_{Ki}) \\ \vdots & & & & \vdots \\ \sum_{i=1}^{n}(X_{Ki}) & \sum_{i=1}^{n}(X_{Ki}X_{1i}) & \sum_{i=1}^{n}(X_{Ki}X_{2i}) & \cdots & \sum_{i=1}^{n}(X_{Ki})^2 \end{pmatrix} \cdot \begin{pmatrix} A_0 \\ A_1 \\ A_2 \\ \vdots \\ A_K \end{pmatrix} = \begin{pmatrix} \sum_{i=1}^{n}(Y_i) \\ \sum_{i=1}^{n}(X_{1i}Y_i) \\ \sum_{i=1}^{n}(X_{2i}Y_i) \\ \vdots \\ \sum_{i=1}^{n}(X_{Ki}Y_i) \end{pmatrix} \tag{4.5-5}$$

where n is the number of samples of experimental data, and i is the index for the data points.

B. Translating the Multiple Linear Regression Model to Go through the Mean

Utilizing the technique of *going through the mean* can also simplify the computations for the coefficients of the multiple regression model. For computations using a digital computer, translating the coordinates by *going through the mean* can reduce the chance of a poorly behaved matrix "blowing up" during the matrix inversion process. Comparing Equation 4.5-5 and Equation 4.5-9 shows that translating the coordinates to *go through the mean* is equivalent to eliminating the first row and column of the matrix in Equation

4.5-5. This in effect takes the term n out of the main diagonal, which can usually make the matrix better behaved if n is much smaller than the other terms in the row.

Translating the coordinates to *go through the mean* is given by the following relationships:

$$X'_{Ki} = X_{Ki} - \overline{X}_K \tag{4.5-6}$$

$$Y'_i = Y_i - \overline{Y} \tag{4.5-7}$$

The multiple linear regression model which *goes through the mean* is given by

$$\hat{Y}' = \sum_{j=1}^{K} A_j X'_j \tag{4.5-8}$$

The coefficients A_j are the same as those in Equation 4.5-5. The equations in matrix notation to solve for A_j are as follows:

$$
\begin{pmatrix}
\sum\limits_{i=1}^{n}(X'_{1i})^2 & \sum\limits_{i=1}^{n}(X'_{1i}X'_{2i}) & \sum\limits_{i=1}^{n}(X'_{1i}X'_{3i}) & \cdots & \sum\limits_{i=1}^{n}(X'_{1i}X'_{Ki}) \\
\sum\limits_{i=1}^{n}(X'_{2i}X'_{1i}) & \sum\limits_{i=1}^{n}(X'_{2i})^2 & \sum\limits_{i=1}^{n}(X'_{2i}X'_{3i}) & \cdots & \sum\limits_{i=1}^{n}(X'_{2i}X'_{Ki}) \\
\sum\limits_{i=1}^{n}(X'_{3i}X'_{1i}) & \sum\limits_{i=1}^{n}(X'_{3i}X'_{2i}) & \sum\limits_{i=1}^{n}(X'_{3i})^2 & \cdots & \sum\limits_{i=1}^{n}(X'_{3i}X'_{Ki}) \\
\vdots & & & & \vdots \\
\sum\limits_{i=1}^{n}(X'_{Ki}X'_{1i}) & \sum\limits_{i=1}^{n}(X'_{Ki}X'_{2i}) & \sum\limits_{i=1}^{n}(X'_{Ki}X'_{3i}) & \cdots & \sum\limits_{i=1}^{n}(X'_{Ki})^2
\end{pmatrix}
\cdot
\begin{pmatrix}
A_1 \\ A_2 \\ A_3 \\ \vdots \\ A_K
\end{pmatrix}
=
\begin{pmatrix}
\sum\limits_{i=1}^{n}(X'_{1i}Y'_i) \\
\sum\limits_{i=1}^{n}(X'_{2i}Y'_i) \\
\sum\limits_{i=1}^{n}(X'_{3i}Y'_i) \\
\vdots \\
\sum\limits_{i=1}^{n}(X'_{Ki}Y'_i)
\end{pmatrix}
\tag{4.5-9}
$$

$$A_0 = \overline{Y} - (A_1\overline{X}_1 + A_2\overline{X}_2 + A_3\overline{X}_3 + \cdots + A_K\overline{X}_K) \tag{4.5-10}$$

where n is the number of samples of experimental data and i is the index for the data points.

C. Example of Multiple Linear Regression Model

A five-step procedure will be used to develop multiple linear regression models of an X-ray emission gauge (XEG) used for on-line analysis of different materials of a finely ground stream of material in a cement plant. These five steps are the following:

1. Description of XEG and measurements.
2. Description of regression models.

3. Equations in matrix form to solve for the coefficients of regression models.

4. Experimental data.

5. Solution.

1. Description of XEG and Measurements

Figure 4.5-1 shows an XEG being fed a continuous sample of the finely powdered material. There are four channels in the XEG for measuring silicon oxide SiO_2, aluminum oxide Al_2O_3, iron oxide Fe_2O_3, and calcium oxide CaO. The problem arises because of the "matrix effect" in which the presence of one material affects the readings of the other materials. Experiments can be conducted to actually take readings of each of the four channels for different samples of material. The "true contents" of the materials may then be determined from wet chemical analysis of the same samples. The regression analysis technique can then be applied to develop the coefficients which relate the "true contents" of each material to the readings of the four channels.

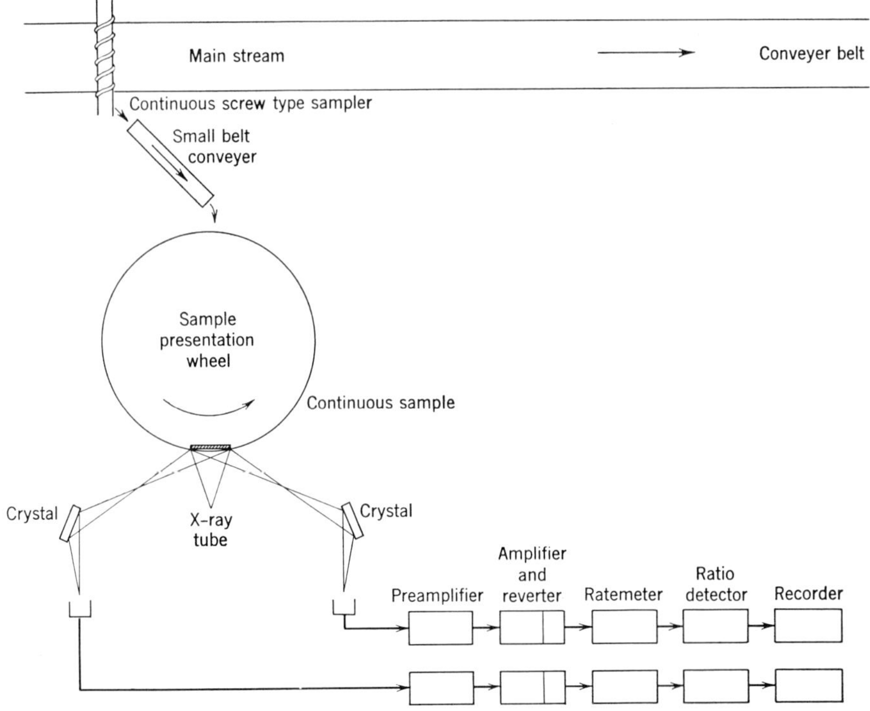

Fig. 4.5-1   Application of X-ray emission gauge in a cement plant.

2. Description of Regression Models

Multiple regression models are required in this example because of the number of relationships to be developed. The regression models are assumed to be linear.

The relationships to be developed are the "true contents" of each material as a function of all the readings of the XEG channels. Let the following symbols* be assigned:

S_n = true SiO_2 contents of nth sample from wet chemical analysis.
A_n = true Al_2O_3 contents of nth sample from wet chemical analysis.
F_n = true Fe_2O_3 contents of nth sample from wet chemical analysis.
C_n = true CaO contents of nth sample from wet chemical analysis.
S_{xn} = XEG reading of nth sample for silicon oxide channel.
A_{xn} = XEG reading of nth sample for aluminum oxide channel.
F_{xn} = XEG reading of nth sample for iron oxide channel.
C_{xn} = XEG reading of nth sample for calcium oxide channel.
N = number of samples taken.
$S_R, B_{SS}, B_{SA}, B_{SF}, B_{SC}$ = coefficients for $\hat{S}$.
$A_R, B_{AS}, B_{AA}, B_{AF}, B_{AC}$ = coefficients for $\hat{A}$.
$F_R, B_{FS}, B_{FA}, B_{FF}, B_{FC}$ = coefficients for $\hat{F}$.
$C_R, B_{CS}, B_{CA}, B_{CF}, B_{CC}$ = coefficients for $\hat{C}$.

The multiple regression models are given in Equations 4.5-11 through 4.5-14:

$$\hat{S} = S_R + B_{SS} S_x + B_{SA} A_x + B_{SF} F_x + B_{SC} C_x \qquad (4.5\text{-}11)$$

$$\hat{A} = A_R + B_{AS} S_x + B_{AA} A_x + B_{AF} F_x + B_{AC} C_x \qquad (4.5\text{-}12)$$

$$\hat{F} = F_R + B_{FS} S_x + B_{FA} A_x + B_{FF} F_x + B_{FC} C_x \qquad (4.5\text{-}13)$$

$$\hat{C} = C_R + B_{CS} S_x + B_{CA} A_x + B_{CF} F_x + B_{CC} C_x \qquad (4.5\text{-}14)$$

Equations 4.5-11 through 4.5-14 are independent. The problem is to derive the coefficients of the equations from a set of experimental data shown in Table 4.5-1.

3. Equations in Matrix Notation to Solve for the Coefficients of Regression Models

The coefficients of silicon oxide can be solved by using Equation 4.5-15, which is in matrix notation. This is derived by using the multiple linear regression model developed in Equation 4.5-5.

---

*n is used as subscript here instead of i which was used in the derivations to avoid confusion with Si.

Table 4.5-1 Experimental Data for X-ray Emission Gauge

Sample Number	Silicon Oxide		Aluminum Oxide		Iron Oxide		Calcium Oxide	
	S_n (Chem)	S_{xn} (XEG)	A_n (Chem)	A_{xn} (XEG)	F_n (Chem)	F_{xn} (XEG)	C_n (Chem)	C_{xn} (XEG)
1	16.59	598.95	1.45	37.45	0.53	335.65	44.09	7185.3
2	32.13	1051.10	1.00	15.20	0.38	211.20	34.41	5739.8
3	1.88	115.50	0.19	8.68	0.18	105.50	50.54	8547.0
4	0.73	28.40	0.30	8.50	0.23	155.60	43.28	7598.5
5	28.39	910.60	6.19	183.70	1.05	738.45	32.97	5205.5
6	33.84	1102.50	6.53	205.58	1.31	999.90	30.31	4680.3
7	19.23	577.80	1.80	41.15	0.49	301.35	43.08	7317.8
8	17.36	611.50	5.11	149.85	0.95	644.85	35.97	5830.3
9	9.59	366.25	3.05	82.90	0.52	358.35	47.21	7729.5
10	15.70	605.40	3.65	96.75	0.63	472.35	41.26	6576.8
11	16.00	617.75	1.31	20.55	0.32	204.35	42.54	6965.3
12	14.17	519.20	3.28	100.02	0.68	457.45	41.42	6750.5
13	16.55	589.80	3.82	112.80	0.69	496.70	41.87	6626.0

$$\begin{pmatrix} N & \sum\limits_{n=1}^{N} S_{xn} & \sum\limits_{n=1}^{N} A_{xn} & \sum\limits_{n=1}^{N} F_{xn} & \sum\limits_{n=1}^{N} C_{xn} \\ \sum\limits_{n=1}^{N} S_{xn} & \sum\limits_{n=1}^{N}(S_{xn}^2) & \sum\limits_{n=1}^{N}(S_{xn}A_{xn}) & \sum\limits_{n=1}^{N}(S_{sn}F_{xn}) & \sum\limits_{n=1}^{N}(S_{xn}C_{xn}) \\ \sum\limits_{n=1}^{N} A_{xn} & \sum\limits_{n=1}^{N}(A_{xn}S_{xn}) & \sum\limits_{n=1}^{N}(A_{xn}^2) & \sum\limits_{n=1}^{N}(A_{xn}F_{xn}) & \sum\limits_{n=1}^{N}(A_{xn}C_{xn}) \\ \sum\limits_{n=1}^{N} F_{xn} & \sum\limits_{n=1}^{N}(F_{xn}S_{xn}) & \sum\limits_{n=1}^{N}(F_{xn}A_{xn}) & \sum\limits_{n=1}^{N}(F_{xn}^2) & \sum\limits_{n=1}^{N}(F_{xn}C_{xn}) \\ \sum\limits_{n=1}^{N} C_{xn} & \sum\limits_{n=1}^{N}(C_{xn}S_{xn}) & \sum\limits_{n=1}^{N}(C_{xn}A_{xn}) & \sum\limits_{n=1}^{N}(C_{xn}A_{xn}) & \sum\limits_{n=1}^{N}(C_{xn}^2) \end{pmatrix} \begin{pmatrix} S_R \\ B_{SS} \\ B_{SA} \\ B_{SF} \\ B_{SC} \end{pmatrix} = \begin{pmatrix} \sum\limits_{n=1}^{N} S_n \\ \sum\limits_{n=1}^{N}(S_{xn}S_n) \\ \sum\limits_{n=1}^{N}(A_{xn}S_n) \\ \sum\limits_{n=1}^{N}(F_{xn}S_n) \\ \sum\limits_{n=1}^{N}(C_{xn}S_n) \end{pmatrix}$$

$$(4.5\text{-}15)$$

where N is the number of samples in the experimental data, and n is the index for the data points.

Similar equations may be developed from Equation 4.5-5 to solve for the following coefficients:

$$A_R,\ B_{AS},\ B_{AA},\ B_{AF},\ B_{AC}$$
$$F_R,\ B_{FS},\ B_{FA},\ B_{FF},\ B_{FC}$$
$$C_R,\ B_{CS},\ B_{CA},\ B_{CF},\ B_{CC}$$

4. Experimental Data

Table 4.5-1 shows 13 sets of experimental data taken by passing selected samples of material through an XEG. The chemical composition is in percent composition by weight, and the XEG readings are in counts per second.

5. Solution

Only the equations for solution of the silicon oxide regression model will be formulated. The equations for the other materials can be formulated in a similar manner. From Equation 4.5-15 the equation (divided by 1000) in matrix notation for the silicon oxide is

$$\begin{pmatrix} 0.013 & 7.69 & 1.06 & 5.48 & 86.75 \\ 7.69 & 5729.88 & 769.08 & 3856.81 & 47761.54 \\ 1.06 & 769.08 & 141.32 & 649.65 & 6447.53 \\ 5.48 & 3856.81 & 649.64 & 3093.62 & 33875.68 \\ 86.75 & 47761.54 & 6447.53 & 33875.68 & 592895.29 \end{pmatrix} \cdot \begin{pmatrix} S_R \\ B_{SS} \\ B_{SA} \\ B_{SF} \\ B_{SC} \end{pmatrix} = \begin{pmatrix} 0.222 \\ 168.852 \\ 22.624 \\ 113.282 \\ 1367.481 \end{pmatrix}$$

$$(4.5\text{-}16)$$

The matrix in Equation 4.5-16 is called the reference matrix, which can be inverted to solve for the regression coefficients.

Instead of inverting the reference matrix, i.e., solving Equation 4.5-16 directly, the regression model will be translated to *go through the mean* by using the techniques discussed in Section 4.5B. The mean values (divided by 1000) are

$$\bar{S} = 0.01708923$$

$$\overline{S_x} = 0.59190384$$

$$\overline{A_x} = 0.08177923 \qquad (4.5\text{-}17)$$

$$\overline{F_x} = 0.42166923$$

$$\overline{C_x} = 6.67327690$$

The set of equations to solve for S_R, B_{SS}, B_{SA}, B_{SF}, B_{SC} may now be translated to *go through the mean* by using Equations 4.5-9 and 4.5-10:

$$
\begin{pmatrix}
\sum_{n=1}^{N}(S'_{xn})^2 & \sum_{n=1}^{N}(S'_{xn}A'_{xn}) & \sum_{n=1}^{N}(S'_{xn}F'_{xn}) & \sum_{n=1}^{N}(S'_{xn}C'_{xn}) \\
\sum_{n=1}^{N}(A'_{xn}S'_{xn}) & \sum_{n=1}^{N}(A'_{xn})^2 & \sum_{n=1}^{N}(A'_{xn}F'_{xn}) & \sum_{n=1}^{N}(A'_{xn}C'_{xn}) \\
\sum_{n=1}^{N}(F'_{xn}S'_{xn}) & \sum_{n=1}^{N}(F'_{xn}A'_{xn}) & \sum_{n=1}^{N}(F'_{xn})^2 & \sum_{n=1}^{N}(F'_{xn}C'_{xn}) \\
\sum_{n=1}^{N}(C'_{xn}S'_{xn}) & \sum_{n=1}^{N}(C'_{xn}A'_{xn}) & \sum_{n=1}^{N}(C'_{xn}F'_{xn}) & \sum_{n=1}^{N}(C'_{xn})^2
\end{pmatrix}
\cdot
\begin{pmatrix}
B_{SS} \\ B_{SA} \\ B_{SF} \\ B_{SC}
\end{pmatrix}
=
\begin{pmatrix}
\sum_{n=1}^{N}(S'_{xn}S'_n) \\
\sum_{n=1}^{N}(A'_{xn}S'_n) \\
\sum_{n=1}^{N}(F'_{xn}S'_n) \\
\sum_{n=1}^{N}(X'_{xn}S'_n)
\end{pmatrix}
$$

$$(4.5\text{-}18)$$

and

$$S_R = \bar{S} - (B_{SS}\,\overline{S_x} + B_{SA}\,\overline{A_x} + B_{SF}\,\overline{F_x} + B_{SC}\,\overline{S_x}) \qquad (4.5\text{-}19)$$

Computing the coefficients for Equations 4.5-18 and Equation 4.5-19 and dividing by $N \times 10^3$ or 13×10^3:

$$
\begin{pmatrix}
90.409745 & 10.754514 & 47.089868 & -275.97365 \\
10.754514 & 4.183036 & 15.489375 & -49.771323 \\
47.089868 & 15.489375 & 60.165775 & -208.09434 \\
-275.97365 & -49.771323 & -208.09434 & 1{,}074.7047
\end{pmatrix}
\cdot
\begin{pmatrix}
B_{SS} \\ B_{SA} \\ B_{SF} \\ B_{SC}
\end{pmatrix}
=
\begin{pmatrix}
2.873506 \\ 0.342754 \\ 1.507959 \\ -8.850299
\end{pmatrix}
$$

$$(4.5\text{-}20)$$

and

$$S_R = 0.01708923 - [(0.59190384)B_{SS} + (0.08177923)B_{SA}$$

$$+ (0.42166923)B_{SF} + (6.6732769)B_{SC}] \qquad (4.5\text{-}21)$$

A computer program was used to solve this example by working on Equations 4.5-20 and 4.5-21. The regression model developed for the silicon oxide composition $\hat{S}$, the coefficient of determination R^2, and the standard deviation s are

$$\hat{S} = 1.627 + (0.03045)S_x - (0.01130)A_x + (0.002701)F_x - (0.000416)C_x$$

$$(4.5\text{-}22)$$

$$R^2 \text{ (maximum likelihood)} = 0.98460426 \qquad (4.5\text{-}23)$$

$$R^2 \text{ (based on unbiased variances)} = 0.97690639 \qquad (4.5\text{-}24)$$

$$s = 1.5237925 \qquad (4.5\text{-}25)$$

Actually, the computer program developed the regression model for the other materials when all the experimental data in Table 4.5-1 was read in. Thus for aluminum oxide

$$\hat{A} = 3.620 - (0.00005656)S_x + (0.03935)A_x$$
$$- (0.003332)F_x - (0.0003748)C_x \qquad (4.5\text{-}26)$$

$$R^2 \text{ (maximum likelihood)} = 0.99661902 \qquad (4.5\text{-}27)$$

$$R^2 \text{ (based on unbiased variances)} = 0.99492853 \qquad (4.5\text{-}28)$$

$$s = 0.15109005 \qquad (4.5\text{-}29)$$

for iron oxide

$$\hat{F} = 0.1733 + (0.00005428)S_x + (0.0008421)A_x$$
$$+ (0.0009856)F_x - (0.00001163)C_x \qquad (4.5\text{-}30)$$

$$R^2 \text{ (maximum likelihood)} = 0.99196610 \qquad (4.5\text{-}31)$$

$$R^2 \text{ (based on unbiased variances)} = 0.98794915 \qquad (4.5\text{-}32)$$

$$s = 0.036215532 \qquad (4.5\text{-}33)$$

and for calcium oxide

$$\hat{C} = -13.31 + (0.003149)S_x - (0.0396)A_x$$
$$+ (0.01389)F_x + (0.007314)C_x \qquad (4.5\text{-}34)$$

$$R^2 \text{ (maximum likelihood)} = 0.98807807 \qquad (4.5\text{-}35)$$

$$R^2 \text{ (based on unbiased variances)} = 0.98211711 \qquad (4.5\text{-}36)$$

$$s = 0.7703776 \qquad (4.5\text{-}37)$$

D. Higher Order Multiple Regression Models

The types of higher order multiple regression models have been selected for discussion because they can easily be transformed into linear models so that the standard matrix operations can be applied to derive the coefficients.

The selection of the type of regression models for actual applications should be made on the basis of the physical laws described in Sections 4.1 and 4.2. If the physical laws indicate that the variables have a linear relationship, using a second-order polynomial regression model would only tend to distort the model, especially when it is used near the end points or outside of the data. Similarly, if the physical laws indicate a cubic relationship, a third-order polynomial regression model would be more appropriate than a linear model. A word of caution should be sounded on the use of higher order polynomials as regression models, especially from fourth-order on. While the use of higher order polynomials may produce a model with a better fit by reducing the error, the shape of the curve may be misleading. Higher order polynomials will tend to oscillate, i.e., have several valleys and peaks. In practice, a polynomial higher than second-order should be used only when the physical laws so indicate.

Care should be taken not to use the regression model beyond the region of the data, or erroneous conclusions may result. Also note that higher order polynomial regression models do not necessarily *go through the mean*.

Second-Order Polynomial Model

Let the following symbols be assigned:

$\hat{Y}$ = predicted state variable.
X_j = control variables, $(j = 1)$ to $(j = K)$.
$\beta_j, \beta_{jm}, \beta_{mm}$ = regression coefficients, $(j = 1)$ to $(j = K)$ and $(m = j + 1)$ to $(m = K)$.

The second-order polynomial multiple regression model has the general form:

$$\hat{Y} = \sum_{j=0}^{K} \beta_j X_j + \sum_{j=1}^{K} \sum_{m=j+1}^{K} \beta_{jm}(X_j X_m) + \sum_{j=1}^{K} \beta_{jj}(X_j)^2 \qquad (4.5\text{-}38)$$

To transform the second-order polynomial model into a linear one let

$$Z_j = X_j \qquad (4.5\text{-}39)$$

$$Z_{jm} = X_j X_m \qquad (4.5\text{-}40)$$

$$Z_{jj} = (X_j)^2 \qquad (4.5\text{-}41)$$

Substitution of Equations 4.5-39 through 4.5-41 into Equation 4.5-38 produces a multiple linear regression model:

$$\hat{Y} = \sum_{j=0}^{K} \beta_j Z_j + \sum_{j=1}^{K} \sum_{m=1}^{K} \beta_{jm} Z_{jm} \tag{4.5-42}$$

For example, if there are two control variables X_1 and X_2, Equation 4.5-38 becomes

$$\hat{Y} = \beta_1 X_1 + \beta_2 X_2 + \beta_{12} X_1 X_2 + \beta_{11} X_1^2 + \beta_{22} X_2^2 \tag{4.5-43}$$

and Equation 4.5-42 becomes

$$\hat{Y} = \beta_1 Z_1 + \beta_2 Z_2 + \beta_{12} Z_{12} + \beta_{11} Z_{11} + \beta_{22} Z_{22} \tag{4.5-44}$$

Multiplicative Model

The multiplicative regression model for K control variables is

$$\hat{Y} = \beta_0 (X_1)^{\beta_1} \cdot (X_2)^{\beta_2} \cdots (X_K)^{\beta_K} \tag{4.5-45}$$

or

$$\hat{Y} = \beta_0 \prod_{j=1}^{K} (X_j)^{\beta_j} \tag{4.5-46}$$

The multiplicative model may be transformed into a first-order linear model by taking the natural logarithm of both sides in Equation 4.5-46 and letting

$$\ln \hat{Y} = \hat{W} \tag{4.5-47}$$

$$\ln X_j = Z_j \tag{4.5-48}$$

$$\ln \beta_0 = B_0 \tag{4.5-49}$$

Thus

$$(\ln \hat{Y}) = (\ln \beta_0) + \sum_{j=1}^{K} \beta_j (\ln X_j) \tag{4.5-50}$$

$$\hat{W} = B_0 + \sum_{j=1}^{K} \beta_j Z_j \tag{4.5-51}$$

Exponential Model

The exponential model for K control variables is

$$\hat{Y} = \exp(\beta_0 + \beta_1 X_1 + \beta_2 X_2 + \cdots + \beta_K X_K) \tag{4.5-52}$$

By taking the natural logarithm of both sides and letting

$$\ln \hat{Y} = \hat{W} \tag{4.5-53}$$

Equation 4.5-52 becomes a linear model:

$$\ln \hat{Y} = \beta_0 + \sum_{j=1}^{K} \beta_j X_j \tag{4.5-54}$$

$$\therefore \quad \hat{W} = \beta_0 + \sum_{j=1}^{K} \beta_j X_j \tag{4.5-55}$$

Reciprocal Model

The reciprocal model for K control variables is

$$\hat{Y} = \left(\beta_0 + \sum_{j=1}^{K} \beta_j X_j \right)^{-1} \tag{4.5-56}$$

By letting $\hat{Y} = 1/\hat{W}$, Equation 4.5-56 becomes a linear model:

$$\hat{W} = \beta_0 + \sum_{j=1}^{K} \beta_j X_j \tag{4.5-57}$$

4.6   DYNAMIC PHYSICAL PROCESS MODELS BY EXPERIMENTAL TECHNIQUES

In classical systems analysis of servomechanisms experimental techniques to develop dynamic process models use either sine waves at various frequencies or step functions to excite the process. The frequency response or step function response as a result of the disturbance can then be analyzed to obtain the coefficients of differential equations.* The type of differential equation assumed to be the physical process model is selected on an *a priori* basis. Unfortunately, frequency response or step function response experiments are difficult to apply for typical fluid, energy, and heat processes for the following reasons:

1. Cost and safety considerations may preclude any experiments which disturb the process or operate the process outside of a normal range.

2. The temporary instrumentation-control equipment needed for the experimentation may mean costly shut-down for installation and removal.

---

* The differential equations are assumed to be linear with coefficients which are not time-dependent.

This section will present a generalized approach to dynamic modeling by experimentation that does not have the objectionable features listed above. An example taken from the cement blending model of Chapter 2 will be used to illustrate the application of this technique. This approach is attributed to Kalman, although the development presented here is not the original approach used by Kalman. Although one control variable is used to keep the derivations simple, the technique is equally applicable to more than one control variable. This approach has the following advantages:

1. The process is not disturbed.
2. No special instrumentation is needed.
3. Only experimental data taken at equally spaced intervals of time are required. The data logging and computations may easily be handled by the computer process control system.

A. The Generalized Approach to Developing Physical Process Models

The generalized approach in developing dynamic physical process models by experimentation uses the following five-step procedure:

1. Assume a difference equation to be the dynamic physical process model.
2. Conduct experiments to record sets of data at equally spaced intervals of time.
3. Arrange the experimental data in sets so that all the past data required to calculate the state variable by the difference equation are within a set.
4. Use regression analysis techniques to solve for the coefficients of the difference equation so that the sum of the error squared between the calculated and measured state variable is a minimum.
5. Try other forms of difference equations and repeat steps 1 through 4 to select the particular difference equation which gives the smallest sum of the error squared.

Each of the steps will be discussed in detail.

STEP 1. Assume a difference equation to be the dynamic physical process model.

If the differential equation describing the process is known, then the difference equation can be found by substituting

$$t = t_i + \Delta t \tag{4.6-1}$$

into the closed form or analytic solution of the differential equation. In this case

t = continuous value of time in the solution of the differential equation.

t_i = ith instant of time in the difference equation.

Δt = interval of time between solutions to difference equation.

Also let

$\hat{Y}_i$ = calculated state variable at ith instant of time.

Y_i = state variable at ith instant of time.

X_i = control variable at ith instant of time.

$A_0, A_1, \ldots A_n, B_0, B_1, \ldots, B_n$ = coefficients of difference equation.

When the form of the differential equation is not known, the general form of the difference equation can be assumed:

$$Y_{(i)} = \theta[X_{(i-1)}, X_{(i-2)}, \ldots, X_{(i-N-1)}; Y_{(i-1)}, Y_{(i-2)}, \ldots, Y_{(i-N-1)}; \Delta t]$$

(4.6-2)

To further simplify matters, Equation 4.6-2 can be assumed to be linear, with coefficients which are not time-dependent.

$$Y_{(i)} = A_0 X_{(i-1)} + A_1 X_{(i-2)} + \cdots + A_N X_{(i-N-1)} + B_0 Y_{(i-1)}$$

$$+ B_1 Y_{(i-2)} + \cdots + B_N Y_{(i-N-1)}$$

(4.6-3)

Equation 4.6-3 is the general form of the $(N+1)$ order difference equation with $(N+1)$ sets of past data.

STEP 2. Conduct experiments to record sets of data at equally spaced intervals of time.

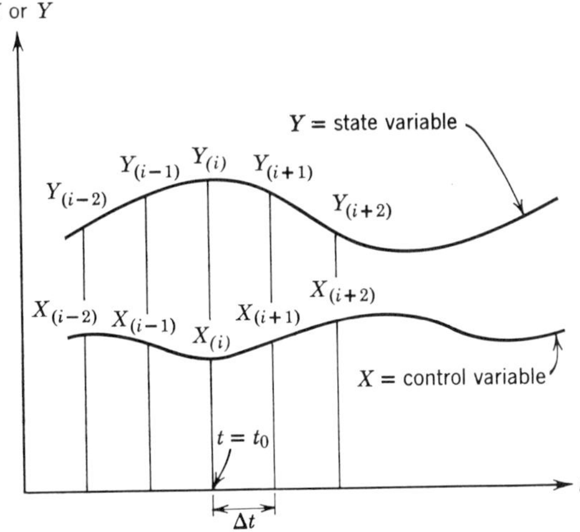

Fig. 4.6-1   Sets of experimental data taken at equally spaced intervals Δt.

Figure 4.6-1 shows the sets of experimental data taken at equally spaced intervals of time t. Thus the values of X and Y at the various intervals are

Time	X	Y
$t = (t_0 - 2\Delta t)$	$X_{(i-2)}$	$Y_{(i-2)}$
$t = (t_0 - \Delta t)$	$X_{(i-1)}$	$Y_{(i-1)}$
$t = t_0$	$X_{(i)}$	$Y_{(i)}$
$t = (t_0 + \Delta t)$	$X_{(i+1)}$	$Y_{(i+1)}$
$t = (t_0 + 2\Delta t)$	$X_{(i+2)}$	$Y_{(i+2)}$
$\vdots$	$\vdots$	$\vdots$

STEP 3. Arrange the experimental values obtained in Step 2 in sets so that all the past data required to calculate the state variable by the difference equation are within a set. $(N + 1)$ is the number of sets of past data collected for each state variable.

Table 4.6-1 shows the experimental data arranged in sets. The calculated state variable $\hat{Y}_{(i)}$ is a function of the values of X and Y at past instants of time, i.e., $X_{(i-1)}$, $X_{(i-2)}$, $\ldots$, $X_{(i-N-1)}$, $Y_{(i-1)}$, $Y_{(i-2)}$, $\ldots$, $Y_{(i-N-1)}$. The calculated state variable is now expressed as a function of the coefficients A_0, A_1, $\ldots$, A_N, B_0, B_1, $\ldots$, B_N after the past values of X and Y are substituted in the equation for $\hat{Y}$. Thus

$$\hat{Y}_{(i)} = \phi[A_0, A_1, \ldots, A_N; B_0, B_1, \ldots, B_N] \tag{4.6-4}$$

STEP 4. Use least squares techniques to solve for the coefficients of the difference equation so that the sum of the error squared between the calculated and measured state variables is minimum.

Let the error function F be defined as

$$F = \sum_{i=1}^{M} (\text{Error})^2$$

$$= \sum_{i=1}^{M} [Y_{(i)} - \hat{Y}_{(i)}]^2 \tag{4.6-5}$$

where M is the number of values of state variable calculated.
Using the expression of $\hat{Y}_{(i)}$ from Equation 4.6-3,

$$F = \sum_{i=1}^{M} [Y_{(i)} - (A_0 X_{(i-1)} + A_1 X_{(i-2)} + \cdots + A_N X_{(i-N-1)}$$

$$+ B_0 Y_{(i-1)} + B_1 Y_{(i-2)} + \cdots + B_N Y_{(i-N-1)})]^2 \tag{4.6-6}$$

Table 4.6-1 Experimental and Calculated Data Arranged in Sets

Set No.	Experimental Data	Calculated $\hat{Y}$
1	$Y_{(i-1)}, Y_{(i-2)}, \ldots, Y_{(i-N-2)}$ $X_{(i-2)}, X_{(i-3)}, \ldots, X_{(i-N-2)}$	$\hat{Y}_{(i-1)} = A_0 X_{(i-2)} + A_1 X_{(i-3)} + \cdots + A_N X_{(i-N-2)}$ $\qquad + B_0 Y_{(i-2)} + B_1 Y_{(i-3)} + \cdots + B_N Y_{(i-N-2)}$
2	$Y_{(i)}, Y_{(i-1)}, \ldots, Y_{(i-N-1)}$ $X_{(i-1)}, X_{(i-2)}, \ldots, X_{(i-N-1)}$	$\hat{Y}_{(i)} = A_0 X_{(i-1)} + A_1 X_{(i-2)} + \cdots + A_N X_{(i-N-1)}$ $\qquad + B_0 Y_{(i-1)} + B_1 Y_{(i-2)} + \cdots + B_N Y_{(i-N-1)}$
3	$Y_{(i+1)}, Y_{(i)}, Y_{(i-1)}, Y_{(i-N)}$ $X_{(i)}, X_{(i-1)}, \ldots, X_{(i-N)}$	$\hat{Y}_{(i+1)} = A_0 X_{(i)} + A_i X_{(i-1)} + \cdots + A_N X_{(i-N)}$ $\qquad + B_0 Y_{(i)} + B_1 Y_{(i-1)} + \cdots + B_N Y_{(i-N)}$
etc.		

F will be at a minimum by setting

$$\frac{\partial F}{\partial A_0} = \frac{\partial F}{\partial A_1} = \cdots = \frac{\partial F}{\partial A_N} = 0 \tag{4.6-7}$$

and

$$\frac{\partial F}{\partial B_0} = \frac{\partial F}{\partial B_1} = \cdots = \frac{\partial F}{\partial B_N} = 0 \tag{4.6-8}$$

Equations 4.6-7 and 4.6-8 result in a set of $(2N + 2)$ simultaneous linear equations which can be used to solve for the coefficients $A_0, A_1, \ldots, A_N$; $B_0, B_1, \ldots, B_N$.

STEP 5. Use other forms of the difference equation and repeat steps 3 and 4 to select the particular difference equation which gives the smallest practical sum of the error squared.

Generally, using more terms of the variables at past instants of time to calculate the state variable will tend to reduce the error. For example, Equation 4.6-9 uses only the set of variables one interval past to calculate the state variable $\hat{Y}$, but Equation 4.6-10 uses the sets of variables at one and two intervals past to calculate the state variable. Thus the difference equation in 4.6-10 will generally give a better fit than that in 4.6-9.

$$\hat{Y}_{(i)} = A_0 X_{(i-1)} + B_0 Y_{(i-1)} \tag{4.6-9}$$

$$\hat{Y}_{(i)} = A_0 X_{(i-1)} + A_1 X_{(i-2)} + B_0 Y_{(i-1)} + B_1 Y_{(i-2)} \tag{4.6-10}$$

Equation 4.6-3 is the general form of difference equation using $(N + 1)$ sets of past readings. Naturally additional accuracy would not be obtained if the order of the difference equation is greater than that of its physical process.

B. Cement Blending Example

Section 2.4 developed the differential equation relating the *composition of output material flow from a homogenizer* and *the composition of input material flow into a homogenizer*. Let the following symbols be assigned:

Y $= \%$ of CaO in homogenizer.
Y_1 $= \%$ of CaO in output stream. If perfect mixing is assumed, then $Y = Y_1$.
X $= \%$ of CaO in input stream of homogenizer.
t_0 $=$ an instant of time.
Δt $=$ interval of time between readings of data X, Y.
$R/T =$ ratio of throughput to capacity of homogenizer, a constant.

The differential equation relating Y and X was given in Equation 2.3-9:

$$\frac{dY}{dt} = \left(\frac{R}{T}\right)(X - Y) \tag{4.6-11}$$

The closed form solution to Equation 4.6-11 is given by Equation 2.3-14:

$$Y_{(t)} = \left\{1 - \exp\left[\frac{-R}{T}(t - t_0)\right]\right\}X_{(t_0)} + \left\{\exp\left[\frac{-R}{T}(t - t_0)\right]\right\}Y_{(t_0)} \tag{4.6-12}$$

The difference equation may be derived by substituting $(t = t_0 + \Delta t)$ in Equation 4.6-12:

$$Y_{(t_0 + \Delta t)} = \left[1 - \exp\left(\frac{-R}{T}\Delta t\right)\right]X_{(t_0)} + \left[\exp\left(\frac{-R}{T}\Delta t\right)\right]Y_{(t_0)} \tag{4.6-13}$$

In the experimental approach, however, the closed form solution of the differential equation is generally not known. Thus it is not possible to substitute $(t = t_0 + \Delta t)$ to get the difference equation. The following is the five step generalized approach to develop the dynamic process model for relating the composition of output material flow to input material flow composition.

STEP 1. Assume the difference equation to be

$$Y_{(t_0 + \Delta t)} = A_0 X_{(t_0)} + B_0 Y_{(t_0)} \tag{4.6-14}$$

A_0 and B_0 are the coefficients which are to be solved to obtain the best fit for the model in Equation 4.6-14.

STEP 2. Gather the set of experimental data, as shown in Table 4.6-2. The first set of data is taken at time $t = t_0$, the next set at $(t = t_0 + \Delta t)$, etc. ....

Table 4.6-2 N Sets of Experimental Data for X and Y Taken at Equal Intervals of Time Δt

Time	X (% CaO Input)	Y (% CaO Output)
t_0	X_0	Y_0
$t_0 + \Delta t$	X_1	Y_1
$t_0 + (2)(\Delta t)$	X_2	Y_2
$\vdots$	$\vdots$	$\vdots$
$t_0 + (i)(\Delta t)$	X_i	Y_i
$\vdots$	$\vdots$	$\vdots$
$t_0 + (N)(\Delta t)$	X_N	Y_N

STEP 3. Arrange the experimental and calculated data in sets as shown in Table 4.6-3. Each set of experimental data consists of the current value of Y_i as well as the values of X and Y one interval past, i.e., $Y_{(i-1)}$ and $X_{(i-1)}$. The calculated value of Y_i is computed by using the assumed difference equation given in Equation 4.6-14. It utilizes the values of X and Y obtained at one interval past.

Table 4.6-3   Experimental and Calculated Data Arranged in Sets

Set No.	Measured Variables	Calculated $\hat{Y}$
1	Y_1, Y_0, X_0	$\hat{Y}_1 = A_0 X_{(0)} + B_0 Y_0$
2	Y_2, Y_1, X_1	$\hat{Y}_2 = A_0 X_{(1)} + B_0 Y_1$
3	Y_3, Y_2, X_2	$\hat{Y}_3 = A_0 X_{(2)} + B_0 Y_2$
$\vdots$	$\vdots$	$\vdots$
i	$Y_i, Y_{(i-1)}, X_{(i-1)}$	$\hat{Y}_i = A_0 X_{(i-1)} + B_0 Y_{(i-1)}$
$i+1$	$Y_{(i+1)}, Y_i, X_i$	$\hat{Y}_{(i+1)} = A_0 X_i + B_0 Y_i$
$\vdots$	$\vdots$	$\vdots$
M	$Y_M, Y_{(M-1)}, X_{(M-1)}$	$\hat{Y}_M = A_0 X_{(M-1)} + B_0 Y_{(M-1)}$

STEP 4. The error function is given by

$$F = \sum_{i=1}^{M} [Y_{(i)} - \hat{Y}_{(i)}]^2$$

$$= \sum_{i=1}^{M} [Y_{(i)} - A_0 X_{(i-1)} - B_0 Y_{(i-1)}]^2 \tag{4.6-15}$$

Setting $\partial F/\partial A_0 = 0$ and $\partial F/\partial B_0 = 0$,

$$\sum_{i=1}^{M} 2[Y_{(i)} - A_0 X_{(i-1)} - B_0 Y_{(i-1)}] X_{(i-1)} = 0 \tag{4.6-16}$$

$$\sum_{i=1}^{M} 2[Y_{(i)} - A_0 X_{(i-1)} - B_0 Y_{(i-1)}] Y_{(i-1)} = 0 \tag{4.6-17}$$

The simultaneous linear equations in matrix notation that will give the solution of A_0 and B_0 are given by Equation 4.6-18.

$$\begin{pmatrix} \sum\limits_{i=1}^{M} X_{(i-1)}^2 & \sum\limits_{i=1}^{M} Y_{(i-1)} X_{(i-1)} \\ \sum\limits_{i=1}^{M} Y_{(i-1)} X_{(i-1)} & \sum\limits_{i=1}^{M} Y_{(i-1)}^2 \end{pmatrix} \cdot \begin{pmatrix} A_0 \\ B_0 \end{pmatrix} = \begin{pmatrix} \sum\limits_{i=1}^{M} Y_{(i)} X_{(i-1)} \\ \sum\limits_{i=1}^{M} Y_{(i)} Y_{(i-1)} \end{pmatrix} \tag{4.6-18}$$

Solving Equation 4.6-18 for A_0 and B_0:

$$A_0 = \frac{\left(\sum\limits_{i=1}^{M} Y_{(i)} X_{(i-1)}\right)\left(\sum\limits_{i=1}^{M} Y_{(i-1)}^2\right) - \left(\sum\limits_{i=1}^{M} Y_{(i-1)} X_{(i-1)}\right)\left(\sum\limits_{i=1}^{M} Y_{(i)} Y_{(i-1)}\right)}{\left(\sum\limits_{i=1}^{M} X_{(i-1)}^2\right)\left(\sum\limits_{i=1}^{M} Y_{(i-1)}^2\right) - \left(\sum\limits_{i=1}^{M} Y_{(i-1)} X_{(i-1)}\right)^2}$$

$$(4.6\text{-}19)$$

$$B_0 = \frac{\left(\sum\limits_{i=1}^{M} X_{(i-1)}^2\right)\left(\sum\limits_{i=1}^{M} Y_{(i)} Y_{(i-1)}\right) - \left(\sum\limits_{i=1}^{M} Y_{(i-1)} X_{(i-1)}\right)\left(\sum\limits_{i=1}^{M} Y_{(i)} X_{(i-1)}\right)}{\left(\sum\limits_{i=1}^{M} X_{(i-1)}^2\right)\left(\sum\limits_{i=1}^{M} Y_{(i-1)}^2\right) - \left(\sum\limits_{i=1}^{M} Y_{(i-1)} X_{(i-1)}\right)^2}$$

$$(4.6\text{-}20)$$

The standard error of estimate s for M samples is

$$s = \left[\frac{\sum\limits_{i=1}^{M} (\text{Error}_i)^2}{(M-2)}\right]^{1/2} \tag{4.6-21}$$

$$s = \left[\frac{\sum\limits_{i=1}^{M} (Y_{(i)} - A_0 X_{(i-1)} - B_0 Y_{(i-1)})^2}{(M-2)}\right]^{1/2} \tag{4.6-22}$$

STEP 5. The general form of the difference equation for this example is

$$Y_{(i)} = A_0 X_{(i-1)} + A_1 X_{(i-2)} + \cdots + A_N X_{(i-N-1)}$$
$$+ B_0 Y_{(i-1)} + B_1 Y_{(i-2)} + \cdots + B_N Y_{(i-N-1)} \tag{4.6-23}$$

Thus, if the form used in Equation 4.6-14 should produce a poor fit, i.e., an unduly large standard estimate of error, other forms derived from Equation 4.6-23 may be tried to reduce the error. Two approaches may be used to select a better form of difference equation. The first approach is to deduce the form by using known process characteristics. Certain types of process have well-known forms of dynamic models to describe their behavior, and the solutions to the known forms of differential equations may be a good starting point for the form of difference equation. The second approach is to select an arbitrary number of terms from Equation 4.6-23 and observe the resulting fit. Once the desired fit is obtained the terms with coefficients which may be insignificant can be discarded, thus simplifying the dynamic physical process model. Once again the physical laws in Sections 4.1 and 4.2 should be the basis of postulating the dynamic physical process models.

4.7 A DYNAMIC PHYSICAL PROCESS MODEL EXAMPLE

A completely mixed stirred tank chemical reactor will be used as an example to illustrate how a dynamic physical process model can be developed by applying the physical laws discussed in Sections 4.1 and 4.2. A simple case of two input streams, each containing one reactant, will be used. A single exothermal chemical reaction is assumed to take place within the reactor. Cooling water is circulated to remove the heat generated by the reaction, and the liquid in the reactor is maintained at a constant level by a level controller.

Material balance and *heat balance equations* are the physical laws applied to develop a *dynamic function for the concentrations of the products and reactants in the reactor.* These functions consist of a set of nonlinear time-dependent differential equations. The dynamic physical process model expresses the state variables (concentrations and temperature in the reactor) as functions of the control variables (flow rates) and parameters (concentrations and temperature of the feed streams). The dynamic physical process model will be generalized for m state variables and n control variables. Finally the physical process model for discrete-time systems will be developed, i.e., in the difference equation form.

A. The Functional Model

Figure 4.7-1 shows the functional diagram of the completely-mixed stirred tank reactor with two input feed streams and one effluent stream. Each input feed stream carries one reactant and the product is also carried by both feed streams. The purpose of the operation is to maintain certain concentrations

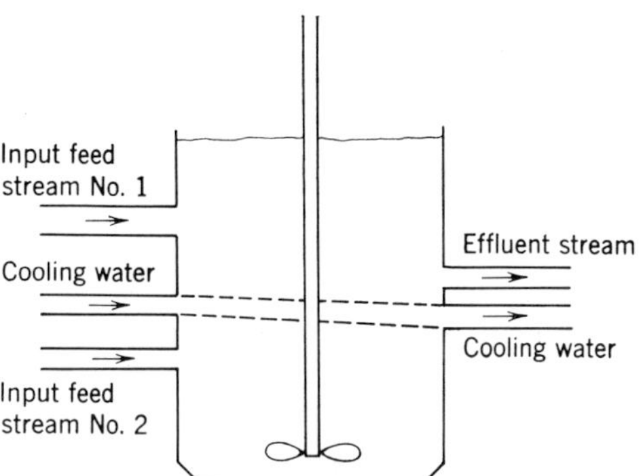

Fig. 4.7-1 A completely-mixed stirred tank chemical reactor with two input feed streams and one effluent stream.

of reactants and products in the effluent stream by controlling the input feed stream flows. The flow of the cooling water controls the rate of heat removed from the reactor.

The state variables are the concentrations of the reactants and products in the reactor and the reactor temperature. Since the reactor is assumed to be completely mixed, the concentrations of the reactants and products and the temperature of the effluent stream are equal to the concentrations and temperature of the reactor. The control variables are the input feed stream flows. The parameters are the input feed stream concentrations and temperatures, and the cooling water temperature.

Let the following symbols be assigned:

F_1, F_2 = flow rates of the input feed streams.

F_e = flow rate of the effluent stream.

F_c = flow rate of the cooling water.

T_1, T_2 = temperatures of the input feed streams.

T = temperature of the reactor and the effluent stream.

T_c, T_o, T_a = cooling water input, output, and average temperatures, respectively.

A, B = volume of reactants A and B in the reactor.

C = volume of product C in the reactor.

Y_A, Y_B, Y_C = concentrations of A, B, and C materials in the reactor.

X_{A1}, X_{C1} = concentrations of A and C materials in the first input feed stream.

X_{B2}, X_{C2} = concentrations of B and C materials in the second input stream.

V = volume of liquid in the reactor, a constant.

K_{AB} = reaction constant for $(A + B \rightarrow C)$. Reaction constant is a function of reactor temperature.

t = time.

B. The Physical Process Model

The physical process model will be developed by using (1) material balances in the reactor to obtain the rate of change in concentrations of products and reactants, (2) the Arrhenius reaction equation to obtain the rates of change in concentrations for the reaction $(A + B \rightarrow C)$, and (3) heat balance in the reactor to obtain the rate of change in reactor temperature.

Material Balance

The total rate of change in volume of a given material in the reactor is equal to (1) the rate of change in volume added by the input feed streams minus (2) the rate of change in volume removed by the effluent stream plus or minus (3) the rate of change in volume due to the chemical reaction.

The rates of change in volume of reactants and products caused by *the input feed streams* are

$$\frac{dA}{dt} = X_{A1}F_1 \tag{4.7-1}$$

$$\frac{dB}{dt} = X_{B2}F_2 \tag{4.7-2}$$

$$\frac{dC}{dt} = X_{C1}F_1 + X_{C2}F_2 \tag{4.7-3}$$

The rates of change in volume of reactants and products caused by *the effluent stream* are

$$\frac{dA}{dt} = -Y_A F_e = -Y_A(F_1 + F_2) \tag{4.7-4}$$

$$\frac{dB}{dt} = -Y_B F_e = -Y_B(F_1 + F_2) \tag{4.7-5}$$

$$\frac{dC}{dt} = -Y_C F_e = -Y_C(F_1 + F_2) \tag{4.7-6}$$

Equations 4.7-4 through 4.7-6 are true because the sum of the input stream flow rates are equal to the effluent stream flow rate for constant volume in the reactor.

The rates of change in the concentrations of the reactants and products are equal to the rates of change of the material volume divided by volume of the reactor. Thus Equations 4.7-1 through 4.7-6 may be summed and converted to rates of change in the concentrations by using the following relationships:

$$\frac{dY_A}{dt} = \frac{dA}{dt}\left(\frac{1}{V}\right) \tag{4.7-7}$$

$$\frac{dY_B}{dt} = \frac{dB}{dt}\left(\frac{1}{V}\right) \tag{4.7-8}$$

$$\frac{dY_C}{dt} = \frac{dC}{dt}\left(\frac{1}{V}\right) \tag{4.7-9}$$

The net rates of change in the concentrations of the reactants and products due to the input feed streams and the effluent stream are

$$\frac{dY_A}{dt} = \left(\frac{1}{V}\right)[X_{A1}F_1 - Y_A(F_1 + F_2)] \tag{4.7-10}$$

$$\frac{dY_B}{dt} = \left(\frac{1}{V}\right)[X_{B1}F_2 - Y_B(F_1 + F_2)] \tag{4.7-11}$$

$$\frac{dY_C}{dt} = \left(\frac{1}{V}\right)[X_{C1}F_1 + X_{C2}F_2 - Y_C(F_1 + F_2)] \tag{4.7-12}$$

For the chemical reaction the reaction dynamics are

$$A + B \xrightarrow{K_{AB}} C \tag{4.7-13}$$

Equation 4.7-13 assumes that a second-order reaction is taking place.

For small or dilute concentrations of the materials in the reactor, the probability of a molecule of material A entering the volume of effective range of intermolecular forces is proportional to the reaction constant K_{AB} and the concentration of A, but it is independent of the concentrations of the other materials. Similarly, the probability of a molecule of material B entering the volume of effective range of intermolecular forces is proportional to the reaction constant K_{AB} and the concentrations of B, but it is independent of the concentrations of the other materials. Since the probabilities of independent events are multiplicative, the rate of change of the concentration of product C is given by

$$\frac{dY_C}{dt} = K_{AB}(Y_A Y_B) \tag{4.7-14}$$

Similarly, the rates of change of the concentrations of reactants A and B are

$$\frac{dY_A}{dt} = -K_{AB}(Y_A Y_B) \tag{4.7-15}$$

$$\frac{dY_B}{dt} = -K_{AB}(Y_A Y_B) \tag{4.7-16}$$

Equation 4.7-14 is the Arrhenius equation, familiar to chemical engineers. The reaction constant K_{AB} is dependent on the temperature of the reactor. Let the following additional symbols be assigned:

α_{AB} = constant for the reaction $(A + B \rightarrow C)$.
E_{AB} = activation energy of the reaction $(A + B \rightarrow C)$.
R = gas constant.

The reaction constant is

$$K_{AB} = \alpha_{AB} \cdot \exp\left[-\left(\frac{E_{AB}}{RT}\right)\right] \qquad (4.7\text{-}17)$$

Summing Equations 4.7-10 through 4.7-12 and Equations 4.7-14 through 4.7-16, the total rates of change in the concentrations of reactants and products in the reactor are

$$\frac{dY_A}{dt} = \left(\frac{1}{V}\right)[X_{A1}F_1 - Y_A(F_1 + F_2)] - K_{AB}(Y_A Y_B) \qquad (4.7\text{-}18)$$

$$\frac{dY_B}{dt} = \left(\frac{1}{V}\right)[X_{B2}F_1 - Y_B(F_1 + F_2)] - K_{AB}(Y_A Y_B) \qquad (4.7\text{-}19)$$

$$\frac{dY_C}{dt} = \left(\frac{1}{V}\right)[X_{C1}F_1 + X_{C2}F_2 - Y_C(F_1 + F_2)] + K_{AB}(Y_A Y_B) \qquad (4.7\text{-}20)$$

where K_{AB} is given by Equation 4.7-17.

The heat balance equation may be used to develop the expression for the rate of change in the reactor temperature due to (1) the input feed streams, (2) the chemical reaction, (3) the effluent stream, and (4) the cooling water. Let the following additional symbols be assigned:

ΔH_{AB} = heat of the reaction ($A + B \rightarrow C$).
ρ = density of the reactor solution.
C_H = heat capacity of the reactor solution.
U = heat transfer coefficient of the cooling water.

The rate of change in the reactor temperature is

$$\frac{dT}{dt} = \frac{1}{V}[T_1 F_1 + T_2 F_2 - T(F_1 + F_1)]$$

$$+ K_{AB}(Y_A Y_B)\left(\frac{\Delta H_{AB}}{\rho C_H}\right) - \frac{U(T - T_a)}{V \rho C_H} \qquad (4.7\text{-}21)$$

Equations 4.7-18 through 4.7-21 completely describe the dynamics of the system, and these *nonlinear differential equations constitute the physical process model of the completely mixed stirred tank reactor.*

C. General Form of Dynamic Physical Process Model

By extension from the physical process model given in Equations 4.7-18 through 4.7-21, the general form of a dynamic physical process model may be presented. This form of dynamic model is quite common among processes.

Let the following symbols be assigned:

Y_j = state variables, $(j = 1)$ to $(j = m)$.
X_i = control variables, $(i = 1)$ to $(i = n)$.
K_p = parameters, $(p = 1)$ to $(p = k)$.
t_0, t_q = initial and present value of time, respectively.

The general form of the physical process model is

$$\frac{dY_j}{dt} = H_j(Y_j, X_i, K_p, t) \tag{4.7-22}$$

Equation 4.7-22 is the *state equation* of the process. This represents a set of nonlinear ordinary differential equations which expresses the rates of change of the state variables as a function of state variables, control variables, parameters, and time.

The functions H_j may be divided into two sets of functions, G_j and J_j:

$$\frac{dY_j}{dt} = G_j(Y_1, Y_2, \ldots, Y_m; t) + J_j(X_1, X_2, \ldots, X_n; K_1, K_2, \ldots, K_k; t)$$

$$\tag{4.7-23}$$

The functions G_j represent the functional dependence on the state variables, and the functions J_j represent the functional dependence on the control variables and parameters.

Two conditions must be satisfied in order to have a unique solution to Equation 4.7-23:

1. The initial conditions of the state variables $Y_j(t_0)$ are given.
2. The functions J_j are given for all values of time from $(t = t_0)$ to $(t = t_q)$.

D. Solution to the General Form State Equations

Any function which satisfies Equation 4.7-23 will be a solution to the general form of the state equations. Assume that such a solution exists and is given by

$$\psi_j[t; Y_1(t_0), Y_2(t_0), \ldots, Y_m(t_0); J_1(t), J_2(t), \ldots, J_m(t); t_0] \tag{4.7-24}$$

The solution is a function of the initial values of the state variables $Y_j(t_0)$, the J_j functions, and time.

The value of any state variable at $t = t_q$ can be obtained by making $t = t_q$ in Equation 4.7-24.

$$Y_j(t_q) = \psi_j[t_q; Y_1(t_0), Y_2(t_0), \ldots, Y_m(t_0); J_1(t_q), J_2(t_q), \ldots, J_m(t_q); t_0] \tag{4.7-25}$$

The set of functions ψ_j in Equation 4.7-25 is called the *state-transition functions* because they specify how the physical process evolves dynamically from one instant of time to another.

E. State Transition Functions in Difference Equation Form

If the assumption can be made that during the sampling interval Δt the external influences are small, the continuous-time state-transition equations can be rewritten as discrete-time difference equations. In other words, if the sampling period is small compared to the process setting times, then the effect of the external influences may be ignored. Thus

$$J_j(t_q) \simeq J_j(t_0) \tag{4.7-26}$$

for small Δt.
   Let

$$t_q = t_0 + \Delta t \tag{4.7-27}$$

By substituting Equation 4.7-26 as an *equality* into Equation 4.7-25 the general difference equation form for a dynamic process is

$$Y_j(t_0 + \Delta t) = \psi_j[(t_0 + \Delta t);\ Y_1(t_0),\ Y_2(t_0),\ \ldots,\ Y_m(t_0);$$
$$J_1(t_0),\ J_2(t_0),\ \ldots,\ J_m(t_0);\ t_0] \tag{4.7-28}$$

4.8   SUMMARY

This chapter has presented some analytical and experimental techniques for developing physical process models. There should be no conflict between the analytical and the experimental approaches; for example, physical process models which are developed by purely analytical approaches must be tested experimentally to validate the physical process model initially. Experiments must be conducted periodically to update the coefficients. On the other hand, analytical approaches can provide many clues to the forms that physical process models should assume for fitting experimental data, as well as guidance for the design of experiments. Thus, in practice, the two approaches complement each other. Above all, common sense, judgment, and balance should prevail in the development of physical process models.
   Chapter 5 will present the techniques of developing economic models for plant optimization.

5

Techniques for Developing Economic Models

5.0 INTRODUCTION

Economic models have traditionally been used in the domain of accounting, finance, and economics. More recently, techniques developed by econometrics and operations research have been applied to develop quantitative models of markets, business operations, and production processes. This chapter will *apply quantitative techniques to describe some of the basic accounting rules and the relationships between business variables.* The result is a concept of practical business models which can be utilized by anyone with the understanding of business operations and simple mathematics.

The chapter will first present *planning economic models* used for evaluating long term investment opportunities. The basic laws and the techniques for developing the models will be presented. The same treatment will be applied to *operating economic models* used for short term profit optimization. Planning economic models may be used for economic feasibility studies of computer process control applications. Alternative proposals with varying degrees, scope, and sophistication of automation may be evaluated by using planning economic models. In a broad sense, the economic feasibility of the entire plant investment proposal may also be evaluated. Operating economic models are the objective functions for process optimization. Thus the development of operating economic models is an integral part of the synthesis of control and optimization systems for the plant. Finally an example of optimal pricing to maximize profits will be developed.

Figure 5.0-1 shows how planning and operating economic models evolve in the systems design of a computer process control system. After developing the physical process and the operating economic models, various methods

143

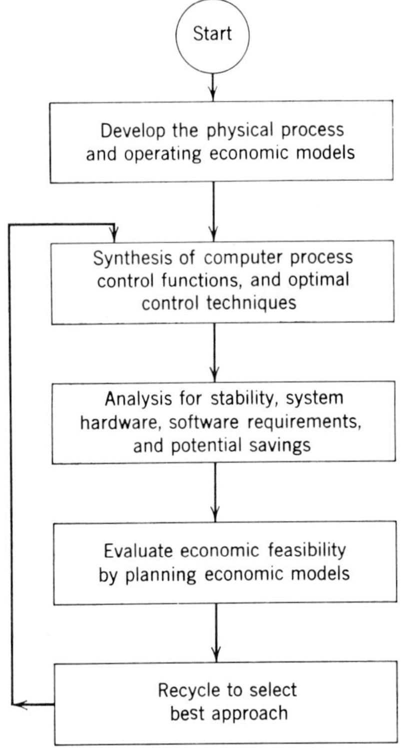

Fig. 5.0-1 Procedure for developing the design of a computer process control system.

of control and optimization systems are synthesized. These are analyzed for stability, response time, and potential economic savings. The computer functions are then translated into hardware, software, and input-output devices. At this point, the economic feasibility of the specific approach may then be analyzed through the use of planning economic models. A recycling process is used to select the best control and optimization system to stay within cost and other constraints.

5.1 PLANNING ECONOMIC MODELS

Planning economic models are used to evaluate long-range investment opportunities. The choices of "do nothing," "modernize," "add a new plant," "introduce a new product line," etc., are typical business decisions that can be made with the assistance of planning economic models. Time periods of planning models may be from 3 to 25 years. Sections 5.1 through 5.3

will first define the variables commonly used in planning economic models, then introduce the basic laws governing these variables, and conclude by presenting techniques for developing planning economic models. Let the following symbols* be assigned:

REV_i = gross revenue for the ith year, in dollars.
EXP_i = expenses for the ith year, in dollars.
PBT_i = profit-before-taxes for the ith year, in dollars.
ICT_i = income tax rate for the ith year, in percent.
NET_i = net income for the ith year, in dollars.
INI_i = initial value of investment made in the ith year, in dollars.
ADF_j = annual depreciation factor for the jth year after the initial investment was made, in percent.
RES_j = annual residual value factor for the jth year after the initial investment was made, in percent.
DEP_i = depreciation expense for the ith year, in dollars.
RSV_i = residual value of investment for the ith year, in dollars.
INV_i = inventory value at end of the ith year, in dollars.
ACR_i = accounts receivable value at end of the ith year, in dollars.
ACP_i = accounts payable value at end of the ith year, in dollars.
NIV_i = net investment value at end of the ith year, in dollars.
CFL_i = cash flow generated for the ith year, in dollars.
DCR = rate for discounting cash flow generated, in percent.
DCF_i = discounted cash flow generated for the ith year, in dollars.
ROS_i = return on sales for the ith year, in percent.
ROI_i = return on investment for the ith year, in percent.
ITO_i = investment turnover ratio for the ith year, in units.
URR = uniform rate of return of cash flow schedule, in percent.

5.2 BASIC LAWS FOR PLANNING ECONOMIC MODELS

Planning economic models for a real business should conform to the established rules and procedures for reporting the results of current operations and the condition of the business. Two types of statement summarize the financial operation and the condition of a business. The first is the *operating statement*, which summarizes the financial results (profit/loss) of a business for a period of time; the second is the *statement of financial condition*,† which summarizes the worth and liabilities of a business at a point of time. The following laws

---

* Three letter symbols are used for clarity.
† Also called the balance sheet.

are essentially those that are used in a business to prepare the operating statement and the statement of financial condition. Only one product and one investment is used for sake of simplicity.

A. Profit-Before-Taxes

Annual profit-before-taxes is equal to the difference between annual gross revenue and expenses:

$$PBT_i = REV_i - EXP_i \qquad (5.2\text{-}1)$$

B. Net Income

Annual net income is equal to annual profit-before-taxes minus federal income tax:

$$NET_i = PBT_i[1 - ICT_i/100] \qquad (5.2\text{-}2)$$

C. Depreciation Expense and Residual Value of Investment

Long-term capital investments for plant, machinery, and equipment cannot be charged off as expense during the year that the investment is made but must be depreciated over a number of years. Tax laws generally allow different schedules of depreciation factors for various types of investments. The total period of depreciation also varies. The depreciation expense for any given year is charged to expense for the year, while the capital value of the investment at year-end is reduced by the same amount. Since the cash was spent during the year the investment was made, no further cash is spent and, therefore, depreciation adds to cash flow. For an investment of INI_i made in the ith year with an annual depreciation factor of ADF_j, the depreciation expense j years after the initial investment is

$$DEP_{i+j} = (INI_i)(ADF_j)/100 \qquad (5.2\text{-}3)$$

The residual value at the end of k years after the initial investment is

$$RSV_{i+k} = [INI_i]\left[1 - \sum_{j=1}^{k} ADF_j/100\right] \qquad (5.2\text{-}4)$$

D. Typical Depreciation Formulas

Figure 5.2-1 shows the application of three typical depreciation formulas. The *straight-line depreciation formula* allows an equal amount of depreciation for each year in the depreciation period. Let N be the number of years in the

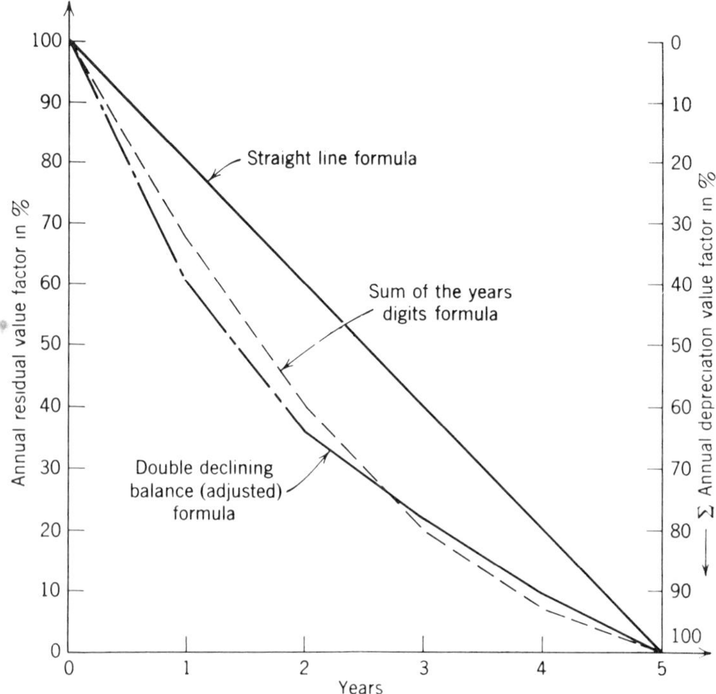

Fig. 5.2-1 Typical depreciation and residual value factors for a 5-year investment period.

depreciation period, the annual depreciation factor ADF in percent for the straight-line formula is

$$\text{ADF}_j = \frac{100}{N} \tag{5.2-5}$$

The *sum of the years' digits formula* computes the annual depreciation factor by using the sum of the years' digits for the denominator and the number of years from the initial investment in reverse sequence for the numerator. The annual depreciation factor ADF which must be computed annually, in percent, for the sum of the years' digits formula is

$$\text{ADF}_j = \frac{(N - j + 1)}{\sum\limits_{K=1}^{N} (K)} \times 100 \tag{5.2-6}$$

For example, the annual depreciation factors and annual residual value factors for a 5-year period, $N = 5$, are as follows:

Years (j) ⟶	1	2	3	4	5
ADF_j	$\frac{5}{15} \times 100$ $= 33$	$\frac{4}{15} \times 100$ $= 27$	$\frac{3}{15} \times 100$ $= 20$	$\frac{2}{15} \times 100$ $= 13$	$\frac{1}{15} \times 100$ $= 7$
RES_j	67	40	20	7	0

The *double declining balance formula* uses annual depreciation factors that are double the factors used in the straight line formula, but the factors operate on the prior years' residual value instead of the initial value. The annual depreciation factor ADF in percent for the double declining balance formula is

$$ADF_j = \left[100 - \sum_{K=0}^{j-1} ADF_K \right]\left(\frac{2}{N}\right) \tag{5.2-7}$$

The formula in Equation 5.2-7 will never reduce the residual value to zero no matter how many years are applied to the formula. An adjustment is usually made in the factors of the last two years, e.g., splitting the total residual value factor for the last two years on a 60/40 basis. For example, the annual depreciation factors and annual residual factors for a five-year period are given for the unadjusted and adjusted cases:

Years (j) ⟶	1	2	3	4	5
ADF_j	40	24	14.4	8.6	5.2
RES_j	60	36	21.6	13.0	7.8
Adj. ADF_j	40	24	14.4	13.0	8.6
Adj. RES_j	60	36	21.6	8.6	0

E. Net Investment at Year-End

In addition to long-term capital investment in plant, buildings, and equipment, *working capital* is required to keep the business in operation. Working capital is short term assets which can be converted to cash in less than a year. In oversimplified terms, net working capital is made up of inventory plus accounts receivable less accounts payable. Inventory generally includes raw materials, work in process, and finished goods. Accounts receivable are the bills which have been sent to customers but are as yet unpaid. Accounts payable are bills received from vendors and suppliers but are as yet unpaid. Generally, the value of inventory may be 15 to 25% of the value of annual gross revenue; bills paid by customers may be 60 to 90 days behind billing, thus accounts receivable may be 15 to 25% of annual gross revenue; accounts payable may be 5 to 12% of annual gross revenue. The total *net investment** of a business at the end of a year is equal to the net capital investment value plus net working capital:

$$NIV_i = RSV_i + INV_i + ACR_i - ACP_i \qquad (5.2\text{-}8)$$

F. Cash Flow Generated

The annual cash flow generated is derived from net income, depreciation, and any decrease in the requirement of net working capital for the year. Conversely, new investment and any increase in the requirement for net working capital for the year will decrease cash flow. In general, omitting the effect of dividends and interest, cash flow generated is equal to "net income" minus "the difference between net investment at this year-end and last year-end."

$$CFL_i = NET_i - [NIV_i - NIV_{i-1}] \qquad (5.2\text{-}9)$$

G. Discounted Cash Flow and Uniform Rate of Return

Discounted cash flow takes into account the reality that cash today is of greater value than cash in the future. It is an important concept to use in evaluating the cash flow schedule of investment opportunities. For example, consider the two simple cash flow alternatives ($ in millions) in Table 5.2-1.

If discounting is ignored, Case 1 in Table 5.2-1 produced $10 million more in cumulative cash flow. If the discount rate is 5%, Case 2 produces $15.8

Table 5.2-1   Alternative Investment Opportunity Evaluated with and without Discounting ($ in millions)

Cash Flow	Years → 1	2	3	4	5	6	Cumulative
Case 1—undiscounted	−100	0	0	0	0	150	50
Case 2—undiscounted	−100	140	0	0	0	0	40
Case 1—discounted	−100	0	0	0	0	117.5	17.5
Case 2—discounted	−100	133.3	0	0	0	0	33.3

Note: This assumes that the cash flow for Case 2 cannot be reinvested at a yield higher than 5% after the second year.

* Or *net worth.*

million more than Case 1 in cumulative discounted cash flow. Thus wrong conclusions can result if the interest rate reflected by discounting is ignored.

The discounted cash flow at n years from the reference is

$$DCF_n = \frac{CFL_n}{(1 + DCR/100)^n} \qquad (5.2\text{-}10)$$

where DCR is the rate for discounting cash flow generated. The uniform rate of return of the cash flow schedule is the discount rate which makes the cumulative value of cash flow equal to zero. The URR may be computed by trial and error.

$$\sum_{n=1}^{K} \frac{CFL_n}{[1 + URR/100]^n} = 0 \qquad (5.2\text{-}11)$$

The uniform rate of return is a true measure of profitability because the cash requirement of a proposed investment could have been invested in a bank to earn interest. The difference between URR and the prevailing bank interest rate is the true profitability of the investment opportunity.

H. Other Measurements of a Business Operation

Other measurements of a business operation include *return on sales, return on investment,* and *investment turnover.* Return on sales is a measure of the amount of annual net income produced by a given amount of annual gross revenue. This is a comparative measurement and is useful in comparing a proposed business opportunity against experience in similar ventures. The return on sales ROS in percent is equal to annual net income divided by gross revenue.

$$ROS_i = \frac{NET_i}{REV_i} \cdot 100 \qquad (5.2\text{-}12)$$

Return on investment compares the net income produced against the required investment. It is the amount of net income that can be produced by a given investment. Generally, the higher the risk and longer the payoff period, the higher the return on investment should be in order to make the investment attractive. The return on investment ROI in percent is equal to annual net income divided by net investment.

$$ROI_i = \frac{NET_i}{NIV_i} \cdot 100 \qquad (5.2\text{-}13)$$

The ratio of investment turnover is a measure of the annual gross revenue that can be produced by a given investment. It is also a comparative measurement. Generally, the higher the turnover ratio, the better the investment opportunity. The investment turnover ratio is equal to annual gross revenue divided by the net investment.

$$\text{ITO}_i = \frac{\text{REV}_i}{\text{NIV}_i} \tag{5.2-14}$$

Cumulative return on sales ROS and the average return on investment for the program life ROI may be computed by using cumulative net income, cumulative revenue, and cumulative net investment in Equations 5.2-12 and 5.2-13.

5.3 TECHNIQUES FOR DEVELOPING PLANNING ECONOMIC MODELS

Planning economic models may be developed by simulating the financial operating results of a proposed investment opportunity. The simulation is merely a set of computational procedures to compute the financial *operating statement* (profit and loss) and *statement of financial condition* (balance sheet) using the laws given in the previous paragraphs. The assumptions and input data for financial simulation falls into the following categories: *sales forecast*, *investments*, and *expenses*.

A. Sales Forecast

Total volume, in units and dollars.
Product line mix, in units.
Product mix within product line, in units and subunits.
Prices for each product, in dollars per unit.
Pricing trends, in percent of change per year.
Associated products, such as replacement parts, accessories, and service, in dollars.
Reasonable product calendar, from birth to obsolescence, for each of the products, in text form.
Markets for each product line, and strategies for penetrating each market, in text form.

B. Investments

Capital investments required, in dollars.
Depreciation factors for the different types of investments, in percent.
Inventory values in dollars, or ratios to revenue in percent.
Accounts receivable and payable in dollars, or ratios to gross revenue in percent.

C. Expenses

Engineering and development expenses, including prototypes and pilot plants, in dollars.

Royalties, warranties, complaints, and other reserves, in dollars or ratios to gross revenue in percent.

Material cost, direct labor, and overhead involved in manufacturing, in ratio to gross revenue in percent.

Sales and market development expenses, in dollars.

Sales support and service costs in ratio to gross revenue, in percent.

Start-up costs for initial hiring and training, in dollars.

Transportation and sales tax in ratio to gross revenue, in percent.

Federal income tax rate, in percent.

General overhead and administrative expenses, in dollars, or ratio to gross revenue in percent.

Other allocated fixed expenses, e.g., support of laboratory and corporate advertising, in dollars.

Given these inputs and a computational procedure that computes the annual financial operating statement and the statement of financial condition, it is possible to evaluate alternate sales volumes, product mix, and cost projections for any proposed investment opportunity.

D. A Planning Model Example

For example, consider an investment opportunity which requires new investment, and is expected to yield the revenue and expenses shown in Table 5.3-1.

The financial measurements in Table 5.3-2 are computed by using a 50% income tax rate and 5% discount rate for cash flow.

Summarizing the results given in Table 5.3-2,

from Equation 5.2-12

$$\text{the cumulative return on sales} = \frac{7.25}{63.0} = 11.5\%;$$

from Equation 5.2-13

$$\text{the average return on investment} = \frac{7.25}{67} = 10.8\%;$$

from Equation 5.2-14

$$\text{the cumulative investment turnover ratio} = \frac{63}{67} = 0.94;$$

from Equation 5.2-11

the uniform rate of return of the cash flow schedule $= 7.8\%$.

Table 5.3-1 Example of Proposed Investment Opportunity ($ in millions)

Years⟶	1	2	3	4	5	6	7	8	9	10	11
	$	$	$	$	$	$	$	$	$	$	$
Revenue	0	0	5.0	7.5	10.0	12.5	12.5	7.5	5.0	3.0	0
Expense	2.0	3.0	6.0	7.5	6.0	8.0	7.0	4.0	3.0	2.0	0
Net investment	5.0	10.0	12.0	10.0	9.0	8.0	6.0	4.0	2.0	1.0	0

Table 5.3-2 Financial Measurements of Proposed Investment Opportunity Given in Table 5.3-1 ($ in millions)

Years ⟶	1	2	3	4	5	6	7	8	9	10	11
	$	$	$	$	$	$	$	$	$	$	$
Net income	−1.0	−1.5	−0.5	0	2.0	2.25	2.75	1.75	1.0	0.5	0
Cumulative net income	−1.0	−2.5	−3.0	−3.0	−1.0	1.25	4.00	5.75	6.75	7.25	7.25
Cash flow	−6.0	−6.5	−2.5	2.0	3.0	3.25	4.75	3.75	3.00	1.50	1.00
Cumulative cash flow	−6.0	−12.5	−15.0	−13.0	−10.0	−6.75	−2.00	1.75	4.75	6.25	7.25
Discount factors @ 5%	1.0	0.953	0.907	0.864	0.823	0.784	0.746	0.710	0.677	0.645	0.614
Discounted cash flow	−6.0	−6.19	−2.27	1.73	2.47	2.55	3.55	2.67	2.03	0.97	0.61
Cumulative discounted cash flow	−6.0	−12.19	−14.46	−12.73	−10.26	−7.71	−4.16	−1.49	0.54	1.51	2.12

Some of the conclusions that can be drawn by these measurements are the following:

The true earnings after paying an equivalent 5% interest charge are 2.8%, or $2.12 million at the present value.

The break-even year for cumulative net income is the sixth year, and the eighth year for cumulative cash flow. The cumulative discounted cash flow breaks even in the ninth year.

The maximum negative cumulative net income is $3 million, and this occurs at the third and fourth year. The maximum negative cumulative cash flow is $15 million, and that occurs at the third year.

This is a relatively high-risk investment because of the long break-even period of six to eight years.

The investment turnover ratio of 0.94, which is quite low, indicates that a relatively large investment is required for a small amount of sales.

In summary the investment opportunity should probably be rejected because of the slow payoff, low yield on cash flow, and low return on investment.

E. Validation of Planning Economic Models

Planning economic models must be validated after a project is approved and launched. Measurements at the various key stages of a project should be taken to see if the initial plans are realistic. If the actual progress is not on schedule financially, steps must be taken to either abort the project or to revise the projections. At the completion of a project, the final results can be compared with the original plans so that the experience gained may be applied to future planning.

5.4   OPERATING ECONOMIC MODELS

In contrast to planning economic models *operating economic models are used for short-range profit optimization.* The problem to be answered by operating economic models is not to select alternative investment opportunities but how to make the best use of existing facilities and equipment. The time scale of operating economic models is in hours and weeks. Operating economic models are simpler to develop because fixed costs and investments may be ignored. Sections 5.4 through 5.6 will first define the variables used in operating economic models, then introduce the basic laws governing these variables, and conclude by presenting techniques for developing operating economic models. The objective function to be optimized is derived by using the techniques for developing operating economic models. Let the following symbols be assigned:

OVL_n = volume of the nth product, in units.
UPR_n = unit price of the nth product, in dollars per unit.
REV_n = gross revenue of the nth product, in dollars.
FIX_n = fixed costs associated with the nth product, in dollars.
UVC_n = unit variable cost of the nth product, in dollars per unit.
UPC_n = unit profit contribution of the nth product, in dollars per unit.
VAR_n = variable cost for OVL_n units of the nth product, in dollars.
PCN_n = profit contribution for OVL_n units of the nth product, in dollars.
PBT_n = profit-before-taxes for OVL_n units, in dollars.
BEV_n = break-even volume for the nth product, in units.
CAP_n = maximum capacity for the nth product, in units.

5.5   BASIC LAWS FOR OPERATING ECONOMIC MODELS

One simplifying assumption in operating economic models is that expenses may be separated into fixed and variable categories. Fixed expenses are assumed not to vary with volume, although sometimes they can vary in steps within specified volume ranges. Variable expenses may vary linearly with volume or they may be nonlinear. In addition, variable expenses may lead or lag the shipment of a product in time. Figure 5.5-1 shows some typical types of expenses.

The following are some of the basic laws relating the variables of operating economic models. All variable expenses are assumed to vary linearly with volume. By the very nature of operating economic models, only the *operating statement* will be simulated. The *statement of financial condition* can usually be ignored.

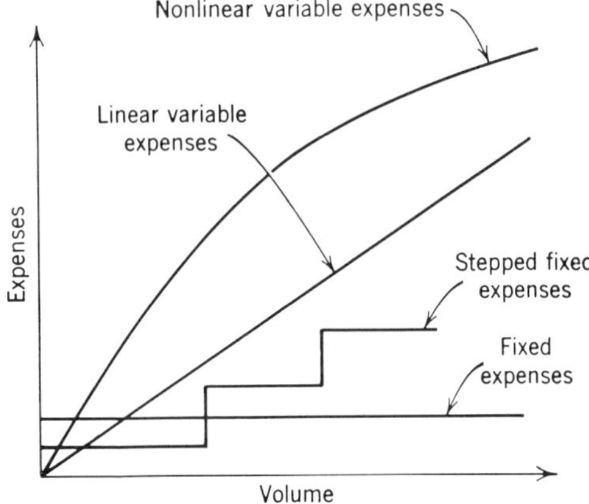

Fig. 5.5-1   Different types of expenses.

A. Gross Revenue

The gross revenue for selling OVL units of products at UPR unit price is equal to the volume times the unit price:

$$REV = (OVL)(UPR) \tag{5.5-1}$$

The total gross revenue resulting from selling K different types of products is

$$\text{total REV} = \sum_{n=1}^{K} (OVL_n \cdot UPR_n) \tag{5.5-2}$$

B. Fixed and Variable Costs

The variable cost for producing OVL units of products at UVC unit variable cost is equal to the volume times the unit variable cost:

$$VAR = (OVL)(UVC) \tag{5.5-3}$$

The total variable cost for producing K types of products is

$$\text{total VAR} = \sum_{n=1}^{K} (OVL_n \cdot UVC_n) \tag{5.5-4}$$

The total fixed costs plus variable costs for producing K types of products is

$$\text{total cost} = \sum_{n=1}^{K} [FIX_n + (OVL_n \cdot UVC_n)] \tag{5.5-5}$$

Fixed costs are usually allocated on a volume, dollar value, or assessed cost basis, and the sum of all the allocated fixed costs must be equal to the total fixed costs of a business:

$$\text{total FIX} = \sum_{n=1}^{K} FIX_n \tag{5.5-6}$$

C. Profit Contribution*

The profit contribution resulting from selling a single unit of a product is equal to the difference between the unit selling price and unit variable cost:

$$UPC_n = UPR_n - UVC_n \tag{5.5-7}$$

The profit contribution resulting from selling K types of products is

$$\text{total PCN} = \sum_{n=1}^{K} (OVL_n \cdot UPC_n) \tag{5.5-8}$$

$$= \sum_{n=1}^{K} (OVL_n) \cdot [UPR_n - UVC_n] \tag{5.5-9}$$

* Also known as marginal income, or incremental profit.

D. Profit-Before-Taxes and Objective Function

The total profit-before-taxes resulting from selling K types of products is equal to total gross revenue less total fixed and variable costs:

$$\text{total PBT} = (\text{total REV}) - (\text{total VAR}) - (\text{total FIX}) \qquad (5.5\text{-}10)$$

$$= \sum_{n=1}^{K} [(\text{OVL}_n)(\text{UPR}_n - \text{UVC}_n) - \text{FIX}_n] \qquad (5.5\text{-}11)$$

$$= \sum_{n=1}^{K} [\text{PCN}_n - \text{FIX}_n] \qquad (5.5\text{-}12)$$

Thus the total profit-before-taxes is equal to the total profit contributions minus the total fixed costs. It is very easy to deduce the objective function for optimization from Equation 5.5-12. Since the fixed cost terms FIX_n are constants, they become zero when the derivative of the profit equation is taken. Thus the objective function for operating economic models is equal to the total profit contribution:

$$\text{objective function} = \sum_{n=1}^{K} \text{PCN}_n \qquad (5.5\text{-}13)$$

E. Break-Even Volume

The break-even volume may be deduced mathematically or by the geometric relationships of Figure 5.5-2. For any given product, the break-even volume is the volume which produces zero profit-before-taxes. Thus from Equation 5.5-12

$$0 = \text{PCN}_n - \text{FIX}_n \qquad (5.5\text{-}14)$$

$$0 = \text{BEV}_n \cdot [\text{UPR}_n - \text{UVC}_n] - \text{FIX}_n \qquad (5.5\text{-}15)$$

$$\text{BEV}_n = \frac{\text{FIX}_n}{\text{UPR}_n - \text{UVC}_n} \qquad (5.5\text{-}16)$$

$$\text{BEV}_n = \frac{\text{FIX}_n}{\text{UPC}_n} \qquad (5.5\text{-}17)$$

Thus the break-even volume is equal to the fixed costs divided by the unit profit contribution. From Figure 5.5-2,

$$\tan \theta = \frac{\text{FIX}_n}{\text{BEV}_n} \qquad (5.5\text{-}18)$$

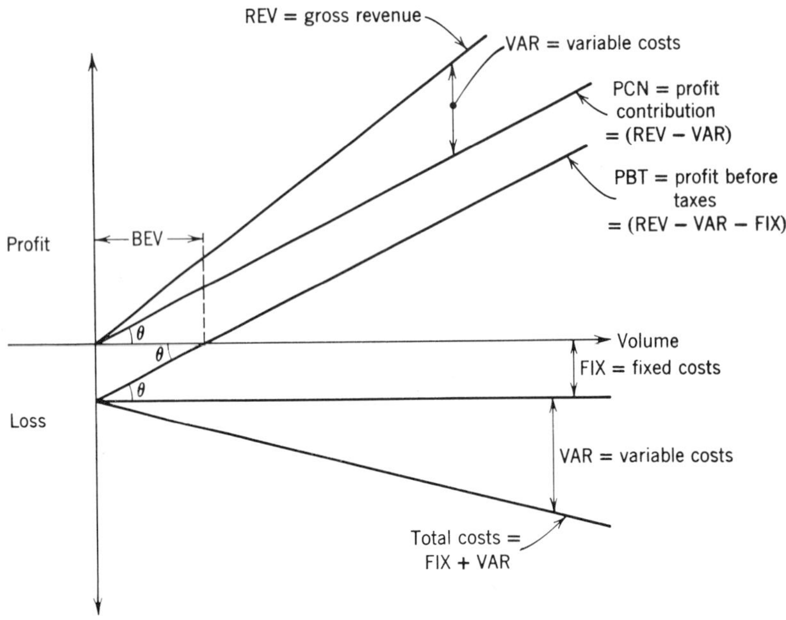

Fig. 5.5-2   Relationship between gross revenue, profit contribution, and fixed costs.

($\tan \theta$) is also the slope of the profit contribution line and is equal to the unit profit contribution. Thus

$$\tan \theta = \mathrm{UPC}_n \tag{5.5-19}$$

Therefore the break-even volume is equal to

$$\mathrm{BEV}_n = \frac{\mathrm{FIX}_n}{\mathrm{UPC}_n} \tag{5.5-20}$$

5.6   TECHNIQUES FOR DEVELOPING OPERATING ECONOMIC MODELS

Operating economic models consist of two parts: *constraints and objective function. Constraints* specify the level of total production, define the physical capacity limitations, and define the product-mix requirements. Other constraints may be due to the restrictions of linearity or other assumptions inherent in the characteristics of the physical process. The *objective function* to be maximized (or minimized) is equal to the profit contribution (or variable costs), as given by Equation 5.5-13 (or Equation 5.5-9).

Let the following additional symbols be assigned:

IVL_n = volume rate of the nth input material, in units per hour.

VLR_m = volume rate of the mth output product, in units per hour.

$A_{m,n}$ = mix coefficient of the nth input material contained in the mth output product, in units.

$B_{m,p}$ = work process requirement for the pth process required to produce a unit of the mth product, in man-hours per unit.

LCS_p = cost of labor at the pth process, in dollars per man-hour.

PRR_p = production rate of the pth process, in units per hour.

UMC_n = unit input material cost, in dollars per unit.

STR_q = storage area or volume of the qth storage location, in units.

UVC_m = unit variable cost for the mth output product, in dollars per unit.

$K_1, K_2, \ldots$ = constants.

The first type of constraint specifies the production level, either in terms of input or output volume rate. For a material flow process

$$\sum_{n=1}^{Q} \text{IVL}_n = K_1 \tag{5.6-1}$$

$$\sum_{m=1}^{T} \text{VLR}_m = K_2 \tag{5.6-2}$$

Most process production levels are specified by using Equation 5.6-2, which specifies the production level in terms of the output. Then the production level can be related to output in dollars. A normalized solution is said to result if the production level is set to equal one, i.e., K_1 or K_2 is equal to one. The normalized approach is feasible only for linear models. For energy processes, the total energy produced is specified by

$$\sum (\text{power generated}) = K_3 \tag{5.6-3}$$

The second type of constraint defines the physical capacity limitations, which include the input material flow rates, production rates of processes or work stations, capacity of storage locations, and output product flow rates. Since the model is for a real physical system, the variables should not be negative or imaginary numbers. Thus:

$$0 \le \text{IVL}_n \le K_4 \tag{5.6-4}$$

$$0 \le \text{VLR}_m \le K_5 \tag{5.6-5}$$

$$0 \le \text{PRR}_p \le K_6 \tag{5.6-6}$$

$$0 \le \text{STR}_q \le K_7 \tag{5.6-7}$$

The third type of constraint specifies the product-mix requirements. The content of each input material for a unit of output is specified by constraint statements through the use of the mix coefficients $A_{m,n}$. The product mix constraints are

$$\text{VLR}_m \le \text{IVL}_1 \cdot A_{m,1}$$

$$\text{VLR}_m \le \text{IVL}_2 \cdot A_{m,2}$$

$$\vdots$$

$$\text{VLR}_m \le \text{IVL}_n \cdot A_{m,n} \tag{5.6-8}$$

or they may be

$$\text{VLR}_m \ge \text{IVL}_1 \cdot A_{m,1}$$

$$\text{VLR}_m \ge \text{IVL}_2 \cdot A_{m,2}$$

$$\vdots$$

$$\text{VLR}_m \ge \text{IVL}_n \cdot A_{m,n} \tag{5.6-9}$$

If the symbol $(\gtrless)$ is used to signify either "*greater than or equal to* or *less than or equal to*," then Equations 5.6-8 and 5.6-9 become

$$\text{VLR}_m \gtrless \text{IVL}_1 \cdot A_{m,1}$$

$$\text{VLR}_m \gtrless \text{IVL}_2 \cdot A_{m,2}$$

$$\vdots$$

$$\text{VLR}_m \gtrless \text{IVL}_n \cdot A_{m,n} \tag{5.6-10}$$

The fourth type of constraint refers to linearity or other assumptions inherent in the model. These may specify that the model is valid only when certain variables are within specified ranges. The specification of this class of constraint is similar to that of capacity constraints. In case the physical process model should specify that different sets of coefficients be selected depending on the range of a given variable, the constraint would be

$$\text{if } K_{n,1} \le \text{IVL}_n \le K_{n,2}, \text{ use } j\text{th set of coefficients};$$

$$\text{if } K_{n,3} \le \text{IVL}_n \le K_{n,4}, \text{ use } (j+1)\text{th set of coefficients.} \tag{5.6-11}$$

$$\vdots$$

The linearity constraints may also restrict the solution to be within the constraint boundaries.

For operating economic models the objective function is equal to the total profit contribution, which is equal to the total gross sales revenue minus the total variable costs:

$$\text{objective function} = (\text{total gross sales revenue rate})$$
$$- (\text{total material cost rate})$$
$$- (\text{total labor cost rate})$$
$$- (\text{total fuel cost rate})$$
$$- (\text{other variable cost rate}) \qquad (5.6\text{-}12)$$

Total gross sales revenue rate is equal to the sum of the product volume rate times the unit price of each product:

$$\text{total gross sales revenue rate} = \sum_{m=1}^{T} \text{VLR}_m \cdot \text{UPR}_m \qquad (5.6\text{-}13)$$

The volume/price relationship generally falls into three categories: *normal elastic, inelastic,* and *abnormal elastic.* Figure 5.6-1 shows *the normal elastic curve* Ⓐ *where volume varies* inversely with price, the abnormal elastic curve Ⓑ where volume varies directly with price,* and the inelastic curve Ⓒ where volume does not vary with price, i.e., price is constant. In the normal elastic case, the total gross sales revenue is a function of price only, since the

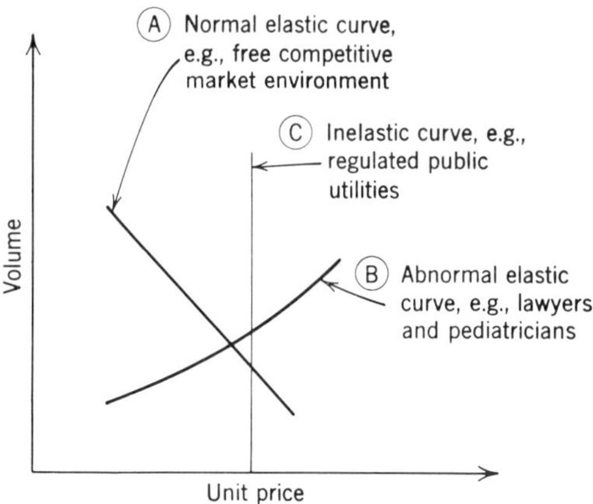

Fig. 5.6-1   Elastic and inelastic volume/price relationships.

* The abnormal elastic curve is found in luxury or conspicuous consumption items, or in services where the customer equates quality with price.

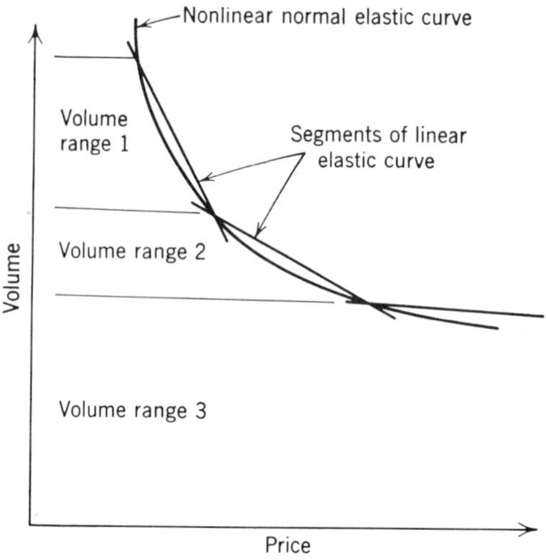

Fig. 5.6-2   Nonlinear normal elastic curve being approximated by linear segments.

volume is now a function of price. Thus

$$\text{total gross sales revenue} = G(\text{UPR}_m) \qquad (5.6\text{-}14)$$

In many cases the normal elastic curve may be approximated by a linear relationship. Figure 5.6-2 shows a nonlinear normal elastic curve being approximated by several linear segments.

Total *material cost rate* is equal to the sum of the input material volume rate times the unit cost of each material:

$$\text{total material cost rate} = \sum_{n=1}^{Q} \text{IVL}_n \cdot \text{UMC}_n \qquad (5.6\text{-}15)$$

Total *labor cost rate* is equal to the sum of the labor cost at each of the processes. The labor required at each process is equal to the sum of the work process requirement times the volume rate and the labor cost is found by multiplying the labor required by the cost of labor. The labor required at each process to produce a unit of a given product is given by the $\mathbf{B}_{m,p}$ matrix, as shown in Figure 5.6-3. Thus

$$\text{total labor cost rate} = \sum_{m=1}^{T} \text{VLR}_m \left(\sum_{p=1}^{s} B_{m,p} \cdot \text{LCS}_p \right) \qquad (5.6\text{-}16)$$

Total *fuel cost* may either be related to the total output volume or input volume. Other variable costs which can be related to output volume include overhead on direct labor, spoilage, waste, royalties, transportation, distribution, installation, service, warranty, etc.

$B_{(1,1)}$	$B_{(1,2)}$	$\cdots$	$B_{(1,p)}$
$B_{(2,1)}$	$B_{(2,2)}$	$\cdots$	$B_{(2,p)}$
$B_{(3,1)}$	$B_{(3,2)}$	$\cdots$	$B_{(3,p)}$
$\vdots$		$\cdots$	
$B_{(m,1)}$	$B_{(m,2)}$	$\cdots$	$B_{(m,p)}$

Fig. 5.6-3   Matrix of work process requirements, listing the labor required at the pth process to produce a unit of the mth product.

5.7   PRICING FOR MAXIMUM PROFIT

Two examples for determining the optimal price which will yield the maximum profit will be used to illustrate the applications of economic models discussed in this chapter.

A. Analytical Solution for Linear Relationships

The first example assumes that the elastic curve between volume and price, and between the profit-before-taxes and volume are both linear. In real life, these relationships will generally not be linear, but the same technique will still be applicable. *The problem is to analytically solve for the optimal price which will maximize the profit-before-tax dollars.* Obviously, a price that is too high will yield a smaller volume, and a price that is too low will yield lower profits per unit of sales.

Let the following symbols be assigned:

OVL   = volume, in units.
UPR   = unit price, in dollars per unit.
UPC   = unit profit contribution, in dollars per unit.
UVC   = unit variable costs, in dollars per unit.
FIX   = fixed costs, in dollars.
PBT   = profit-before-taxes, in dollars.
K_1, K_2 = constants $(K_1 > 0)$, $(K_2 < 0)$.

Figure 5.7-1 shows the elastic curve relating volume OVL to unit price UPR. Figure 5.7-2 shows a plot of profit-before-taxes PBT as a function of

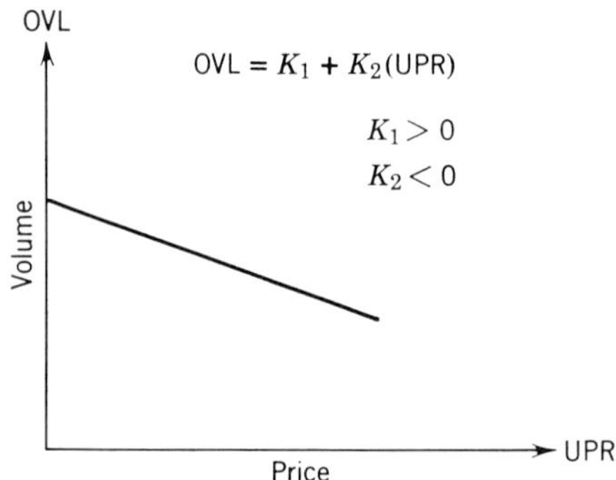

Fig. 5.7-1   Linear elastic curve.

volume for a given price. Profit-before-taxes PBT is equal to fixed costs FIX at zero volume, and the slope of the curve is equal to unit profit contribution UPC. From Section 5.5, the following relationships are true.

$$PBT = (OVL)(UPC) - FIX \qquad (5.7\text{-}1)$$

$$PBT = OVL(UPR - UVC) - FIX \qquad (5.7\text{-}2)$$

Let the elastic curve be

$$OVL = K_1 + K_2(UPR) \qquad (5.7\text{-}3)$$

Substituting Equation 5.7-3 into Equation 5.7-2,

$$PBT = [K_1 + K_2(UPR)][UPR - UVC] - FIX$$
$$= K_2(UPR)^2 + [K_1 - K_2(UVC)](UPR) - [K_1(UVC) + FIX] \qquad (5.7\text{-}4)$$

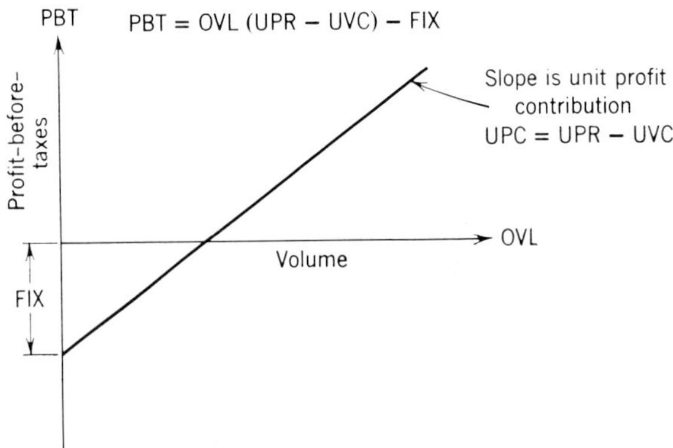

Fig. 5.7-2   Linear profit-before-taxes as a function of volume.

From calculus the profit will be maximum if

$$\frac{d(\text{PBT})}{d(\text{UPR})} = 0 \qquad (5.7\text{-}5)$$

and

$$\frac{d^2(\text{PBT})}{d(\text{UPR})^2} < 0 \qquad (5.7\text{-}6)$$

Thus:

$$\frac{d(\text{PBT})}{d(\text{UPR})} = 2K_2(\text{UPR}) + [K_1 - K_2(\text{UVC})] = 0 \qquad (5.7\text{-}7)$$

and

$$\frac{d^2(\text{PBT})}{d(\text{UPR})^2} = 2K_2 \qquad (5.7\text{-}8)$$

Solving for the optimal price,

$$\text{UPR} = \frac{K_2(\text{UVC}) - K_1}{2K_2} = \frac{\text{UVC}}{2} - \frac{K_1}{2K_2} \qquad (5.7\text{-}9)$$

Since

$$K_1 > 0 \quad \text{and} \quad K_2 < 0 \qquad (5.7\text{-}10)$$

$$\therefore \quad \frac{d^2(\text{PBT})}{d(\text{UPR})^2} < 0 \qquad (5.7\text{-}11)$$

The optimal price that will produce the maximum profit is given by Equation 5.7-9. The resulting volume at the optimal price is obtained by substituting Equation 5.7-9 into Equation 5.7-3.

$$\text{OVL} = K_1 + K_2\left[\frac{\text{UVC}}{2} - \frac{K_1}{2K_2}\right]$$

$$= \frac{K_1}{2} + \frac{K_2(\text{UVC})}{2} \qquad (5.7\text{-}12)$$

The profit-before-taxes is given by Equation 5.7-2.

Note that the optimal price and the resulting volume are only dependent on the unit variable cost UVC and the slope K_2 and intercept K_1 of the elastic curve. The optimal price is independent of the fixed costs FIX.

As specific examples, let the following values be assigned to these variables:

Case 1

$K_1 = 1000$ units.
$K_2 = -20$ units per dollar.
$\text{UVC} = 10$ dollars per unit.
$\text{FIX} = \$4000$ (for all cases).

The optimal price is

$$\text{UPR} = \left[\frac{10}{2} - \frac{1000}{(2)(-20)}\right] = 5 + 25 = 30 \text{ dollars per unit} \qquad (5.7\text{-}13)$$

the resulting volume is

$$\text{OVL} = \left[\frac{1000}{2} - \frac{(20)(10)}{2}\right] = 500 - 100 = 400 \text{ units} \qquad (5.7\text{-}14)$$

and the profit-before-taxes is

$$\text{PBT} = 400[30 - 10] - 4000 = \$4000 \qquad (5.7\text{-}15)$$

Case 2

$K_1 = 700$ units.
$K_2 = -10$ units per dollar.
$\text{UVC} = 10$ dollars per unit.

The optimal price is

$$\text{UPR} = \left[\frac{10}{2} + \frac{700}{(2)(10)}\right] = 5 + 35 = 40 \text{ dollars per unit} \qquad (5.7\text{-}16)$$

the resulting volume is

$$\text{OVL} = \left[\frac{700}{2} - \frac{(10)(10)}{2}\right] = 350 - 50 = 300 \text{ units} \qquad (5.7\text{-}17)$$

and the profit-before-taxes is

$$\text{PBT} = 300[40 - 10] - 4000 = \$5000 \qquad (5.7\text{-}18)$$

Case 3

$K_1 = 1000$ units.
$K_2 = -20$ units per dollar.
$\text{UVC} = 20$ dollars per unit.

The optimal price is

$$\text{UPR} = \left[\frac{20}{2} + \frac{1000}{(2)(20)}\right] = 10 + 25 = 35 \text{ dollars per unit} \qquad (5.7\text{-}19)$$

the resulting volume is

$$\text{OVL} = \left[\frac{1000}{2} - \frac{(20)(20)}{2}\right] = 500 - 200 = 300 \text{ units} \qquad (5.7\text{-}20)$$

and the profit-before-taxes is

$$\text{PBT} = 300(35 - 20) - 4000 = \$500 \qquad (5.7\text{-}21)$$

By comparing Cases 1 and 2 it is possible to conclude that as the slope of the elastic curve decreases (smaller K_2), the optimal price will be increased. The resulting profit will increase although the volume will decrease. By comparing Cases 1 and 3 it is possible to conclude that as the variable unit cost is increased, the optimal price will be increased. The resulting volume and the profit will decrease. These conclusions are justified by examining the equations for optimal price, resulting volume, and profit-before-taxes given by Equation 5.7-9, 5.7-12, and 5.7-2.

B. Numerical Solution for Nonlinear Relationships

The analytical solution for optimal price discussed in Section 5.7A is only valid for linear relationships between volume and price, and between total costs and volume. For nonlinear elastic and profit curves, it is often simpler to directly solve for optimal price by numerical methods. The procedure is to vary the price by small increments and look-up the volume from the

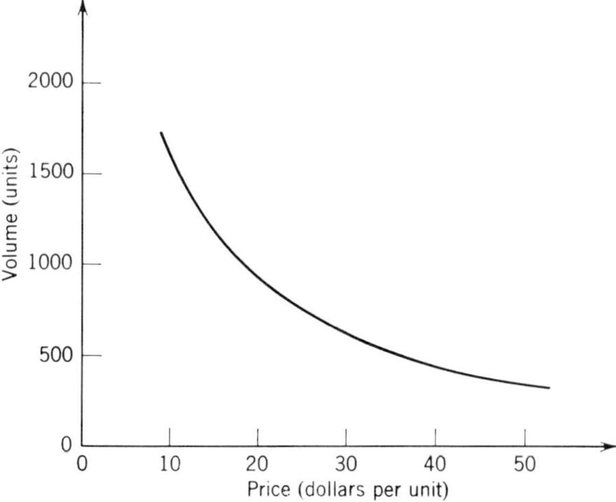

Fig. 5.7-3   Nonlinear elastic curve.

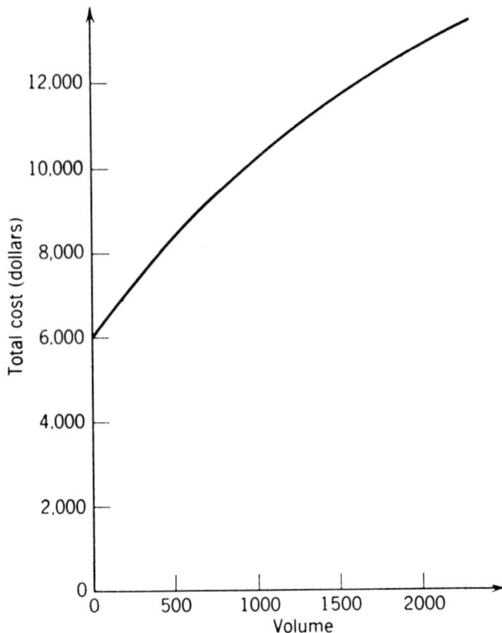

Fig. 5.7-4 Nonlinear total costs versus volume.

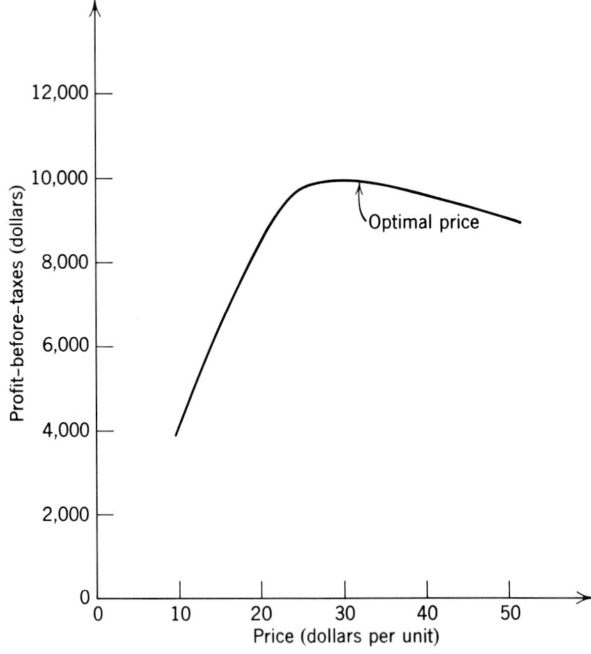

Fig. 5.7-5 Profit-before-taxes versus price.

elastic curve, then the volume is used to look-up the total costs. The profit-before-taxes is obtained by subtracting the total costs from volume times price. The optimal price is that which will produce the maximum profit-before-taxes.

Figure 5.7-3 shows the nonlinear elastic curve, and Figure 5.7-4 shows the nonlinear relationship between total cost and volume. Table 5.7-1 shows the computation of volume and profit-before-taxes as price is varied. Figure 5.7-5 shows the profit-before-taxes plotted against price. The optimal price is $32 per unit, and the corresponding profit-before-taxes is $10,000.

Table 5.7-1   Direct Numerical Solution for Optimal Price

① Price	② Volume Use ① To Look-up Figure 5.7-3	③ Revenue ① × ②	④ Total Cost Use ② To Look-up Figure 5.7-4	⑤ Profit- Before-Taxes ③ − ④	
10	1,600	16,000	12,000	4,000	
12	1,400	16,800	11,600	5,200	
14	1,250	17,500	11,100	6,400	
16	1,100	17,600	10,700	6,900	
18	1,000	18,000	10,400	7,600	
20	920	18,400	10,000	8,400	
22	850	18,700	9,200	9,500	
24	800	19,200	9,600	9,600	
26	740	19,240	9,400	9,840	
28	680	19,040	9,200	9,840	
30	630	18,900	9,000	9,900	
32	590	18,800	8,800	10,000	← Optimal price
34	540	18,300	8,600	9,700	
36	500	18,000	8,400	9,600	
38	475	18,050	8,300	9,750	
40	450	18,000	8,200	9,800	
42	420	17,640	8,100	9,540	
44	390	17,160	7,900	9,260	
46	380	17,480	7,850	9,630	
48	360	17,280	7,800	9,480	
50	340	17,000	7,750	9,250	

5.8   SUMMARY

This chapter has presented the fundamental principles of economic models and the techniques for developing economic models. Planning economic models are used in the section of investment opportunities, including the functions and equipment of the computer process control system. Operating economic models are used to develop the objective function to be optimized.

Students and other readers who may not have had exposure to economic models are encouraged to make a start in this field. Any quantitative analysis in this area is better than the "seat of the pants" approach.

This chapter marks the end of the first subject of this book—*modeling*. Chapter 6 will begin the second subject of this book—*optimization*.

6

Introduction to the Optimization Problem

6.0 INTRODUCTION

Chapters 2 through 5 were devoted to the techniques of developing *functional, physical,* and *economic models.* Chapter 6 marks the beginning of the second topic of this book—*optimization.* The mathematical derivations of the various optimization techniques will be covered in Chapters 7 through 10. This chapter presents an introductory survey of the overall optimization problem. It is descriptive rather than mathematical in nature, the purpose being to present in one place the overall view of optimization so that the reader will not "lose sight of the forest because of the trees."

This chapter will first present the *major elements of the optimization problem:*

The physical process model.
The objective function.
Constraints on state and control variables.

Particular emphasis will be devoted to the place of each of these major elements in the overall optimization procedure and the various forms of these major elements. Chapter 6 will include a summary of the various practical approaches to optimal control implementation, which will be discussed in Chapters 7 through 10.

6.1 THE PHYSICAL PROCESS MODEL

The physical process model is a mathematical representation of the physical process under study. More specifically, the physical process model relates the state variables to the control variables of the process. As pointed out in

171

Chapter 4, a great deal of intuition and judgment as well as the applications of the laws of physics and chemistry are involved in developing practical physical process models. Also physical process models are subject to inaccuracies, even when the assumptions used to develop them are valid. Thus, on one hand, the optimization problem solution is only as good as the physical process model; on the other hand, the expenditure of more effort and cost may not necessarily result in a model that will provide significantly better results than the simple model that is adequate for the purpose. The key factors in developing physical process models are the following:

(1) Identification of the variables of interest.

(2) Determination of the extent of information available, either from existing documents or planned tests.

(3) Derivation of a physical process model that is adequate for the purpose, no more and no less.

This section will first discuss the place of physical process models in the optimization procedure and then present a survey of the various forms of physical process models which will be used in Chapters 7 through 10.

A. The Place of the Physical Process Model

The place of the physical process model in the optimization problem is to relate the state variables to the control variables of the process. Note that the physical process model itself is never optimized. It is always the objective function which is optimized. The objective function and constraints will generally be functions of both the state and the control variables. In some cases, a direct substitution of the physical process model can be made that results in the *objective function and constraints being functions of the control variables only.* In other cases, the objective function may be modified by adding the physical process model multiplied by a constant, i.e., the Lagrange multiplier. In this case, the physical process model may be considered as a constraint on the objective function.

Let the following symbols be assigned:

Y_j = state variables, $(j = 1)$ to $(j = m)$.
X_i = control variables, $(i = 1)$ to $(i = n)$.
f_j = physical process model functions.
F = objective function expressed in terms of Y_j and X_i.
G = objective function expressed in terms of X_i only.
R_p = constraint function expressed in terms of Y_j and X_i, $(p = 1)$ to $(p = v)$.
Q_p = constraint function expressed in terms of X_i only.

The most desirable form of the physical process model is *one in which each state variable is expressed as a function of the control variables*. This is the form most convenient for direct substitution into the objective function and constraints to eliminate the state variables. The physical process model in this form is

$$Y_j = f_j(X_1, X_2, \ldots, X_n) \tag{6.1-1}$$

The objective function in *F*-form is a function of the state and control variables.

$$F = F(Y_1, Y_2, \ldots, Y_m; X_1, X_2, \ldots, X_n) \tag{6.1-2}$$

By substituting Equation 6.1-1 into the objective function in Equation 6.1-2, the state variables may be eliminated and the result is the objective function in *G* form:

$$G = G(X_1, X_2, \ldots, X_n) \tag{6.1-3}$$

where $F = G$. Similarly, the constraint functions R_p may have the state variables eliminated by substituting Equation 6.1-1 into R_p:

$$R_p(Y_1, Y_2, \ldots, Y_m; X_1, X_2, \ldots, X_n) = Q_p(X_1, X_2, \ldots, X_n) \tag{6.1-4}$$

At this point the optimization problem is to optimize the objective function

$$G = G(X_1, X_2, \ldots, X_n) \tag{6.1-5}$$

subject to the constraints:

$$Q_p(X_1, X_2, \ldots, X_n) = 0 \tag{6.1-6}$$

B. Linear Algebraic Model

The simplest and most convenient form of physical process model is the linear algebraic model. This can readily be developed by regression analysis techniques discussed in Sections 4.3 through 4.5. The application of linear algebraic physical process models to optimization problems will be discussed in Sections 7.1, 7.2, 7.3, and 7.5.

The general form of the linear algebraic model for *m* state variables and *n* control variables is

$$Y_j = \sum_{i=0}^{n} (A_{ji})(X_i) \tag{6.1-7}$$

where $X_0 = 1$.

In matrix notation, the linear algebraic model is

$$\mathbf{Y} = \mathbf{A}\mathbf{X} \tag{6.1-8}$$

where $\mathbf{Y}$ is a (1 by m) column vector, $\mathbf{X}$ is a $[1$ by $(1 + n)]$ column vector, and $\mathbf{A}$ is an $[m$ by $(1 + n)]$ matrix. Thus:

$$
\begin{pmatrix} Y_1 \\ Y_2 \\ \vdots \\ Y_m \end{pmatrix} = \begin{pmatrix} A_{10} & A_{11} & A_{12} & \cdots & A_{1n} \\ A_{20} & A_{21} & A_{22} & \cdots & A_{2n} \\ \vdots & & & & \vdots \\ A_{m0} & A_{m1} & A_{m2} & \cdots & A_{mn} \end{pmatrix} \cdot \begin{pmatrix} X_0 \\ X_1 \\ X_2 \\ \vdots \\ X_n \end{pmatrix} \qquad (6.1\text{-}9)
$$

C. Second-Order Polynomial Model

The physical process model in second-order polynomial form may provide a better fit for certain processes known to be nonlinear, e.g., fluid flow processes. The application of second-order polynomial physical process models to the optimization problems will be discussed in Sections 7.3 and 7.4. The general form of the second-order polynomial physical process model for m state variables and n control variables is

$$
Y_j = A_j + \sum_{i=1}^{n} (B_{ji})(X_i) + \sum_{k=i+1}^{n} \sum_{i=1}^{n} (C_{ji})(X_i)(X_k) + \sum_{i=1}^{n} (D_{ji})(X_i)^2 \qquad (6.1\text{-}10)
$$

For example, a physical process model with two state variables and two control variables is

$$
Y_1 = A_1 + B_{11}(X_1) + B_{12}(X_2) + C_{12}(X_1)(X_2) \qquad (6.1\text{-}11)
$$
$$
+ D_{11}(X_1)^2 + D_{12}(X_2)^2
$$
$$
Y_2 = A_2 + B_{21}(X_1) + B_{22}(X_2) + C_{21}(X_1)(X_2) \qquad (6.1\text{-}12)
$$
$$
+ D_{21}(X_1)^2 + D_{22}(X_2)^2
$$

D. Incremental Linear Algebraic Model

Incremental linear algebraic physical process models *relate the changes of the state variables to the changes of the control variables.* Incremental physical process models will be applied to the optimization problem in Sections 8.2, 8.3, and 8.4. The incremental physical process model corresponding to the linear algebraic physical process model given in Equation 6.1-7 is

$$
\Delta Y_j = \sum_{i=1}^{n} (A_{ji})(\Delta X_i) \qquad (6.1\text{-}13)
$$

The changes ΔY_j and ΔX_i are in turn defined by

$$
\Delta Y_j = Y_j^{t+1} - Y_j^t \qquad (6.1\text{-}14)
$$
$$
\Delta X_i = X_i^{t+1} - X_i^t \qquad (6.1\text{-}15)
$$

where the variables Y_j^{t+1} and X_i^{t+1} are measured at time interval $(t + 1)$, and Y_j^t and X_i^t are measured at the time interval t. Since these are steady-state measurements, the time intervals should be far enough apart to allow process transients to settle.

E. No Physical Process Model

For poorly defined processes without a physical process model, Chapter 9 presents the evolutionary optimization approach, which seeks to improve the objective function by direct experimentation on the process. In a sense, physical process models are not absolutely essential to the optimization problem.

F. Dynamic Models in Continuous-Time Form

Dynamic physical process models in continuous-time form express the derivatives of the state variables as functions of all the state variables, control variables, and time. This form of model is applied to the optimization problem in Sections 10.1 and 10.2. The general form of the continuous-time dynamic physical process model is

$$\frac{dY_j}{dt} = H_j(Y_1, Y_2, \ldots, Y_m; X_1, X_2, \ldots, X_n; t) \qquad (6.1\text{-}16)$$

G. Dynamic Model in Discrete-Time Form

Dynamic physical process models in discrete-time form express the value of the state variable at the next time interval $Y_j(t_0 + \Delta t)$ as a function of the following:

(1) The state variables at past intervals $Y_j(t_0)$.
(2) The functions which represent the effect of the control variable at past intervals $J_j(t_0)$.
(3) The instants of time: t_0 and $(t_0 + \Delta t)$.
The general form discrete-time dynamic physical process model is

$$Y_j(t_0 + \Delta t) = \psi_j[(t_0 + \Delta t); Y_1(t_0), Y_2(t_0), \ldots, Y_m(t_0);$$
$$J_1(t_0), J_2(t_0), \ldots, J_m(t_0); t_0] \qquad (6.1\text{-}17)$$

The difference equation given in Equation 6.1-17 will be used for the dynamic programming approach discussed in Sections 10.3, 10.4, and 10.5. Note that it is a recurrence relationship, i.e., the same equation holds for all the time-intervals, $t_0, (t_0 + \Delta t), (t_0 + 2\Delta t), \ldots.$

6.2 THE OBJECTIVE FUNCTION

The objective function is the mathematical expression of process operation to be maximized or minimized, i.e., optimized. It is also called the criterion function or the performance index. *The definition of the objective function and the solution for the optimum of the objective function is the heart of the optimization problem.* In contrast to physical process models, objective functions usually express nonphysical quantities such as profit, cost, quality, and time.

This section will first discuss some of the basic principles for defining objective functions and then describe some of the more useful forms of objective functions used in Chapters 7 through 10.

A. Basic Principles for Defining Objective Functions

The basic principles presented for defining objective functions are pragmatic guides developed from experience. They should not be regarded as scientific laws, and there are many situations where some of these basic principles should probably not be used.

The *principle of uniqueness* states that one and only one objective function should be maximized or minimized. In situations where two objective functions P and W are to be optimized, they may be combined into a single objective function by means of a linear combination. The composite objective function is

$$F = \psi_1(P) + \psi_2(W) \qquad (6.2\text{-}1)$$

where ψ_1 and ψ_2 are weighting constants. If P is to be maximized and W is to be minimized, then a function V may be substituted for W if V is equal to the inverse of W, so that the maximum of V will occur at the same point as the minimum of W. Thus if

$$W \text{ is a minimum when } V \text{ is a maximum at } (X_1, X_2, \ldots, X_n) \qquad (6.2\text{-}2)$$

the composite objective function to be maximized is

$$F = \psi_3(P) + \psi_4(V) \qquad (6.2\text{-}3)$$

where ψ_3 and ψ_4 are weighting constants.

The *principle of accountability* states that the objective function should be closely related to the accountability to which the management of the process is held responsible. In other words, the scope and nature of the objective function should be defined so that optimal operation of the process will make the management of the process *most successful.*

The *principle of controllability* states that the objective function should be expressed in terms of the variables that are controllable by the process operator or by the computer process control system. Objective functions which are expressed in variables beyond the scope of the control variables of the process are exercises in futility.

The *principle of profit orientation* states that objective functions should generally express profits or quantities which are closely related to profits, e.g., cost and quality. If it is required to select one objective function from among two possible objective functions, the one which has a larger impact on profits should be selected.

The *principle of proper form* states that it is desirable to have an objective function which has a maximum or minimum. Objective functions which do not have maxima or minima will require constraints to provide meaningful solutions, e.g., linear programming. Figure 6.2-1 shows that constants and linear functions without constraints have no minima or maxima. Other undesirable forms of the objective function include functions which have discontinuities, functions which have local maxima or minima, and multiple-valued functions, as shown in Figure 6.2-2. One of the most desirable forms of objective function is the quadratic form which will be used for quality control in Chapters 7, 8, and 10.

The *principle of optimal control equations* states that the optimization procedure should develop *optimal control equations* which can be used to solve for the control variables that will optimize the objective function. In feedforward control, the optimal control equations are derived by setting the objective function to the conditions of optimum and then solving the conditions of optimum for the control variables. The equations that express the optimal control variables in terms of the desired operating conditions and parameters are called the *optimal control equations*.

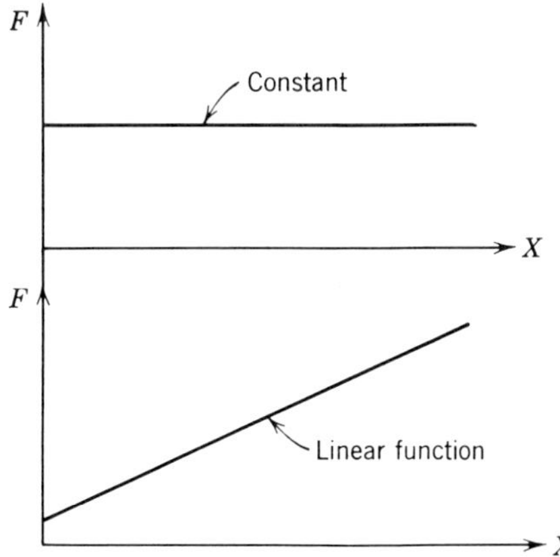

Fig. 6.2-1 Constant and linear objective functions without constraints have no maxima or minima.

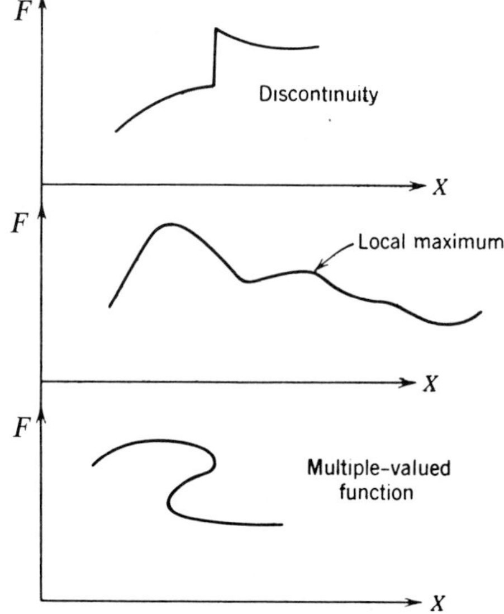

Fig. 6.2-2 Some undesirable forms of the objective function.

B. Profit Objective Function

The profit objective function expresses the profit contribution that can be obtained from operating the process. Section 5.4, 5.5, and 5.6 defined profit contribution as the *difference between gross revenue and variable costs.* Fixed costs can be ignored for process control applications because these are constant terms in the objective function which do not affect the location of the optimum. For example, the objective function which will be used in the linear programming problem of Section 7.5 uses profit contribution:

$$F = \sum (\text{price})(\text{output volume})$$

$$- \sum (\text{cost})(\text{raw material volume}) \qquad (6.2\text{-}4)$$

C. Cost Objective Function

The cost objective function expresses the costs associated with operating the process. Usually only the variable costs which are controllable are included in the cost objective function, e.g., material costs and fuel costs. Fixed costs and other noncontrollable costs should not be included, e.g., fixed investment costs, overhead burden, and other indirect expenses.

Section 7.4 will show an example of steady-state blending of styrene from several reactors which will have the objective function as the sum of the costs for producing the styrene from each reactor. Let $C_i(P_i)$ be the cost of producing the volume P_i from the ith reactor, then the objective function to be minimized is

$$F = \sum_i C_i(P_i) \qquad (6.2\text{-}5)$$

Section 10.1 will show that dynamic optimization problems have objective functions which are integrals of the cost-rate function over time. The cost-rate function F corresponds to the steady-state objective function, and is the expression of the rate of cost expenditure in dollars per hour. The dynamic objective function to be minimized is

$$\phi = \int_{t_0}^{t_f} F(Y_1, Y_2, \ldots, Y_m; X_1, X_2, \ldots, X_n; t)\, dt \qquad (6.2\text{-}6)$$

This is called the optimal trajectory problem because the problem is to select a path or trajectory of the control variables between two boundary conditions which will minimize ϕ.

D. Quality Objective Function

There are many methods to optimize the quality objective function. The first method is to consider that the process is at an optimum if the value of the quality objective function is greater than or equal to a pre-set number. The second method is to minimize the weighted deviation from the ideal state, i.e.. minimize the objective function which is made up of the weighted sum of the deviations from the desired state or quality variables. A useful and convenient form of objective function is the "weighted quadratic form." Such a form defines optimal quality process operation as the condition where the sum of the squares of the differences between the desired state variables $\overline{Y}_j$ and the actual state variables Y_j is a minimum. The state variables may be either measured or indirectly calculated.

The steady-state quadratic form quality objective function will be used extensively in Chapters 7 and 8 and is given by

$$F = \sum_{j=1}^{m} \psi_j(\overline{Y}_j - Y_j)^2 \qquad (6.2\text{-}7)$$

where ψ_j are positive weighting constants.

The dynamic quadratic form quality objective function in continuous-time form can be stated as

$$\phi[Y_1(t_0), Y_2(t_0), \ldots, Y_m(t_0)] = \int_{t_0}^{\infty} \sum_{j=1}^{m} \psi_j[\,\overline{Y}_j - Y_j(t)]^2 \exp[-\alpha(t - t_0)]\, dt$$

$$(6.2\text{-}8)$$

where ψ_j and α are positive constants, and t_0 is the initial time for which the initial states are known. The exponential term is an additional weighting constant which is time-dependent.

The dynamic quadratic form quality objective function in discrete-time form is

$$\phi[Y_1(t_0), Y_2(t_0), \ldots, Y_m(t_0)] = \sum_{K=1}^{\infty} \left\{ \sum_{j=1}^{m} \psi_j [\overline{Y}_j - Y_j(t_0 + K\Delta t)]^2 \right\}$$
$$\exp[-\alpha K(\Delta t)]$$

$$(6.2\text{-}9)$$

Equations 6.2-8 and 6.2-9 will be used in the dynamic optimization discussions of Chapter 10. The difference between dynamic and steady-state quality control is that dynamic quality control attempts to bring the process back to the optimum with minimum cost, whereas steady-state quality control ignores the dynamic performance of the process and the speed of arriving at the new operating point.

E. Time Objective Function

The time objective function expresses the elapsed time required for the process to proceed between two fixed boundary conditions, and will be discussed in Section 10.1. The objective function to be minimized is

$$T = \int_{t_0}^{t_f} dt \qquad (6.2\text{-}10)$$

For boundary conditions of

$$Y_j(t_0) = Y_{j0} \qquad (6.2\text{-}11)$$

$$Y_j(t_f) = Y_{jf} \qquad (6.2\text{-}12)$$

where j indexes the specified state variables.

Again the physical process model may be substitued into the objective function in Equation 6.2-10 to make it a function of the state and control variables.

6.3   CONSTRAINTS ON STATE AND CONTROL VARIABLES

Constraints on state and control variables either limit the region of allowable operation for state and control variables or impose further relationships between state and control variables. Constraints are always present in the optimization of real physical processes. In a sense, constraints on state and control variables may be thought of as extensions of the physical process model. This is because constraints define the relationship between state and control variables or the allowed region of process operation.

Constraints play a vital part in the optimization procedure. For example, the linear objective function used in linear programming would not have an optimum without the constraints. Also, the effect of constraints is generally to degrade the quality of optimal process operation. This section will first describe the different types of constraints, and then discuss the role of constraints in the optimization problem.

A. Hard Constraints on the Limits of Control Variables

Hard constraints on the limits of control variables prescribe specific upper and/or lower limits for a control variable. Hard constraints should never be violated, otherwise disasters or nonsensical solutions may result. Examples are safety constraint limits and negative values for physical variables. Let the following symbols be defined:

R_{iL} = lower constraint limit for the ith control variable.
R_{iU} = upper constraint limit for the ith control variable.
Z_{iL}, Z_{iU} = slack variables, $Z_{iL} \geq 0$ and $Z_{iU} \geq 0$.

The general form for hard constraints on the limit of control variables is given by the inequalities:

$$X_i \geq R_{iL} \tag{6.3-1}$$

$$X_i \leq R_{iU} \tag{6.3-2}$$

By introducing *positive slack variables*, the constraints may be expressed as equalities.

$$X_i - Z_{iL} = R_{iL} \tag{6.3-3}$$

$$X_i + Z_{iU} = R_{iU} \tag{6.3-4}$$

The optimization problem is to optimize F subject to the constraints of Equations 6.3-3 and 6.3-4.

Hard constraints on the limits of control variables will be discussed in Sections 7.5, 8.3, and 9.4.

B. Soft Constraints on the Limit of Control Variables

Soft constraints do not prohibit the control variables to exceed the constraint limit. However, the quality of process operation will deteriorate very rapidly if the constraint limit is exceeded by any significant extent. Soft constraints may be implemented indirectly, i.e., by modifying the objective function to reflect the penalty imposed by significant excursions of a control variable beyond the constraint limit. A possible penalty function to be added to the objective function is

$$P_1 = K_i \left(\frac{X_i}{R_{iU}} \right)^{M_i} \tag{6.3-5}$$

where K_i is a positive constant, M_i is a positive integer, and R_{iU} is the soft constraint limit on the control variable X_i.

Another possible penalty function is

$$P_2 = K_i \exp[M_i(X_i - R_{iU})] \tag{6.3-6}$$

The modified objective function to be *maximized* will be $(F - P_1)$ or $(F - P_2)$. The modified objective function to be *minimized* will be $(F + P_1)$ or $(F + P_2)$.

C. Constraints on State and Control Variables Expressed Only as Functions of Control Variables

Constraints on state and control variables have the general form:

$$S_p(Y_1, Y_2, \ldots, Y_m; X_1, X_2, \ldots, X_n) \leq S_{pU} \tag{6.3-7}$$

or

$$S_p(Y_1, Y_2, \ldots, Y_m; X_1, X_2, \ldots, X_n) \geq S_{pL} \tag{6.3-8}$$

where $(p = 1)$ to $(p = v)$ and S_{pU} and S_{pL} are positive constraint limits. By introducing positive slack variables Z_p and moving S_{pU} and S_{pL} to the left-hand side of Equations 6.3-7 and 6.3-8, the general form of constraints on state and control variables is

$$R_p(Y_1, Y_2, \ldots, Y_m; X_1, X_2, \ldots, X_n, Z_p) = 0 \tag{6.3-9}$$

To express constraints on state and control variables only as a function of control variables, it may be possible to substitute the physical process model into Equation 6.3-9 to eliminate the state variables. Of course, the operation is not necessary if the constraints originally contain only the control variables. The general form of the constraint on state and control variables expressed only as functions of control variables is

$$R_p(X_1, X_2, \ldots, X_n, Z_p) = 0 \tag{6.3-10}$$

Each constraint given in Equation 6.3-10 effectively reduces the number of independent control variables by one.

D. Constraints on State and Control Variables

For constraints on state and control variables where the state variables cannot be eliminated, the objective function to be optimized may be modified by the Lagrange multiplier technique, as will be shown in Section 7.4. If λ_p is the Lagrange multiplier, the modified objective function subject to the constraint of Equation 6.3-7 or Equation 6.3-8 is

$$\phi = F + \sum_{p=1}^{v} (\lambda_p)[R_p(Y_1, Y_2, \ldots, Y_m; X_1, X_2, \ldots, X_n)] \quad (6.3\text{-}11)$$

E. Initial and Terminal Boundary Values for Dynamic Optimization

Dynamic optimization problems which attempt to minimize the cost of proceeding between two points are called optimal trajectory problems, as will be discussed in Section 10.1. The initial and terminal boundary conditions for the specified state variables are given by

$$Y_j(t_0) = Y_{j0} \quad (6.3\text{-}12)$$

$$Y_j(t_f) = Y_{jf} \quad (6.3\text{-}13)$$

where j is the index for the specified state variables; t_0 and t_f are initial and final times; Y_{j0} and Y_{jf} are constants.

In some of the dynamic optimization techniques which will be discussed in Chapter 10, the actual measured value of the state variables at each step may be used as the initial boundary value to improve the accuracy of control.

6.4   PRACTICAL APPROACHES TO OPTIMAL CONTROL IMPLEMENTATION

Optimal control implementation is the application of optimization problem solutions to control the physical process. This section will describe the various practical approaches to optimal control implementation that will be discussed in detail in Chapters 7 through 10. This summary will also serve to present the overall picture of optimal control implementation and the relationship between the major elements of the optimization problem.

A. Feedforward Optimal Control with the Physical Process Model

Feedforward optimal control requires a physical process model and is particularly suited to well-defined steady-state processes, as will be shown in Chapter 7. Figure 6.4-1 shows the feedforward optimal control approach. The

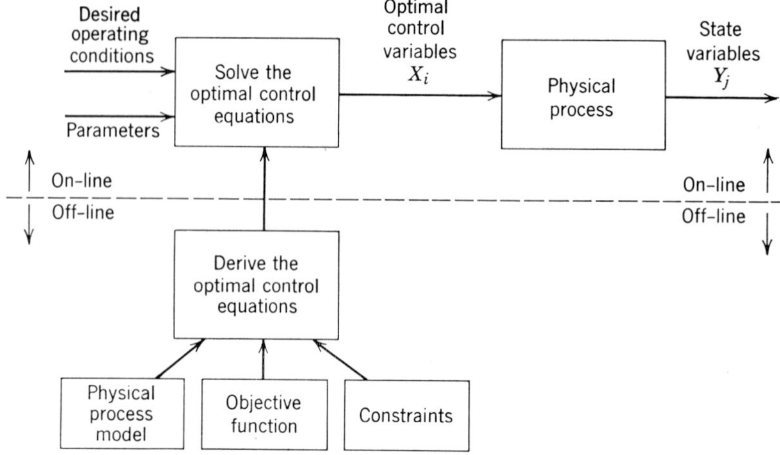

Fig. 6.4-1   Feedforward optimal control.

optimal equations, which are solved on-line by the computer process control system, yield the optimal control variables for the desired operating conditions and parameters. The optimal control equations are derived, however, in an off-line fashion by using the physical process model, the objective function, and the constraints.

B. Feedforward Optimal Control with the Updated Physical Process Model

Section 8.1 will show how the coefficients of the physical process model can be updated by using the computer process control system to perform on-line experimentation. Figure 6.4-2 shows the feedforward optimal control approach using the updated physical process model. Note that while the derivation of the optimal control equations is done initially in an off-line fashion, the updating of the coefficients of the optimal control equation must be done on-line. Thus updating the coefficients of the physical process model and the coefficients of the optimal control equations are the two elements added to the computer process control functions.

C. Feedback Optimal Control with the Incremental Physical Process Model

Sections 8.2, 8.3, and 8.4 will show how feedback optimal control may be applied by using an incremental physical process model. In contrast to physical process models which relate state variables to control variables, *incremental physical process models relate the changes in the state variables to the changes in the control variables.* Figure 6.4-3 shows the feedback optimal control approach using the incremental physical process model. In using this

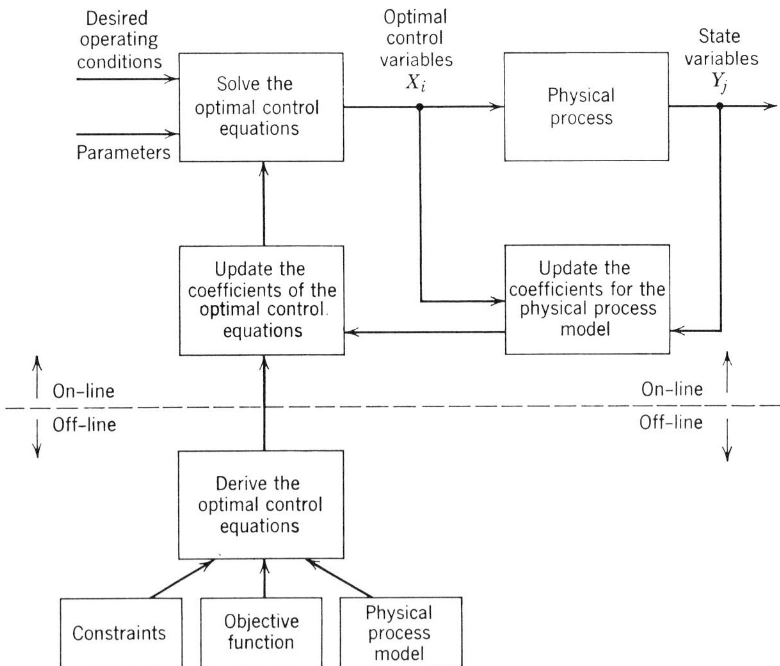

Fig. 6.4-2 Feedforward optimal control with on-line updating of the physical process model.

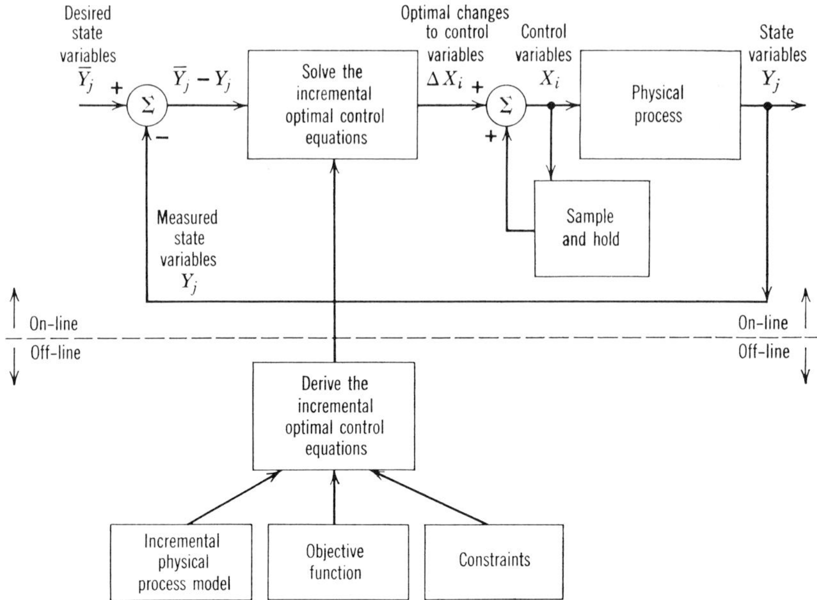

Fig. 6.4-3 Feedback optimal control using an incremental physical process model.

approach, successive incremental adjustments are made to the control variables so that errors are reduced in turn. Therefore, the accuracy of the incremental physical process model does not affect the accuracy of optimal control, but the accuracy does affect the speed of arriving at the optimum. Again the optimal control equations, which yield the optimal changes of the control variables, are solved on-line but derived off-line.

D. Evolutionary Optimization using the Process for On-Line Experimentation

For poorly defined processes which do not have a physical process model, or where direct optimization cannot be applied, the evolutionary optimization approach, EVOP, which is discussed in Chapter 9, can be used. The evolutionary optimization approach actually uses the physical process itself instead of the physical process model to obtain the relationship between the state and control variables. This method is also referred to as the

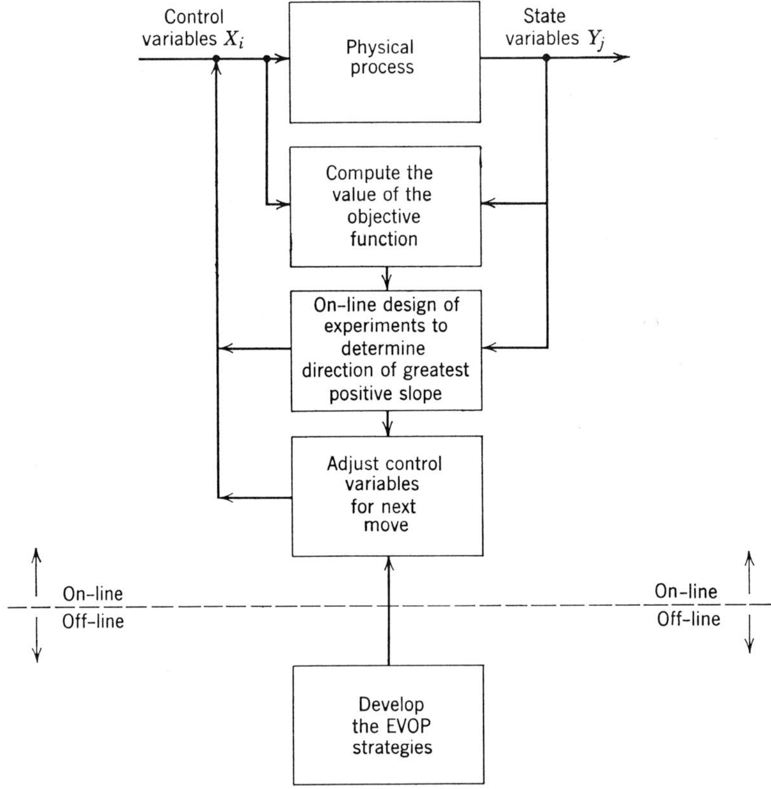

Fig. 6.4-4   Evolutionary optimization approach.

"hill-climbing" technique. The computer process control system can be used to conduct on-line design of experiments to determine the direction of the greatest positive (or negative) slope of the objective function. For maximization problems, the control variables may be adjusted to move the objective function in the direction of the greatest positive slope; for minimization problems, the control variables may be adjusted to move the objective function in the direction of the greatest negative slope. This approach is continued until the optimum is reached. Figure 6.4-4 shows the evolutionary optimization approach applied to a poorly defined steady-state process.

E. Combined Feedforward and Feedback Optimal Control Using the EVOP Technique with the On-Line Simulated Physical Process Model

Another way of using the EVOP technique is to use the simulated physical process model instead of the actual physical process to obtain the relationship between the state and control variables, as shown in Figure 6.4-5. The state and control variables are measured to determine the operating region. Then

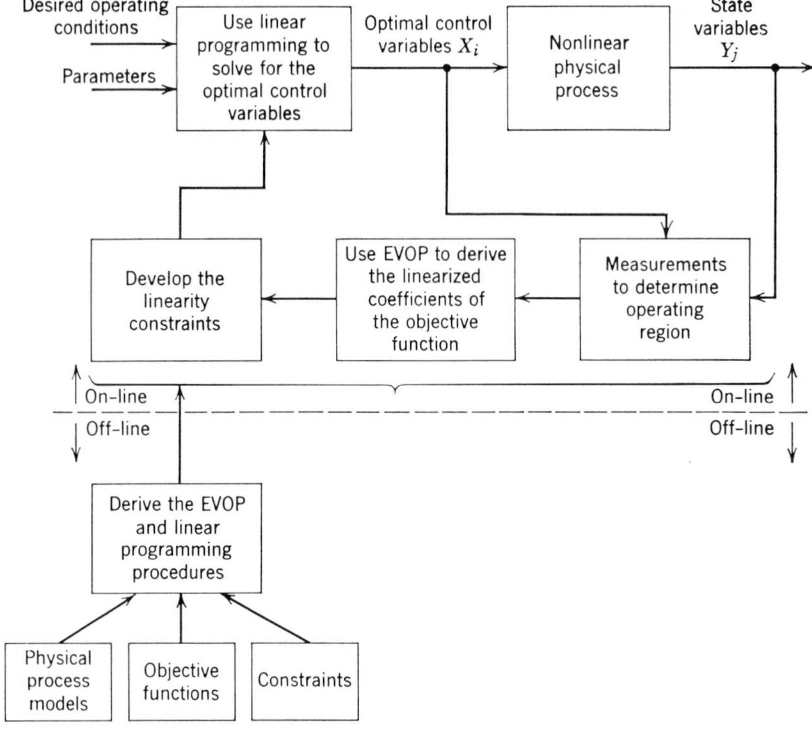

Fig. 6.4-5 Combined feedforward and feedback optimal control.

the EVOP technique is used on the simulated physical process model to derive the linearized coefficients of the objective function and the linearity restrictions. Then linear programming techniques may be used to solve for the optimal objective function. This approach is naturally much faster than using EVOP on the physical process itself. The nonlinearity or complexity of the model and the objective function should not add to the difficulty of this approach, since the simulation approach circumvents the problems of direct analytical solutions.

F. Dynamic Optimal Control by Feedback

Dynamic optimal control by feedback uses the measured state variables at each time interval as initial conditions for the solutions of the optimal control equations. This has the advantage of taking into account the inaccuracies of the physical process model and the effect of uncontrolled variables. Figure 6.4-6 shows the dynamic optimal control by feedback,

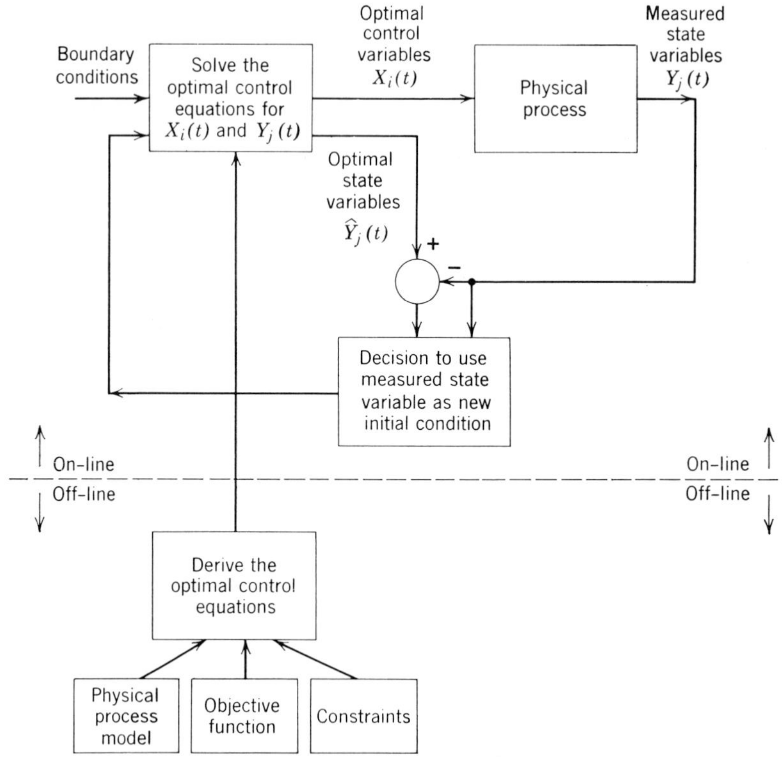

Fig. 6.4-6   Dynamic optimization by feedback.

which will be discussed in Section 10.2. The solution of the optimal control equations yields the optimal state and control variables. The calculated optimal state variables are compared against the measured state variables of the corresponding time interval. A decision can be made to determine whether the difference is significant enough to warrant using the measured state variables as the initial conditions of the time interval. Again the derivation of the optimal control equations are shown as off-line functions.

G. Direct Implementation of the Euler-Lagrange Equation for Optimal Control

The Euler-Lagrange equation is the equation that holds true at optimal conditions, as will be discussed in Section 10.2. It corresponds to the conditions of optimum for steady-state processes. Instead of using the conditions of optimum (i.e., the Euler-Lagrange equation) to solve for the optimal control variables, the Euler-Lagrange equations may be implemented directly. As an example, Figure 6.4-7 shows the feedback control loop to implement

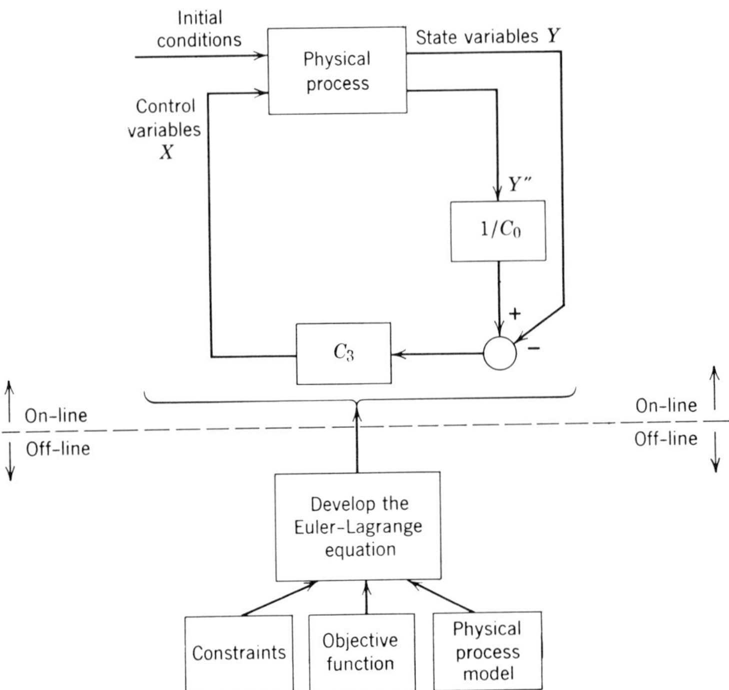

Fig. 6.4-7   Direct implementation of the Euler-Lagrange equation.

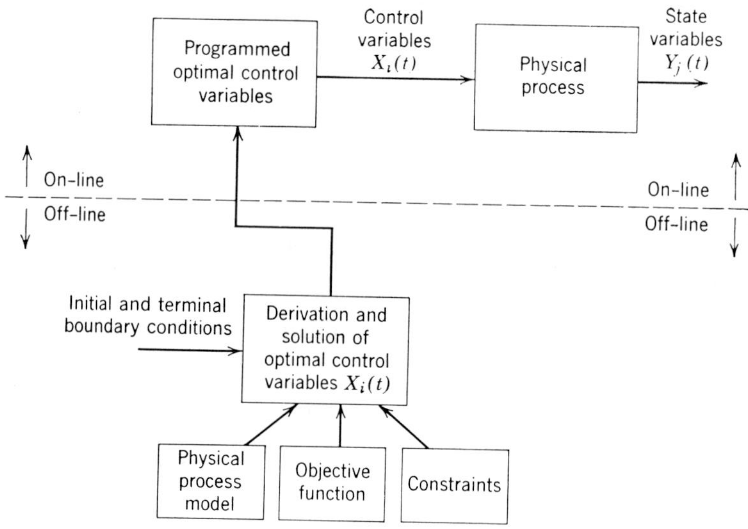

Fig. 6.4-8 Feedforward optimal control by dynamic programming.

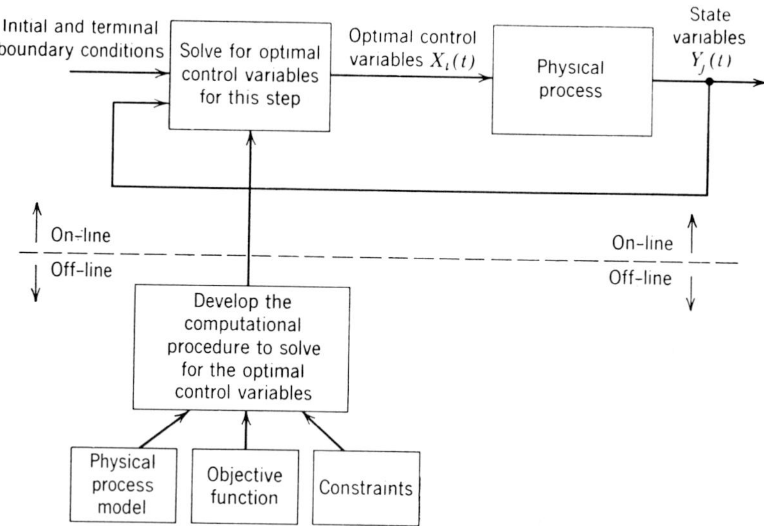

Fig. 6.4-9 Feedback optimal control by dynamic programming.

the following Euler-Lagrange equation:

$$\frac{Y''}{C_0} - Y = 0 \tag{6.4-1}$$

where Y'' is the second derivative of the state variable Y with respect to time, and C_0 is a constant. The feedback loop shown in Figure 6.4-7 ignores the terminal boundary condition, which can be controlled by superimposing another loop on the system. The control variable X is adjusted to make the difference given in Equation 6.4-1 equal to zero.

H. Feedforward or Feedback Optimal Control by Dynamic Programming

Dynamic programming techniques can be used in a feedforward or feedback optimal control fashion, as will be discussed in Section 10.5. In feedforward optimal control using dynamic programming techniques, the optimal control variables $X_i(t)$ are solved ahead of time and used to "program" the control variables with time. Figure 6.4-8 shows feedforward optimal control by dynamic programming. The derivations and the solutions of the optimal control variables may be done in an off-line fashion.

In feedback optimal control using dynamic programming techniques to compute the optimal control variables $X_i(t)$, the solution of the control equations for each time interval uses the measured state variables $Y_j(t)$ as the initial conditions. Figure 6.4-9 shows feedback optimal control by dynamic programming. The solution of the optimal control equation is done on-line while the development of the computational procedure may be done in an off-line fashion.

6.5 SUMMARY

This chapter has surveyed the overall optimization problem and presented a brief summary of Chapters 7 through 10. It is descriptive and narrative in nature, the objective being to familiarize the reader with optimization problems and how optimal control is implemented without confusing the reader with details and mathematical derivations. The various types of physical process models, objective functions, constraints, and approaches to optimal control implementation discussed in this chapter are summarized in Table 6.5-1. The section references are given in parentheses in the first column. If this chapter is a summary of Chapters 7 through 10, then *Table 6.5-1 is a summary of the summary.* Hopefully, this chapter has whetted the reader's appetite and provided him with the proper perspective on the optimization problem so he will proceed to study the remaining chapters.

Table 6.5-1 Summary of the Optimization Problem

Optimal Control Approach (Sections)	Major Elements		
	Physical Process Model	Objective Function	Constraints
Feedforward (7.1, 7.2, 7.3)	Steady-state, well-defined	Steady-state	None
Feedforward (7.3, 7.4, 7.5)	Steady-state, well-defined	Steady-state	Yes
Feedforward with updated model (8.1, 8.4)	Steady-state, moderately well-defined	Steady-state	Yes
Feedback with incremental model (8.2, 8.3, 8.4, 8.5)	Steady-state, moderately well-defined	Steady-state	Yes
Evolutionary optimization (Chapter 9)	Steady-state, poorly defined, no model	Steady-state, may or may not be defined	Yes
Combined feedforward and feedback (Chapter 9)	Steady-state, well-defined, nonlinear	Steady-state	Yes
Dynamic optimization by feedback (10.2)	Dynamic, well-defined	Dynamic	Yes
Direct implementation of Euler-Lagrange equation (10.2)	Dynamic, well-defined	Dynamic	Yes
Feedforward control by dynamic programming (10.3, 10.4, 10.5)	Dynamic, well-defined	Dynamic	Yes
Feedback control by dynamic programming (10.3, 10.4, 10.5)	Dynamic, well-defined	Dynamic	Yes

192

7

Optimal Control of Well-Defined Processes—Steady-State

7.0 INTRODUCTION

The purpose of this chapter is to develop the basic concepts of *optimal control* of processes by limiting the discussion to the techniques that are applicable to well-defined processes. Chapter 8 will present the more practical aspects of optimal control as these techniques are applied to processes with changing characteristics. The mathematical derivations and manipulations used in this chapter can be found in most advanced calculus and operations research texts. This chapter will seek to present the following concepts:

Conditions for optimal process operation.
Feedforward optimal control by solving the optimal control equation.
Lagrange multiplier technique.
Linear programming technique.

By definition, the four characteristics of the objective function, physical process model, and constraints of *a well-defined process* are (1) steady-state, (2) continuous-value, (3) deterministic, and (4) well-behaved. Section 2.5 described information systems that have the first three characteristics. Well-behaved functions are functions which have unique values and whose derivatives exist.

This chapter will first develop the *conditions for optimum* and the *optimal control equations* for the case of one control variable without constraints and then extend the treatment to two control variables. The *Lagrange multiplier technique* is introduced by an example which applies the Lagrange multiplier technique to the optimal control of *styrene manufacture*. Finally, the special

193

case is presented where the objective function and physical process model are both linear and the constraints are linear inequalities. The last case can be recognized as the classical linear programming problem mentioned in Section 2.4.

7.1   NONLINEAR OBJECTIVE FUNCTIONS WITH ONE CONTROL VARIABLE WITHOUT CONSTRAINTS

This section will introduce the basic notions of optimal control by using a nonlinear objective function with a simple physical process having one control variable without constraints.

A. The Objective Function and Physical Process Model

The simplest case of optimal control is the classical calculus problem of solving for the maximum and minimum. The objective function to be maximized, however, is subject to the constraints of the physical process model, i.e., the control and state variables of the objective function must satisfy the physical process model. Let the following symbols be assigned:

X = control variable.
Y = state variable.
f = physical process model function.
F = objective function expressed in terms of X and Y.
G = objective function expressed in terms of X only.
$\overline{X}, \overline{Y}$ = desired operating point.

Figure 7.1-1 shows the physical process with one control variable and one state variable. Note that the physical process model merely relates the state variable to the control variable and *is not the function to be optimized*.

The physical process model of the process is

$$Y = f(X) \tag{7.1-1}$$

Assume that the objective function F to be optimized is

$$F = F(\overline{X}, \overline{Y}, X, Y) \tag{7.1-2}$$

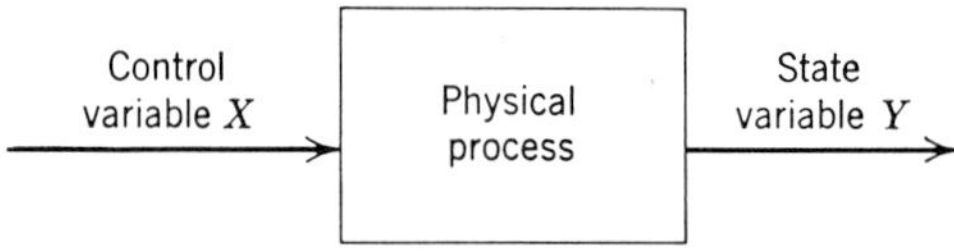

Fig. 7.1-1   Physical process with one control variable X and one state variable Y.

The objective function with Y eliminated by substituting Equation 7.1-1 into Equation 7.1-2 is

$$G = G(\overline{X}, \overline{Y}, X) \tag{7.1-3}$$

B. Conditions for Optimum

The condition for optimum is satisfied when the first derivative of G with respect to X is equal to zero:

$$\frac{dG}{dX} = 0 \tag{7.1-4}$$

In addition, when the *second derivative of G with respect to X is positive*, then the objective function is at a *minimum*. When the *second derivative of G with respect to X is negative*, then the objective function is at a *maximum*. Thus the conditions for the minimum of the objective function are

$$\frac{dG}{dX} = 0 \quad \text{and} \quad \frac{d^2 G}{dX^2} > 0 \tag{7.1-5}$$

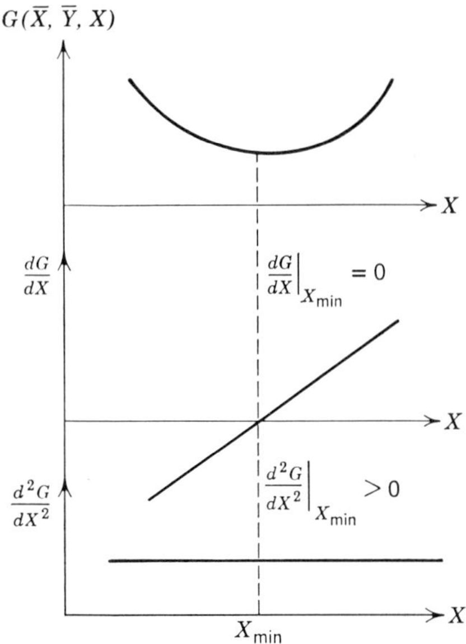

Fig. 7.1-2 Conditions for minimum objective function with one control variable.

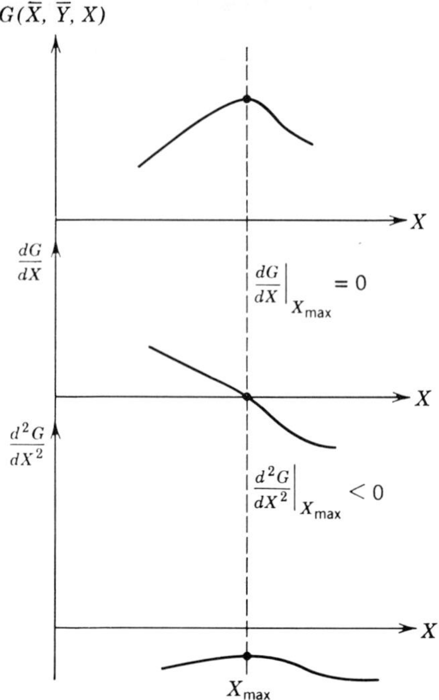

Fig. 7.1-3 Conditions for maximum objective function with one control variable.

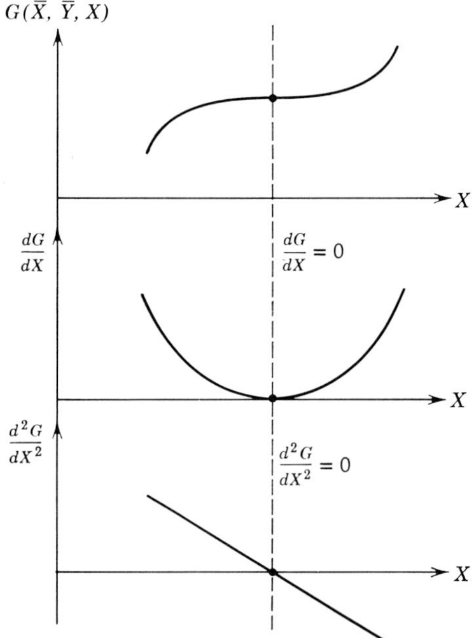

Fig. 7.1-4 No maximum or minimum when first and second derivatives are both equal to zero.

and the conditions for the maximum of the objective function are

$$\frac{dG}{dX} = 0 \quad \text{and} \quad \frac{d^2G}{dX^2} < 0 \tag{7.1-6}$$

When both the first and second derivatives are equal to zero, the objective function is neither at a maximum nor a minimum, or indeterminate. Thus

$$\text{no maximum or minimum at } \frac{dG}{dX} = 0 \quad \text{and} \quad \frac{d^2G}{dX^2} = 0 \tag{7.1-7}$$

Figures 7.1-2 through 7.1-4 provide the geometric interpretations of Equations 7.1-5 through 7.1-7.

C. Example

The physical process model in the example is

$$Y = A_0 + A_1 X \tag{7.1-8}$$

Assume that the objective function to be minimized is

$$F = \psi_1(\overline{Y} - Y)^2 + \psi_2(\overline{X} - X)^2 \tag{7.1-9}$$

where ψ_1 and ψ_2 are positive constants.

The objective function may be expressed in the G form by substituting the physical process model, as given in Equation 7.1-8, into F:

$$G = \psi_1(\overline{Y} - A_0 - A_1 X)^2 + \psi_2(\overline{X} - X)^2 \tag{7.1-10}$$

The *conditions for minimum are*

$$\frac{dG}{dX} = 0 \quad \text{and} \quad \frac{d^2G}{dX^2} > 0 \tag{7.1-11}$$

$$\frac{dG}{dX} = 2\psi_1(\overline{Y} - A_0 - A_1 X)(-A_1) + 2\psi_2(\overline{X} - X)(-1) = 0 \tag{7.1-12}$$

and

$$\frac{d^2G}{dX^2} = 2\psi_1 A_1^2 + 2\psi_2 > 0 \tag{7.1-13}$$

Solving for X in Equation 7.1-12 yields the *optimal control equation*

$$X = \frac{[\overline{X} + A_1(\psi_1/\psi_2)\overline{Y} - A_0 A_1(\psi_1/\psi_2)]}{[1 + A_1^2(\psi_1/\psi_2)]} \tag{7.1-14}$$

Let three constants be assigned:

$$K^1 = \left[1 + A_1^2\left(\frac{\psi_1}{\psi_2}\right)\right]^{-1} \tag{7.1-15}$$

$$K_2 = \frac{A_1(\psi_1/\psi_2)}{[1 + A_1^2(\psi_1/\psi_2)]} \tag{7.1-16}$$

$$K_3 = \frac{-A_0 A_1(\psi_1/\psi_2)}{[1 + A_1^2(\psi_1/\psi_2)]} \tag{7.1-17}$$

The *optimal control equation* is

$$X = K_1\overline{X} + K_2\overline{Y} + K_3 \tag{7.1-18}$$

The block diagram for optimal control of the example is shown in Figure 7.1-5.

The block diagram in Figure 7.1-5 implements the optimal control equation in Equation 7.1-18. The control is the *feedforward* type which utilizes the physical process model to predict the desired control variable setting X that will minimize the objective function G. In this example, X and Y satisfy the physical model given in Equation 7.1-8 and the form of F has been chosen so as to have a minimum at the operating point $(\overline{X}, \overline{Y})$.

Let the following values be assigned to the constants and coefficients of the example

$$\begin{aligned} A_0 &= 1, & A_1 &= 1 \\ \psi_1 &= \tfrac{1}{2}, & \psi_2 &= \tfrac{1}{2} \\ \overline{X} &= 2, & \overline{Y} &= 3 \end{aligned} \tag{7.1-19}$$

Thus

$$K_1 = \tfrac{1}{2}, \qquad K_2 = \tfrac{1}{2}, \qquad K_3 = -\tfrac{1}{2} \tag{7.1-20}$$

$$X = \tfrac{1}{2}\overline{X} + \tfrac{1}{2}\overline{Y} - \tfrac{1}{2} \tag{7.1-21}$$

$$X = 1 + 1.5 - 0.5 = 2 \tag{7.1-22}$$

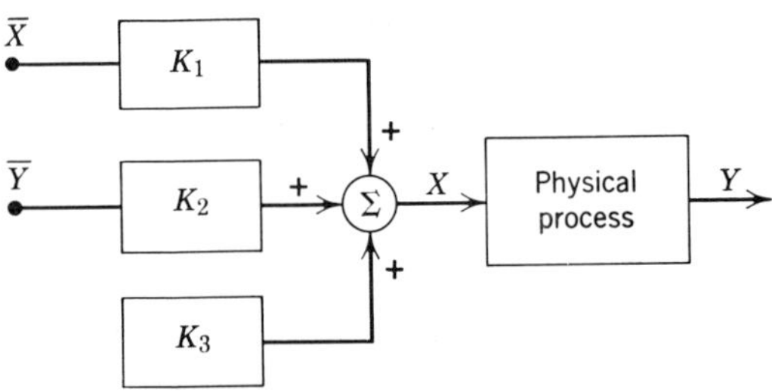

Fig. 7.1-5   Block diagram for optimal control with one control variable.

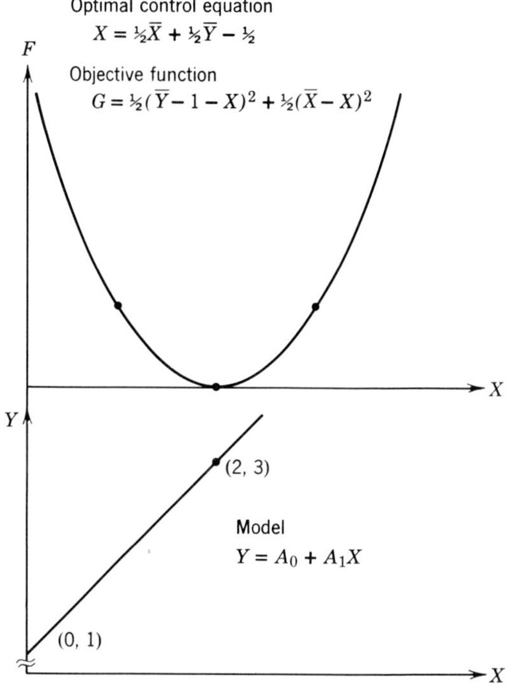

Fig. 7.1-6 Optimal control for objective function with one control variable.

The objective function is

$$G = \tfrac{1}{2}(2 - X)^2 + \tfrac{1}{2}(2 - X)^2 \qquad (7.1\text{-}23)$$
$$= (2 - X)^2$$

Figure 7.1-6 shows the example with the coefficients given in Equation 7.1-19. Notice that for $\overline{X} = 2$ and $\overline{Y} = 3$, the objective function is indeed at a minimum $G = 0$ if X is set at 2.

Note that if the conditions of the physical process model are not met for the desired values $\overline{X}$, $\overline{Y}$, the conditions for a minimum are still satisfied. For example, using the same coefficients in the example and $\overline{Y} = 2.5$, the optimal control equation yields

$$X = \tfrac{1}{2}\overline{X} + \tfrac{1}{2}\overline{Y} - \tfrac{1}{2} = \tfrac{1}{2}(2) + (\tfrac{1}{2})(2.5) - \tfrac{1}{2} = 1.75 \qquad (7.1\text{-}24)$$

At $X = 1.75$

$$G = \tfrac{1}{2}(\overline{Y} - 1 - X)^2 + \tfrac{1}{2}(\overline{X} - X)^2$$
$$= \tfrac{1}{2}(1.5 - X)^2 + \tfrac{1}{2}(2 - X)^2 = 0.0625 \qquad (7.1\text{-}25)$$

Note that G is no longer equal to zero but it is still at a minimum.

7.2 NONLINEAR OBJECTIVE FUNCTIONS WITH TWO CONTROL VARIABLES WITHOUT CONSTRAINTS

This section will extend the development of Section 7.1 by increasing the complexity of the problem to a process with two state variables and two control variables.

A. The Objective Function and Physical Process Model

The next step of complexity is to add another control variable and state variable so that the physical process has two control variables and two state variables, as shown in Figure 7.2-1. Let the following symbols be assigned:

X_1, X_2 = control variables.
Y_1, Y_2 = state variables.
f_1, f_2 = physical process model functions.
F = objective function expressed in terms of X_1, X_2, Y_1, Y_2.
G = objective function expressed in terms of X_1, X_2 only.
$(\overline{X}_1, \overline{X}_2, \overline{Y}_1, \overline{Y}_2)$ = desired operating point.

$$G_{X_1} = \frac{\partial G}{\partial X_1}, \quad G_{X_2} = \frac{\partial G}{\partial X_2}$$

$$G_{X_1 X_1} = \frac{\partial^2 G}{\partial X_1^2}, \quad G_{X_2 X_2} = \frac{\partial^2 G}{\partial X_2^2}, \quad G_{X_1 X_2} = \frac{\partial^2 G}{\partial X_1 \partial X_2}$$

$\psi_1, \psi_2, \psi_3, \psi_4$ = constants.

The physical process model is

$$Y_1 = f_1(X_1, X_2) \tag{7.2-1}$$

$$Y_2 = f_2(X_1, X_2) \tag{7.2-2}$$

The objective function to be optimized is

$$F = F(\overline{X}_1, \overline{X}_2, \overline{Y}_1, \overline{Y}_2, X_1, X_2, Y_1, Y_2) \tag{7.2-3}$$

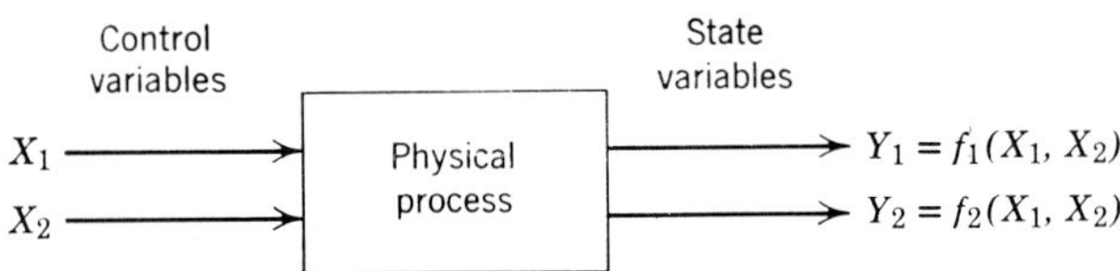

Fig. 7.2-1 Physical process with two control variables and two state variables.

Substituting Equations 7.2-1 and 7.2-2 into Equation 7.2-3, the objective function in G form becomes a function of X_1 and X_2 and the desired operating point:

$$G = G(\overline{X}_1, \overline{X}_2, \overline{Y}_1, \overline{Y}_2, X_1, X_2) \qquad (7.2\text{-}4)$$

B. Conditions for Optimum

To develop the conditions for optimum for $G(X_1, X_2)$, assume that there is a maximum at $X_1 = A$ and $X_2 = B$. Let h and k be positive constants. Thus in the neighborhood of a maximum:

$$\Delta^+ = G(A + h, B + k) - G(A, B) < 0 \qquad (7.2\text{-}5)$$

$$\Delta^- = G(A - h, B - k) - G(A, B) < 0 \qquad (7.2\text{-}6)$$

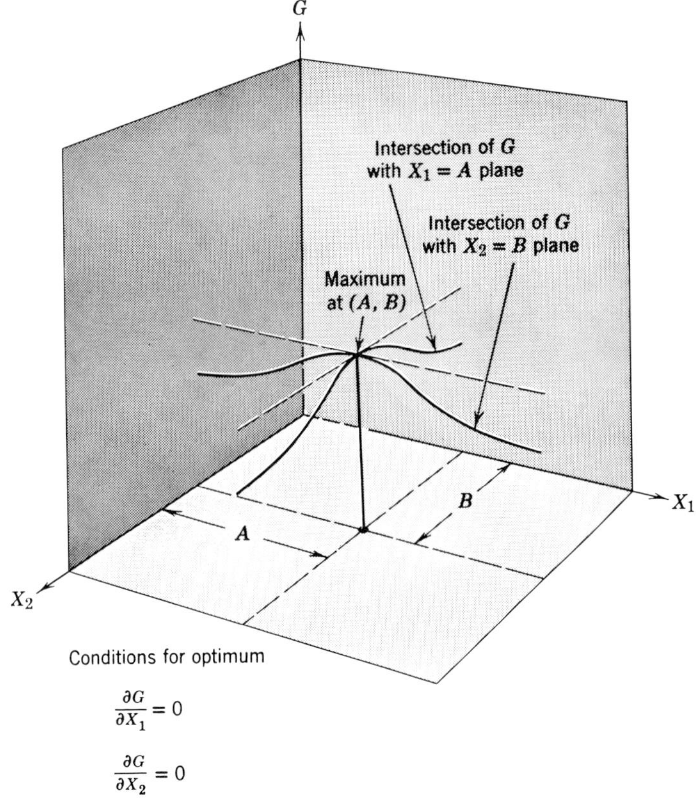

Fig. 7.2-2 Conditions for optimum for objective function with two control variables.

where Δ^+ and Δ^- are as defined above, and h and k are sufficiently small in value. In contrast, $(\Delta^+ > 0)$ and $(\Delta^- > 0)$ in the neighborhood of a minimum.

The intersection of the planes $X_1 = A$ and $X_2 = B$ with the surface $G(X_1, X_2)$ will form curves, as shown in Figure 7.2-2. The curve which is the intersection of the $X_1 = A$ plane and the surface G will have a maximum at (A, B), and the curve which is the intersection of the $X_2 = B$ plane and the surface G will also have a maximum at (A, B). The conditions for maximum can be derived by using the condition for maximum for one control variable on each of these planes. Thus

$$\frac{\partial G}{\partial X_1} = 0 \tag{7.2-7}$$

$$\frac{\partial G}{\partial X_2} = 0 \tag{7.2-8}$$

The same holds true for the conditions for minimum. Equations 7.2-7 and 7.2-8 are the conditions for optimum (maximum or minimum) for an objective function with two control variables.

C. Tests for Maximum or Minimum

The tests to determine whether the point is a maximum or a minimum can be derived by expanding the functions Δ^+ and Δ^- by the Taylor Series. The series will be expanded so as to include the second partial derivative terms. Higher order terms will not be included. Thus expanding about (A, B)

$$\Delta^+ = G(A + h, B + k) - G(A, B) = h\left(\frac{\partial G(A, B)}{\partial X_1}\right) + k\left(\frac{\partial G(A, B)}{\partial X_2}\right)$$

$$+ \tfrac{1}{2}\left[h^2\left(\frac{\partial^2 G(A, B)}{\partial^2 X_1}\right) + 2hk\left(\frac{\partial^2 G(A, B)}{\partial X_1 \partial X_2}\right) + k^2\left(\frac{\partial^2 G(A, B)}{\partial^2 X_2}\right) \right] \tag{7.2-9}$$

and

$$\Delta^- = G(A - h, B - k) - G(A, B) = -h\left(\frac{\partial G(A, B)}{\partial X_1}\right) - k\left(\frac{\partial G(A, B)}{\partial X_2}\right)$$

$$+ \tfrac{1}{2}\left[h^2\left(\frac{\partial^2 G(A, B)}{\partial^2 X_1}\right) + 2hk\left(\frac{\partial^2 G(A, B)}{\partial X_1 \partial X_2}\right) + k^2\left(\frac{\partial^2 G(A, B)}{\partial^2 X_2}\right) \right] \tag{7.2-10}$$

Using the abbreviations for the partial derivatives defined in Section 7.2A,

$$\Delta^+ = hG_{X_1} + kG_{X_2} + \tfrac{1}{2}[h^2G_{X_1X_1} + 2hkG_{X_1X_2} + k^2G_{X_2X_2}] \qquad (7.2\text{-}11)$$

$$\Delta^- = -hG_{X_1} - kG_{X_2} + \tfrac{1}{2}[h^2G_{X_1X_1} + 2hkG_{X_1X_2} + k^2G_{X_2X_2}] \qquad (7.2\text{-}12)$$

At the optimum

$$G_{X_1} = 0 \quad \text{and} \quad G_{X_2} = 0 \qquad (7.2\text{-}13)$$

$$\therefore \quad 2\Delta^+ = h^2G_{X_1X_1} + 2hkG_{X_1X_2} + k^2G_{X_2X_2} \qquad (7.2\text{-}14)$$

and

$$2\Delta^- = h^2G_{X_1X_1} + 2hkG_{X_1X_2} + k^2G_{X_2X_2} \qquad (7.2\text{-}15)$$

From geometry it can be deduced that *both Δ^+ and Δ^- will be negative in the neighborhood of the maximum*, and *both Δ^+ and Δ^- will be positive in the neighborhood of the minimum*. Thus the conditions for maximum and minimum can be deduced by examining Equations 7.2-14 and 7.2-15 in the neighborhood of the optimum (A, B). Equations 7.2-14 and 7.2-15 can be re-written as

$$2\Delta^+ = G_{X_1X_1}\left[h + k\,\frac{G_{X_1X_2}}{G_{X_1X_1}}\right]^2 + \frac{k^2}{G_{X_1X_1}}\,[G_{X_1X_1}G_{X_2X_2} - (G_{X_1X_2})^2]$$

$$= G_{X_2X_2}\left[k + h\,\frac{G_{X_1X_2}}{G_{X_2X_2}}\right]^2 + \frac{h^2}{G_{X_2X_2}}\,[G_{X_1X_1}G_{X_2X_2} - (G_{X_1X_2})^2] \qquad (7.2\text{-}16)$$

Similar expressions can be obtained for $2\Delta^-$. *Thus the conditions for maximum are*

(1) $$[G_{X_1X_1}G_{X_2X_2} - (G_{X_1X_2})^2] > 0 \qquad (7.2\text{-}17)$$

(2) $$G_{X_1X_1} < 0 \qquad (7.2\text{-}18)$$

(3) $$G_{X_2X_2} < 0 \qquad (7.2\text{-}19)$$

and the conditions for minimum are

(1) $$[G_{X_1X_1}G_{X_2X_2} - (G_{X_1X_2})^2] > 0 \qquad (7.2\text{-}20)$$

(2) $$G_{X_1X_1} > 0 \qquad (7.2\text{-}21)$$

(3) $$G_{X_2X_2} > 0 \qquad (7.2\text{-}22)$$

However if

$$[G_{X_1X_1}G_{X_2X_2} - (G_{X_1,X_2})^2] < 0 \qquad (7.2\text{-}23)$$

and either

$$G_{X_1X_1} \neq 0 \quad \text{or} \quad G_{X_2X_2} \neq 0 \tag{7.2-24}$$

then $2\Delta^+$ and $2\Delta^-$ can be positive or negative depending on the signs of $G_{X_1X_1}$, $G_{X_2X_2}$, and the relative magnitudes of h and k. In this case the point is neither a maximum or a minimum.

If

$$[G_{X_1X_1} G_{X_2X_2} - (G_{X_1X_2})^2] = 0 \tag{7.2-25}$$

the test gives no information at all. The higher order derivatives must be obtained and examined to determine whether the optimum is a maximum or a minimum.

D. Example

The physical process model of the example is

$$Y_1 = A_1 + B_1 X_1 + C_1 X_2 \tag{7.2-26}$$

$$Y_2 = A_2 + B_2 X_1 + C_2 X_2 \tag{7.2-27}$$

The quadratic form objective function F to be minimized is

$$F = \psi_1(\overline{Y}_1 - Y_1)^2 + \psi_2(\overline{Y}_2 - Y_2)^2 + \psi_3(\overline{X}_1 - X_1)^2 + \psi_4(\overline{X}_2 - X_2)^2 \tag{7.2-28}$$

The objective function may be expressed in the G form by substituting Y_1 and Y_2 from the physical process model into Equation 7.2-28:

$$G = \psi_1(\overline{Y}_1 - A_1 - B_1 X_1 - C_1 X_2)^2 + \psi_2(\overline{Y}_2 - A_2 - B_2 X_1 - C_2 X_2)^2$$
$$+ \psi_3(\overline{X}_1 - X_1)^2 + \psi_4(\overline{X}_2 - X_2)^2 \tag{7.2-29}$$

The conditions for optimum are

$$\frac{\partial G}{\partial X_1} = 0 \tag{7.2-30}$$

$$\frac{\partial G}{\partial X_2} = 0 \tag{7.2-31}$$

Substituting Equation 7.2-29 into Equation 7.2-30,

$$2\psi_1(\overline{Y}_1 - A_1 - B_1 X_1 - C_1 X_2)(-B_1) + 2\psi_2(\overline{Y}_2 - A_2 - B_2 X_1 - C_2 X_2)(-B_2)$$
$$+ 2\psi_3(\overline{X}_1 - X_1)(-1) = 0 \tag{7.2-32}$$

Substituting Equation 7.2-29 into Equation 7.2-31,

$$2\psi_1(\overline{Y}_1 - A_1 - B_1 X_1 - C_1 X_2)(-C_1) + 2\psi_2(\overline{Y}_2 - A_2 - B_2 X_1 - C_2 X_2)(-C_2)$$
$$+ 2\psi_4(\overline{X}_2 - X_2)(-1) = 0 \qquad (7.2\text{-}33)$$

Collecting terms for Equations 7.2-32 and 7.2-33,

$$(X_1)(\psi_1 B_1^2 + \psi_2 B_2^2 + \psi_3) + (X_2)(\psi_1 B_1 C_1 + \psi_2 B_2 C_2) = (\overline{Y}_1)(\psi_1 B_1)$$
$$+ (\overline{Y}_2)(\psi_2 B_2) + (\overline{X}_1)(\psi_3) - (\psi_1 A_1 B_1 + \psi_2 A_2 B_2) \qquad (7.2\text{-}34)$$

$$(X_1)(\psi_1 B_1 C_1 + \psi_2 B_2 C_2) + (X_2)(\psi_1 C_1^2 + \psi_2 C_2^2 + \psi_4) = (\overline{Y}_1)(\psi_1 C_1)$$
$$+ (\overline{Y}_2)(\psi_2 C_2) + (\overline{X}_2)(\psi_4) - (\psi_1 A_1 C_1 + \psi_2 A_2 C_2) \qquad (7.2\text{-}35)$$

Equations 7.2-34 and 7.2-35 are the simultaneous equations which may be solved to obtain the optimal X_1 and X_2. Thus

$$X_1 = \frac{1}{D}\,[\{(\overline{Y}_1)(\psi_1 B_1) + (\overline{Y}_2)(\psi_2 B_2) + (\overline{X}_1)(\psi_3) - (\psi_1 A_1 B_1 + \psi_2 A_2 B_2)\}$$

$$\times \{\psi_1 C_1^2 + \psi_2 C_2^2 + \psi_4\}$$

$$- \{(\overline{Y}_1)(\psi_1 C_1) + (\overline{Y}_2)(\psi_2 C_2) + (\overline{X}_2)(\psi_4) - (\psi_1 A_1 C_1 + \psi_2 A_2 C_2)\}$$

$$\times \{\psi_1 B_1 C_1 + \psi_2 B_2 C_2\}] \qquad (7.2\text{-}36)$$

$$X_2 = \frac{1}{D}\,[\{(\overline{Y}_1)(\psi_1 C_1) + (\overline{Y}_2)(\psi_2 C_2) + (\overline{X}_2)(\psi_4) - (\psi_1 A_1 C_1 + \psi_2 A_2 C_2)\}$$

$$\times \{\psi_1 B_1^2 + \psi_2 B_2^2 + \psi_3\}$$

$$- \{(\overline{Y}_1)(\psi_1 B_1) + (\overline{Y}_2)(\psi_2 B_2) + (\overline{X}_1)(\psi_3) - (\psi_1 A_1 B_1 + \psi_2 A_2 B_2)\}$$

$$\times \{\psi_1 B_1 C_1 + \psi_2 B_2 C_2\}] \qquad (7.2\text{-}37)$$

where

$$D = (\psi_1 B_1^2 + \psi_2 B_2^2 + \psi_3)(\psi_1 C_1^2 + \psi_2 C_2^2 + \psi_4) - (\psi_1 B_1 C_1 + \psi_2 B_2 C_2)^2$$
$$(7.2\text{-}38)$$

Collecting terms, the optimal control equations are

$$X_1 = K_1 \overline{Y}_1 + K_2 \overline{Y}_2 + K_3 \overline{X}_1 + K_4 \overline{X}_2 + K_5 \qquad (7.2\text{-}39)$$
$$X_2 = K_6 \overline{Y}_1 + K_7 \overline{Y}_2 + K_8 \overline{X}_1 + K_9 \overline{X}_2 + K_{10} \qquad (7.2\text{-}40)$$

where

$$K_1 = \frac{1}{D}\left[(\psi_1 B_1)(\psi_1 C_1^2 + \psi_2 C_2^2 + \psi_4) - (\psi_1 C_1)(\psi_1 B_1 C_1 + \psi_2 B_2 C_2)\right]$$

(7.2-41)

$$K_2 = \frac{1}{D}\left[(\psi_2 B_2)(\psi_1 C_1^2 + \psi_2 C_2^2 + \psi_4) - (\psi_2 C_2)(\psi_1 B_1 C_1 + \psi_2 B_2 C_2)\right]$$

(7.2-42)

$$K_3 = \frac{1}{D}\left[(\psi_3)(\psi_1 C_1^2 + \psi_2 C_2^2 + \psi_4)\right]$$

(7.2-43)

$$K_4 = \frac{1}{D}\left[(-\psi_4)(\psi_1 B_1 C_1 + \psi_2 B_2 C_2)\right]$$

(7.2-44)

$$K_5 = \frac{1}{D}\left[(\psi_1 A_1 C_1 + \psi_2 A_2 C_2)(\psi_1 B_1 C_1 + \psi_2 B_2 C_2)\right.$$

$$\left. - (\psi_1 A_1 B_1 + \psi_2 A_2 B_2)(\psi_1 C_1^2 + \psi_2 C_2^2 + \psi_4)\right]$$

(7.2-45)

$$K_6 = \frac{1}{D}\left[(\psi_1 C_1)(\psi_1 B_1^2 + \psi_2 B_2^2 + \psi_3) - (\psi_1 B_1)(\psi_1 B_1 C_1 + \psi_2 B_2 C_2)\right]$$

(7.2-46)

$$K_7 = \frac{1}{D}\left[(\psi_2 C_2)(\psi_1 B_1^2 + \psi_2 B_2^2 + \psi_3) - (\psi_2 B_2)(\psi_1 B_1 C_1 + \psi_2 B_2 C_2)\right]$$

(7.2-47)

$$K_8 = \frac{1}{D}\left[(-\psi_3)(\psi_1 B_1 C_1 + \psi_2 B_2 C_2)\right]$$

(7.2-48)

$$K_9 = \frac{1}{D}\left[(\psi_4)(\psi_1 B_1^2 + \psi_2 B_2^2 + \psi_3)\right]$$

(7.2-49)

$$K_{10} = \frac{1}{D}\left[(\psi_1 A_1 B_1 + \psi_2 A_2 B_2)(\psi_1 B_1 C_1 + \psi_2 B_2 C_2)\right.$$

$$\left. - (\psi_1 A_1 C_1 + \psi_2 A_2 C_2)(\psi_1 B_1^2 + \psi_2 B_2^2 + \psi_3)\right]$$

(7.2-50)

Figure 7.2-3 shows the optimal control equations in block diagram form. The control is the feedforward open-loop type, where the optimal control variables X_1 and X_2 will be computed to minimize the objective function G to satisfy the desired operating point $(\overline{X}_1, \overline{X}_2, \overline{Y}_1, \overline{Y}_2)$.

If it is not desirable to maintain a particular variable at a given set point, then the corresponding ψ coefficient in Equation 7.2-28 and Equations 7.2-36 through 7.2-50 may be set to zero. For example, if it is not desirable to maintain X_1 at $\overline{X}_1$ then ψ_3 may be set to zero. Thus the generalized optimal control equations given above may be modified readily to fit special cases.

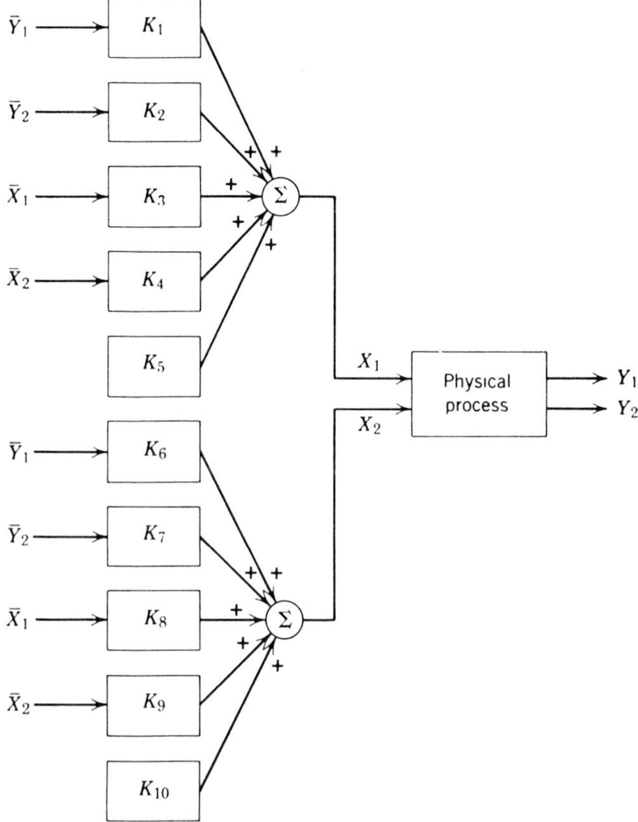

Fig. 7.2-3 Block diagram for feedforward optimal control with two control variables.

E. Example In Matrix Notation

The example in Section 7.2D can be restated in matrix notation. Matrix notation has the advantages of being compact and is convenient for computer solution. From Equations 7.2-26 and 7.2-27, the physical process model in matrix notation is

$$\begin{pmatrix} Y_1 \\ Y_2 \end{pmatrix} = \begin{pmatrix} A_1 \\ A_2 \end{pmatrix} + \begin{pmatrix} B_1 & C_1 \\ B_2 & C_2 \end{pmatrix} \cdot \begin{pmatrix} X_1 \\ X_2 \end{pmatrix} \tag{7.2-51}$$

From Equation 7.2-28, the quadratic form objective function in matrix notation is

$$F = \left((\overline{Y}_1 - Y_1) \quad (\overline{Y}_2 - Y_2) \quad (\overline{X}_1 - X_1) \quad (\overline{X}_2 - X_2) \right)$$

$$\cdot \begin{pmatrix} \psi_1 & 0 & 0 & 0 \\ 0 & \psi_2 & 0 & 0 \\ 0 & 0 & \psi_3 & 0 \\ 0 & 0 & 0 & \psi_4 \end{pmatrix} \cdot \begin{pmatrix} (\overline{Y}_1 - Y_1) \\ (\overline{Y}_2 - Y_2) \\ (\overline{X}_1 - X_1) \\ (\overline{X}_2 - X_2) \end{pmatrix} \tag{7.2-52}$$

From Equations 7.2-34 and 7.2-35, the conditions for optimum in matrix notation are

$$\begin{pmatrix} (\psi_1 B_1^2 + \psi_2 B_2^2 + \psi_3) & (\psi_1 B_1 C_1 + \psi_2 B_2 C_2) \\ (\psi_1 B_1 C_1 + \psi_2 B_2 C_2) & (\psi_1 C_1^2 + \psi_2 C_2^2 + \psi_4) \end{pmatrix} \cdot \begin{pmatrix} X_1 \\ X_2 \end{pmatrix}$$

$$= \begin{pmatrix} \{(\overline{Y}_1)(\psi_1 B_1) + (\overline{Y}_2)(\psi_2 B_2) + (\overline{X}_1)(\psi_3) - (\psi_1 A_1 B_1 + \psi_2 A_2 B_2)\} \\ \{(\overline{Y}_1)(\psi_1 C_1) + (\overline{Y}_2)(\psi_2 C_2) + (\overline{X}_2)(\psi_4) - (\psi_1 A_1 C_1 + \psi_2 A_2 C_2)\} \end{pmatrix}$$

$$(7.2\text{-}53)$$

From Equations 7.2-39 and 7.2-40, the optimal control equations in matrix notation are

$$\begin{pmatrix} X_1 \\ X_2 \end{pmatrix} = \begin{pmatrix} K_1 & K_2 & K_3 & K_4 \\ K_6 & K_7 & K_8 & K_9 \end{pmatrix} \cdot \begin{pmatrix} \overline{Y}_1 \\ \overline{Y}_2 \\ \overline{X}_1 \\ \overline{X}_2 \end{pmatrix} + \begin{pmatrix} K_5 \\ K_{10} \end{pmatrix} \qquad (7.2\text{-}54)$$

where K_1 through K_{10} are defined in Equations 7.2-41 through 7.2-50.

7.3 QUADRATIC OBJECTIVE FUNCTIONS WITH MORE THAN TWO CONTROL VARIABLES WITHOUT CONSTRAINTS

This section will generalize the process to n control variables and m state variables. The objective function considered is assumed to be in the quadratic form which can yield the optimal control equations simply through the solution of simultaneous linear algebraic equations. The special cases of determined, underdetermined and overdetermined systems will also be considered.

A. The Objective Function and the Physical Process Model

Let the following symbols be assigned:

X_i = control variables, $(i = 1)$ to $(i = n)$.
Y_j = state variables, $(j = 1)$ to $(j = m)$.
$\overline{X}_i, \overline{X}_j$ = desired operating point.
f_j = physical process model functions.
F = objective function expressed in terms of X_i and Y_j.
G = objective function expressed in terms of X_i only.
$\theta_n, \psi_m, A_i, B_i, C_{ij}, D$, etc. = constants.

Figure 7.3-1 shows the physical process with n control variables and m state variables.

The physical process model is defined by

$$Y_j = f_j(X_i) \qquad (7.3\text{-}1)$$

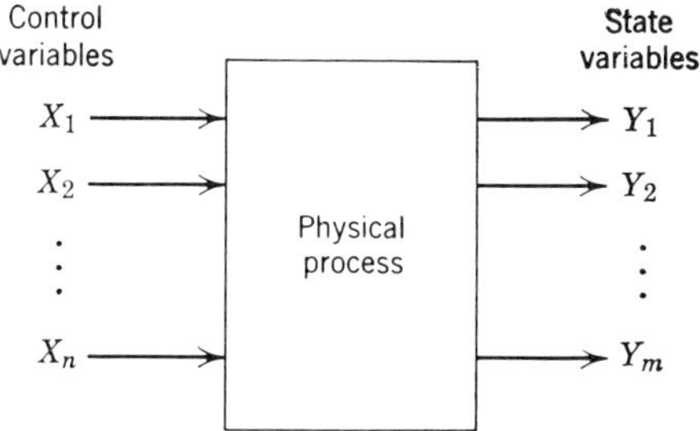

Fig. 7.3-1 Process with n control variables and m state variables.

The objective function F is

$$F = F(\overline{X}_i, \overline{Y}_j, X_i, Y_j) \tag{7.3-2}$$

Substituting the values of Y_j from Equation 7.3-1 into Equation 7.3-2, the objective function in G form is

$$G = G(\overline{X}_i, \overline{Y}_j, X_i) \tag{7.3-3}$$

The quadratic form of the objective function to be optimized is

$$F = \sum_{j=1}^{m} \psi_j(\overline{Y}_j - Y_j)^2 + \sum_{i=1}^{n} \theta_i(\overline{X}_i - X_i)^2 \tag{7.3-4}$$

B. Conditions for Optimum without Constraints

Without constraints, the conditions for the objective function to be optimum is

$$\frac{\partial G}{\partial X_i} = 0 \tag{7.3-5}$$

In the general case in which the physical process model is nonlinear, Equation 7.3-5 becomes a set of simultaneous nonlinear equations. Iterative techniques may be used to solve the nonlinear equations in Equation 7.3-5 for the optimal control variables X_i.

C. The Optimal Control Equation

In the special case in which the physical process model is linear, that is

$$Y_1 = f_1(X_1, X_2, \ldots, X_n) = A_{10} + A_{11}X_1 + A_{12}X_2 + \cdots + A_{1n}X_n$$

$$Y_2 = f_2(X_1, X_2, \ldots, X_n) = A_{20} + A_{21}X_1 + A_{22}X_2 + \cdots + A_{2n}X_n$$
$$\vdots \tag{7.3-6}$$

$$Y_m = f_m(X_1, X_2, \ldots, X_n) = A_{m0} + A_{m1}X_1 + A_{m2}X_2 + \cdots + A_{mn}X_n$$

The optimal control equation for the quadratic form objective function can be derived by extension of Equation 7.2-54. The optimal control equation for a linear physical process model with n control variables and m state variables in matrix notation is

$$
\begin{pmatrix} X_1 \\ X_2 \\ \vdots \\ X_n \end{pmatrix} = \begin{pmatrix} K_{1,1} & K_{1,2} & \cdots & K_{1,n+m} \\ K_{2,1} & K_{2,2} & \cdots & K_{2,n+m} \\ \vdots & & & \\ K_{n,1} & K_{n,2} & \cdots & K_{n,n+m} \end{pmatrix} \cdot \begin{pmatrix} \overline{Y}_1 \\ \overline{Y}_2 \\ \vdots \\ \overline{Y}_m \\ \overline{X}_1 \\ \overline{X}_2 \\ \vdots \\ \overline{X}_n \end{pmatrix} + \begin{pmatrix} C_1 \\ C_2 \\ \vdots \\ C_n \end{pmatrix}
\tag{7.3-7}
$$

where $K_{1,1} \cdots K_{n,n+m}$ and $C_1 \cdots C_n$ are constants.

D. Overdetermined and Underdetermined Systems

For linear physical process models with the quadratic form objective function, there is a special case where the coefficients of the desired control variables $\overline{X}_i$ in Equation 7.3-4 are equal to zero

$$
\theta_1 = \theta_2 = \cdots = \theta_n = 0
\tag{7.3-8}
$$

For this special case, there are three conditions of optimum depending on the number of state variables m relative to the number of control variables n. By definition, *a determined system* is one in which the number of state variables is equal to the number of control variables:

$$
m = n
\tag{7.3-9}
$$

An overdetermined system is one in which the number of control variables is more than the number of state variables:

$$
n > m
\tag{7.3-10}
$$

An underdetermined system is one in which the number of control variables is less than the number of state variables:

$$
n < m
\tag{7.3-11}
$$

Case 1   Determined System

The objective function for the special case with $\theta_i = 0$ is

$$
F = \sum_{j=1}^{m} \psi_j (\overline{Y}_j - Y_j)^2
\tag{7.3-12}
$$

The linear physical process model given in Equation 7.3-6 can be substituted into the objective function in Equation 7.3-12. The conditions of optimum may be derived by taking the partial derivatives of the objective function with respect to the control variables X_i and then setting these to zero. For the determined system where the number of state variables and control variables are equal, $m = n$, the condition of optimum is

$$\begin{pmatrix} A_{11} & A_{12} & \cdots & A_{1n} \\ A_{21} & A_{22} & \cdots & A_{2n} \\ \vdots & \vdots & & \vdots \\ A_{n1} & A_{n2} & \cdots & A_{nn} \end{pmatrix} \cdot \begin{pmatrix} X_1 \\ X_2 \\ \vdots \\ X_n \end{pmatrix} = \begin{pmatrix} (\overline{Y}_1 - A_{10}) \\ (\overline{Y}_2 - A_{20}) \\ \vdots \\ (\overline{Y}_n - A_{n0}) \end{pmatrix} \qquad (7.3\text{-}13)$$

The condition of optimum is always a square matrix which may be inverted directly to solve for the optimal control variables. For example, let ($m = n = 2$). Substituting the physical process model into the objective function:

$$F = \psi_1(\overline{Y}_1 - A_{10} - A_{11}X_1 - A_{12}X_2)^2 + \psi_2(\overline{Y}_2 - A_{20} - A_{21}X_1 - A_{22}X_2)^2 \qquad (7.3\text{-}14)$$

The conditions of optimum are

$$\frac{\partial F}{\partial X_1} = 2\psi_1(-A_{11})(\overline{Y}_1 - A_{10} - A_{11}X_1 - A_{12}X_2)$$

$$+ 2\psi_2(-A_{21})(\overline{Y}_2 - A_{20} - A_{21}X_1 - A_{22}X_2) = 0 \qquad (7.3\text{-}15)$$

$$\frac{\partial F}{\partial X_2} = 2\psi_1(-A_{12})(\overline{Y}_1 - A_{10} - A_{11}X_1 - A_{12}X_2)$$

$$+ 2\psi_2(-A_{22})(\overline{Y}_2 - A_{20} - A_{21}X_1 - A_{22}X_2) = 0 \qquad (7.3\text{-}16)$$

In matrix notation, the condition of optimum is

$$\begin{pmatrix} -2\psi_1 A_{11} & -2\psi_2 A_{21} \\ -2\psi_1 A_{12} & -2\psi_2 A_{22} \end{pmatrix} \cdot \begin{pmatrix} (\overline{Y}_1 - A_{10} - A_{11}X_1 - A_{12}X_2) \\ (\overline{Y}_2 - A_{20} - A_{21}X_1 - A_{22}X_2) \end{pmatrix} = \begin{pmatrix} 0 \\ 0 \end{pmatrix} \qquad (7.3\text{-}17)$$

Equation 7.3-17 may be premultiplied by the inverse of

$$\begin{pmatrix} -2\psi_1 A_{11} & -2\psi_2 A_{21} \\ -2\psi_1 A_{12} & -2\psi_2 A_{22} \end{pmatrix} \qquad (7.3\text{-}18)$$

The result is

$$\begin{pmatrix} (\overline{Y}_1 - A_{10} - A_{11}X_1 - A_{12}X_2) \\ (\overline{Y}_2 - A_{20} - A_{21}X_1 - A_{22}X_2) \end{pmatrix} = \begin{pmatrix} 0 \\ 0 \end{pmatrix} \qquad (7.3\text{-}19)$$

Equation 7.3-19 may be rewritten as

$$\begin{pmatrix} A_{11} & A_{12} \\ A_{21} & A_{22} \end{pmatrix} \cdot \begin{pmatrix} X_1 \\ X_2 \end{pmatrix} = \begin{pmatrix} (\overline{Y}_1 - A_{10}) \\ (\overline{Y}_2 - A_{20}) \end{pmatrix} \tag{7.3-20}$$

Equation 7.3-20 may be recognized as a special case of Equation 7.3-13 with $n = 2$.

Case 2 Overdetermined System

For the overdetermined system in which the number of control variables is more than the number of state variables, $n > m$, the condition of optimum given by Equation 7.3-13 is no longer applicable. In general, the condition of optimum will result in fewer independent equations than unknown control variables. As will be shown in the example to follow, $(n - m)$ control variables can be assigned arbitrary constant values, e.g., set equal to zero. The remaining m control variables can be solved for in terms of the assigned values of the $(m - n)$ constants. Linear programming problems, which will be discussed in Section 7.5, are examples of overdetermined systems.

For example, let $m = 1$ and $n = 2$. The physical process model is

$$Y_1 = A_{10} + A_{11}X_1 + A_{12}X_2 \tag{7.3-21}$$

The objective function is

$$F = \psi_1(\overline{Y}_1 - Y_1)^2 \tag{7.3-22}$$

Substituting Equation 7.3-21 into Equation 7.3-22,

$$F = \psi_1(\overline{Y}_1 - A_{10} - A_{11}X_1 - A_{12}X_2)^2 \tag{7.3-23}$$

The conditions of optimum are

$$\frac{\partial F}{\partial X_1} = 2\psi_1(-A_{11})(\overline{Y}_1 - A_{10} - A_{11}X_1 - A_{12}X_2) = 0 \tag{7.3-24}$$

$$\frac{\partial F}{\partial X_2} = 2\psi_1(-A_{12})(\overline{Y}_1 - A_{10} - A_{11}X_1 - A_{12}X_2) = 0 \tag{7.3-25}$$

Note that Equations 7.3-24 and 7.3-25 are not independent. What results is only one independent equation to solve for X_1. Thus X_1 will be expressed in terms of the arbitrary constant value assigned to X_2.

$$X_1 = \frac{1}{A_{11}}(\overline{Y}_1 - A_{10} - A_{12}X_2) \tag{7.3-26}$$

Note also that the value of the objective function F will be zero at optimum.

Case 3   *Underdetermined System*

For the underdetermined system in which the number of control variables is less than the number of state variables, $n < m$, the condition of optimum given by Equation 7.3-13 is also not applicable. The optimal control variables may be solved for by setting the partial derivatives of the objective function with respect to the control variables equal to zero. The optimal state variables Y_j will generally not be equal to their desired values $\overline{Y}_j$, and the optimal objective function will not be equal to zero.

For example, let $m = 3$ and $n = 2$. The physical process model is

$$Y_1 = A_{10} + A_{11}X_1 + A_{12}X_2 \tag{7.3-27}$$

$$Y_2 = A_{20} + A_{21}X_1 + A_{22}X_2 \tag{7.3-28}$$

$$Y_3 = A_{30} + A_{31}X_1 + A_{32}X_2 \tag{7.3-29}$$

The objective function to be minimized is

$$F = \psi_1(\overline{Y}_1 - A_{10} - A_{11}X_1 - A_{12}X_2)^2 + \psi_2(\overline{Y}_2 - A_{20} - A_{21}X_1 - A_{22}X_2)^2$$
$$+ \psi_3(\overline{Y}_3 - A_{30} - A_{31}X_1 - A_{32}X_2)^2 \tag{7.3-30}$$

The conditions of optimum are

$$\frac{\partial F}{\partial X_1} = 2\psi_1(-A_{11})(\overline{Y}_1 - A_{10} - A_{11}X_1 - A_{12}X_2)$$

$$+ 2\psi_2(-A_{21})(\overline{Y}_2 - A_{20} - A_{21}X_1 - A_{22}X_2)$$
$$+ 2\psi_3(-A_{31})(\overline{Y}_3 - A_{30} - A_{31}X_1 - A_{32}X_2) = 0 \tag{7.3-31}$$

$$\frac{\partial F}{\partial X_2} = 2\psi_1(-A_{12})(\overline{Y}_1 - A_{10} - A_{11}X_1 - A_{12}X_2)$$

$$+ 2\psi_2(-A_{22})(\overline{Y}_2 - A_{20} - A_{21}X_1 - A_{22}X_2)$$
$$+ 2\psi_3(-A_{32})(\overline{Y}_3 - A_{30} - A_{31}X_1 - A_{32}X_2) = 0 \tag{7.3-32}$$

Equations 7.3-31 and 7.3-32 may be expressed in matrix notation:

$$\begin{pmatrix} -2\psi_1 A_{11} & -2\psi_2 A_{21} \\ -2\psi_1 A_{12} & -2\psi_2 A_{22} \end{pmatrix} \cdot \begin{pmatrix} (\overline{Y}_1 - A_{10} - A_{11}X_1 - A_{12}X_2) \\ (\overline{Y}_2 - A_{20} - A_{21}X_1 - A_{22}X_2) \end{pmatrix}$$

$$+ (\overline{Y}_3 - A_{30} - A_{31}X_1 - A_{32}X_2)(-2\psi_3)\begin{pmatrix} A_{31} \\ A_{32} \end{pmatrix} = \begin{pmatrix} 0 \\ 0 \end{pmatrix} \tag{7.3-33}$$

Comparing Equation 7.3-33 with Equation 7.3-17, it is obvious that these two equations will not yield the same solutions for X_1 and X_2. The difference is due to the second term in the left hand side of Equation 7.3-33, which was caused by the additional state variable Y_3.

From a strictly mathematical viewpoint it may seem foolish to try to control m state variables with n control variables with $m > n$, as suggested by Equations 7.3-27 through 7.3-30. In practice, however, circumstances may dictate operation of a plant as an underdetermined system. For example, for a blending operation which is a determined system with $m = n = 3$, the physical facility represented by one of the control variables X_3 may become inoperative, resulting in $X_3 = 0$. Thus the original physical process model is

$$Y_1 = A_{10} + A_{11}X_1 + A_{12}X_2 + A_{13}X_3 \qquad (7.3\text{-}34)$$

$$Y_2 = A_{20} + A_{21}X_1 + A_{22}X_2 + A_{23}X_3 \qquad (7.3\text{-}35)$$

$$Y_3 = A_{30} + A_{31}X_1 + A_{32}X_2 + A_{33}X_3 \qquad (7.3\text{-}36)$$

With $X_3 = 0$, the physical process model becomes the physical process model given by Equations 7.3-27 through 7.3-29. Note that Equations 7.3-34 through 7.3-36 become Equations 7.3-27 through 7.3-29 by setting $A_{13} = A_{23} = A_{33} = 0$. The use of control equations derived by solving Equation 7.3-33 will allow a degree of control to be retained over all three state variables. This is in contrast to an alternative practice of dropping the term involving $\overline{Y}_3$ from Equation 7.3-30 which would result in good control over Y_1 and Y_2 but no deliberate control over Y_3.

A second example of an underdetermined system would be the case in which only two raw materials are used in order to minimize the costs. From Equations 7.3-30 and 7.3-33, it can be seen that good control over Y_1, Y_2, and Y_3 can be obtained if the following condition is true:

$$\overline{Y}_3 - A_{30} - A_{31}X_1 - A_{32}X_2 = 0 \qquad (7.3\text{-}37)$$

Thus, if, on the average, X_1 is set at the average $X_1{}^{\text{av}}$ and X_2 is set at the average value $X_2{}^{\text{av}}$, this condition will be satisfied if the raw materials X_1 and X_2 are available in quantities to satisfy the following condition:

$$A_{31}X_1{}^{\text{av}} + A_{32}X_2{}^{\text{av}} = \overline{Y}_3 - A_{30} \qquad (7.3\text{-}38)$$

Many other examples of plant operation under similar situations can be found.

7.4 NONLINEAR OBJECTIVE FUNCTION WITH CONSTRAINTS, USING THE LAGRANGE MULTIPLIER TECHNIQUE

This section will introduce the case of *nonlinear objective functions with nonlinear constraints*. The constraint may be the physical process model which is not in a form that can be directly substituted into the objective function. The constraint may also be equations relating some of the control variables to the other control variables, which in effect reduces the degree of freedom of the control variables. The second type of constraint may be viewed conceptually as an extension of the physical process model. The approach used in this section will be to use the geometric and then the calculus approaches to derive the conditions of optimum. The conditions of optimum may then be solved by iterative techniques to obtain the numerical solution of the optimum.

A. Geometric Interpretation of the Lagrange Multiplier

The concept of the Lagrange multiplier can be developed geometrically as follows:

STEP 1. At the optimum of the objective function F subject to the constraint $H = 0$, as shown in Figure 7.4-1, both the objective function F and the constraint $H = 0$ are satisfied. Furthermore, since the optimum is an extrema, i.e., either minimum or maximum, the *objective function F must intersect the constraint $H = 0$ at a point at the optimum.* Thus the tangent lines to the objective function and to the constraint $H = 0$ must coincide.

STEP 2. For the example in Figure 7.4-1 the objective function is

$$F = X^2 + Y^2 \qquad (7.4\text{-}1)$$

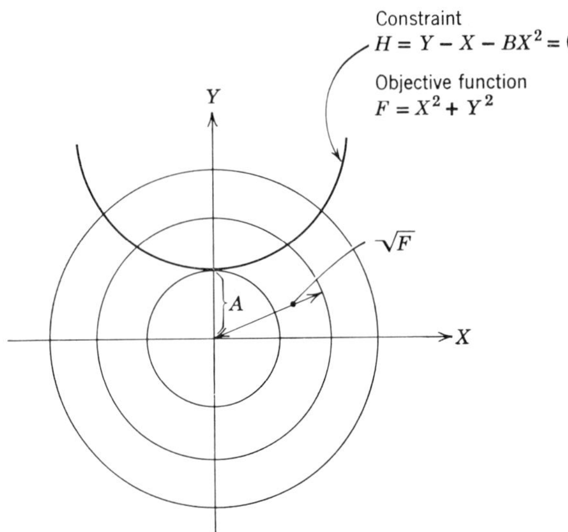

Fig. 7.4-1   Geometric interpretation of minimizing F subject to constraint $H = 0$.

and the constraint is

$$H = Y - A - BX^2 = 0 \tag{7.4-2}$$

The problem is to find the minimum of F subject to the constraint $H = 0$.
The equations for the tangent lines for F and H at the optimum $X = 0$, $Y = A$ are

$$(X - 0)\frac{\partial F(X, Y)}{\partial X}\bigg|_{(0,A)} + (Y - A)\frac{\partial F(X, Y)}{\partial Y}\bigg|_{(0,A)} = 0 \tag{7.4-3}$$

$$(X - 0)\frac{\partial H(X, Y)}{\partial X}\bigg|_{(0,A)} + (Y - A)\frac{\partial H(X, Y)}{\partial Y}\bigg|_{(0,A)} = 0 \tag{7.4-4}$$

Equations 7.4-3 and 7.4-4 may be rearranged:

$$(X - 0)\frac{\partial F(X, Y)}{\partial X}\bigg|_{(0,A)} = (A - Y)\frac{\partial F(X, Y)}{\partial Y}\bigg|_{(0,A)} \tag{7.4-5}$$

$$(X - 0)\frac{\partial H(X, Y)}{\partial X}\bigg|_{(0,A)} = (A - Y)\frac{\partial H(X, Y)}{\partial Y}\bigg|_{(0,A)} \tag{7.4-6}$$

STEP 3. Equation 7.4-5 may be divided by Equation 7.4-6:

$$\frac{\dfrac{\partial F(X, Y)}{\partial X}\bigg|_{(0,A)}}{\dfrac{\partial H(X, Y)}{\partial X}\bigg|_{(0,A)}} = \frac{\dfrac{\partial F(X, Y)}{\partial Y}\bigg|_{(0,A)}}{\dfrac{\partial H(X, Y)}{\partial Y}\bigg|_{(0,A)}} = -\lambda \tag{7.4-7}$$

The ratios of the partial derivatives are $-\lambda$, where λ is the Lagrange multiplier. Equation 7.4-7 can be rearranged into two equations:

$$\frac{\partial F(X, Y)}{\partial X}\bigg|_{(0,A)} + \lambda \frac{\partial H(X, Y)}{\partial X}\bigg|_{(0,A)} = 0 \tag{7.4-8}$$

and

$$\frac{\partial F(X, Y)}{\partial Y}\bigg|_{(0,A)} + \lambda \frac{\partial H(X, Y)}{\partial Y}\bigg|_{(0,A)} = 0 \tag{7.4-9}$$

Equations 7.4-8 and 7.4-9 are the conditions for the optimum of the objective function F subject to the constraint $H = 0$. The constraint $H = 0$ together with Equations 7.4-8 and 7.4-9 may be used to solve for the optimum.

Applying the conditions for optimum to the example in Figure 7.4-1:

$$\frac{\partial F}{\partial X} = 2X, \qquad \frac{\partial F}{\partial Y} = 2Y \tag{7.4-10}$$

$$\frac{\partial H}{\partial X} = -2BX, \qquad \frac{\partial H}{\partial Y} = 1 \tag{7.4-11}$$

$$\therefore \quad 2X - 2B\lambda X = 0 \tag{7.4-12}$$

$$2Y + \lambda = 0 \tag{7.4-13}$$

and the constraint is

$$Y - A - BX^2 = 0 \tag{7.4-14}$$

Solving Equations 7.4-12 through 7.4-14, the optimum is

$$X = 0, \qquad Y = A, \qquad \lambda = -2A \tag{7.4-15}$$

By examining Figure 7.4-1, it can be shown that the addition of the constraint has increased the optimal value of the objective function from $F = 0$ to $F = A$. In general, *the effect of constraints is to degrade the quality of the optimum*. In other words, in comparison with the case without constraints, the value of the minimum will be increased and the value of the maximum will be decreased.

B. Lagrange Multiplier by Calculus

Equations 7.4-8 and 7.4-9 have been developed from a *geometric point of view*. By examining the form of these equations, it can be seen that these can also be derived by taking the partial derivatives of Equation 7.4-16 with respect to X and Y and setting these to zero:

$$F(X, Y) + \lambda H(X, Y) = 0 \tag{7.4-16}$$

Thus using the approach of calculus, the first step in deriving the conditions for optimum is to define a modified objective function:

$$F' = F + \lambda H \tag{7.4-17}$$

The optimization techniques discussed in Sections 7.2 and 7.3 can then be followed to obtain the optimum of F'. This discussion leads directly to Equation 7.4-20.

The general form of the Lagrange multiplier solution may be applied to optimize an objective function with n control variables, m state variables, and p constraints, with $p < n$. The problem is to optimize the objective function

$$F = F(X_i, Y_j) \qquad (7.4\text{-}18)$$

subject to the following constraints

$$H_k = H_k(X_i, Y_j) = 0 \qquad (7.4\text{-}19)$$

where $(k = 1)$ to $(k = p)$.

The modified objective function is

$$F' = F + \sum_{k=1}^{p} \lambda_k H_k \qquad (7.4\text{-}20)$$

and the conditions for optimum are

$$\frac{\partial F'}{\partial X_1} = \frac{\partial F}{\partial X_1} + \sum_{k=1}^{p} \lambda_k \frac{\partial H_k}{\partial X_1}$$

$$\frac{\partial F'}{\partial X_2} = \frac{\partial F}{\partial X_1} + \sum_{k=1}^{p} \lambda_k \frac{\partial H_k}{\partial X_2} \qquad (7.4\text{-}21)$$

$$\vdots$$

$$\frac{\partial F'}{\partial X_n} = \frac{\partial F}{\partial X_n} + \sum_{k=1}^{p} \lambda_k \frac{\partial H_k}{\partial X_n}$$

Equations 7.4-19 and 7.4-21 may be solved iteratively to yield the optimal operating point.

C. Example of Applying the Lagrange Multiplier Technique to Styrene Manufacture

A typical application of the Lagrange multiplier technique is the *optimal control of the styrene manufacturing process*. Figure 7.4-2 shows the functional model of the styrene manufacturing process. To manufacture styrene,

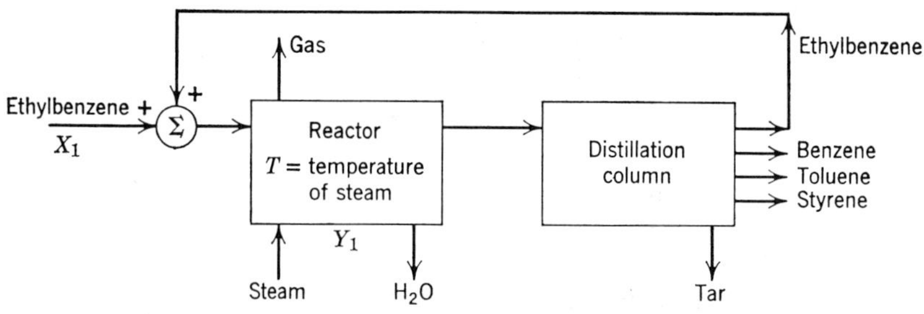

Fig. 7.4-2 Functional model of the single reactor styrene manufacturing process.

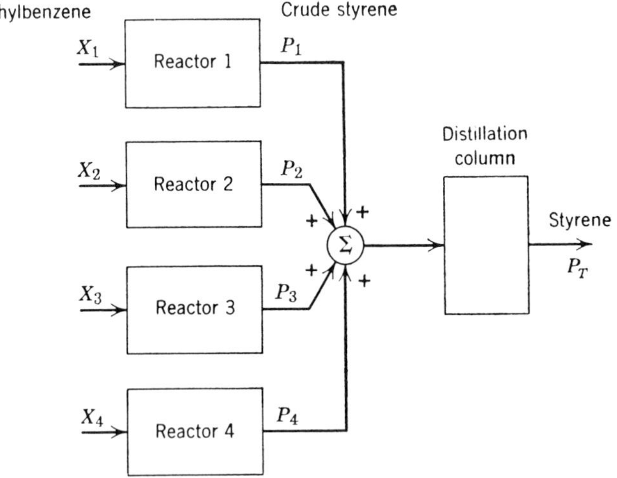

Fig. 7.4-3   Four reactors feeding crude styrene into a distillation column.

ethylbenzene is heated in a reactor in the presence of a catalyst to form crude styrene. The crude styrene output stream from the reactor contains benzene, ethylbenzene, toluene, and styrene. The output stream from the reactor is fed to a distillation column to separate the components listed above. The ethylbenzene output of the distillation column is recycled to the reactor, the benzene is recycled to an ethylbenzene unit, and the styrene is used directly. In an *actual process*, there may be several reactors feeding into a common distillation column, as shown in Figure 7.4-3. The basic reactions occurring in the reactors are quite simple in concept. The reaction rate depends on the following factors:

Temperature.
Catalyst age and history.
Concentration of basic compounds.

The physical process model can be developed experimentally through the use of regression analysis techniques discussed in Chapter 4.

Let the following symbols be assigned:

X_1, X_2, X_3, X_4 = ethylbenzene feed rate into Reactors 1, 2, 3, and 4, respectively.

Y_1, Y_2, Y_3, Y_4 = steam feed rate into Reactors 1, 2, 3, and 4, respectively.

P_1, P_2, P_3, P_4 = crude styrene production rate from Reactors 1, 2, 3, and 4, respectively.

P_T = total styrene produced.

C_1, C_2, C_3, C_4 = relative production cost for Reactors 1, 2, 3, and 4, respectively.

T_1, T_2, T_3, T_4 = steam temperature of Reactors 1, 2, 3, and 4, respectively.

The objective of operation is to operate the four reactors shown in Figure 7.4-3 at minimum cost to meet a specified total production requirement of styrene P_T. The production cost of the nth reactor depends on the following factors:

Ethylbenzene feed rate X_n.
Steam flow Y_n.
Crude styrene production rate P_n.
Steam temperature T_n.

Figure 7.4-4 shows the relative cost contours for a single reactor plotted as a function of X_n, Y_n, P_n, and T_n. For constant steam temperature T_n, the relative unit production cost C_n increases as the ethylbenzene feed rate X_n is increased. For constant steam temperature T_n and constant ethylbenzene feed rate X_n, it is possible to increase crude styrene output rate P_n by increasing catalyst deterioration and steam flow Y_n. This, however, is at the expense of increasing the production cost C_n. The limits on the ethylbenzene flow rate and the upper and lower "conversion" limits are also shown in Figure 7.4-4. The "conversion" limits attempt to force the operation into a reasonable region and to conserve catalyst usage. The most economic cost for any given

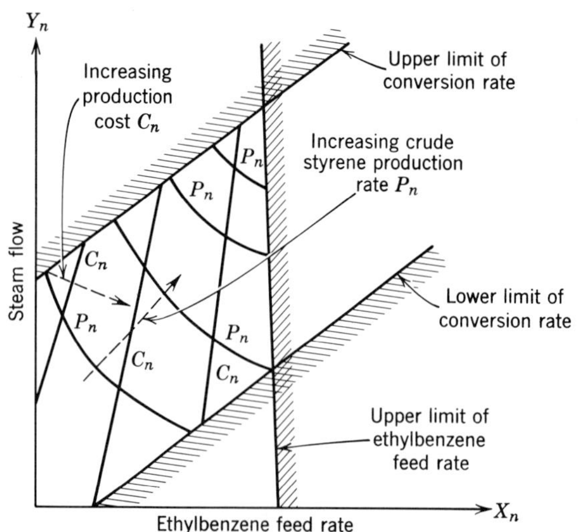

Fig. 7.4-4   Relative production cost of styrene for constant steam temperature.

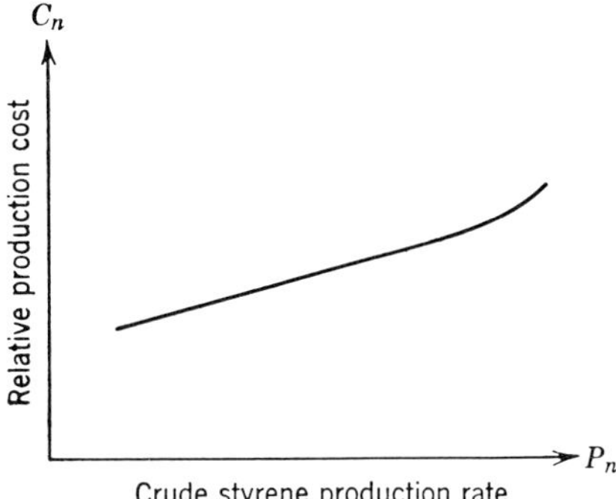

Fig. 7.4-5 "Cost versus production rate" relationship for a given reactor producing styrene.

crude styrene production rate is along the upper conversion limit. Figure 7.4-5 shows the relationship of relative production cost C_n plotted against crude styrene production rate P_n *for operation along the upper conversion limit.* Thus

$$C_n = C_n(P_n) \tag{7.4-22}$$

Each of the four reactors will have different characteristics of "cost versus production rate" curves. The problem is to solve for the minimum of the cost function F which is defined as

$$F = C_1(P_1) + C_2(P_2) + C_3(P_3) + C_4(P_4) \tag{7.4-23}$$

subject to the constraint to meet the given required total production

$$P_T = P_1 + P_2 + P_3 + P_4 \tag{7.4-24}$$

Applying the Lagrange multiplier technique, the modified objective function is

$$F' = C_1(P_1) + C_2(P_2) + C_3(P_3) + C_4(P_4) + \lambda(P_T - P_1 - P_2 - P_3 - P_4) \tag{7.4-25}$$

Note that as long as the constraint equation is satisfied, the last term in Equation 7.4-25 is always zero.

The *conditions for optimum* are satisfied when the partial derivatives of the objective function in Equation 7.4-23 are set to zero:

$$\frac{\partial F'}{\partial P_1} = \frac{\partial C_1(P_1)}{\partial P_1} - \lambda = 0 \tag{7.4-26}$$

$$\frac{\partial F'}{\partial P_2} = \frac{\partial C_2(P_2)}{\partial P_2} - \lambda = 0 \tag{7.4-27}$$

$$\frac{\partial F'}{\partial P_3} = \frac{\partial C_3(P_3)}{\partial P_3} - \lambda = 0 \tag{7.4-28}$$

$$\frac{\partial F'}{\partial P_4} = \frac{\partial C_4(P_4)}{\partial P_4} - \lambda = 0 \tag{7.4-29}$$

Thus the *conditions of optimum* require that

$$\frac{\partial C_1(P_1)}{\partial P_1} = \frac{\partial C_2(P_2)}{\partial P_2} = \frac{\partial C_3(P_3)}{\partial P_3} = \frac{\partial C_4(P_4)}{\partial P_4} = \lambda \tag{7.4-30}$$

Equation 7.4-30 states that *at the optimum, all the reactors should operate at equal incremental costs,* i.e., the slope of all the "costs versus production rate" curves will be equal. Figure 7.4-6 shows a plot of four different "incremental cost" curves. The minimum cost for any given specified total production can be solved iteratively by varying the values of the "incremental cost" λ, so that the sum of the production equals the total required.

Figure 7.4-7 shows a flow chart for the iterative procedure. Note that practical consideration of constraints on production limits, etc., have been ignored in the flow chart. Constraints on the production rate for each reactor can be readily built into the iterative solution by introducing a few tests to ensure that the values P_n are within limits. An alternative method of introducing the production constraints is to modify the incremental cost curves, as shown in Figure 7.4-8. By using an arbitrarily steep rise of the incremental cost curve near the upper production limit, the iterative procedure will generally not exceed the upper limit by any significant extent.

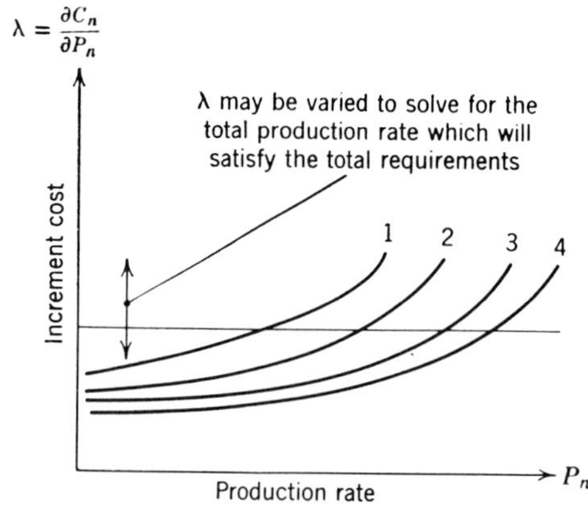

Fig. 7.4-6　Incremental cost curves for four reactors.

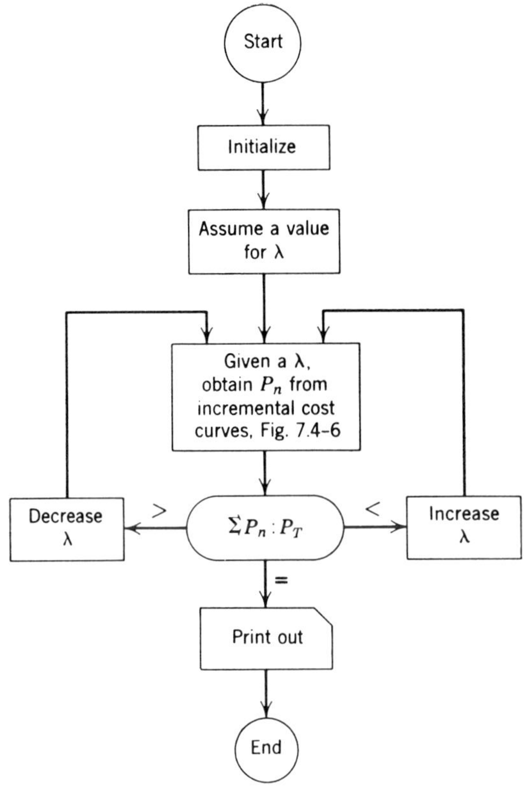

Fig. 7.4-7 Iterative solution for optimal control of styrene manufacturing process.

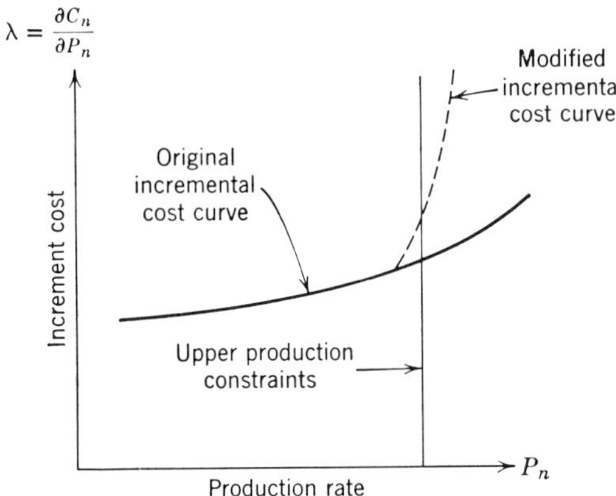

Fig. 7.4-8 Modified incremental cost curve to introduce the upper production constraint.

7.5 LINEAR OBJECTIVE FUNCTIONS WITH LINEAR INEQUALITY CONSTRAINTS, USING THE LINEAR PROGRAMMING TECHNIQUE

This section will present the special case where the *objective function, the physical process model, and constraints are all linear functions.* Optimal control of this class of problems can be solved by classical linear programming techniques. This section will first discuss the conditions for optimum for the linear objective function without constraints, then the constraints for control variables will be introduced, and finally the Simplex method for solving for the optimum will be presented.

Let the following symbols be assigned:

F = objective function expressed in terms of the state and control variables.
G = objective function expressed in terms of the control variables only.
H_k = constraint functions for control variables, $(k = 1)$ to $(k = p)$.
X_i = control variables, $(i = 1)$ to $(i = n)$.
Y_j = state variables, $(j = 1)$ to $(j = m)$.
R_{iU}, R_{iL} = upper and lower limits for control variables.
Q_{jU}, Q_{jL} = upper and lower limits for state variables.
Z_i = slack variables.
W_{kU}, W_{kL} = slack variables.
A, B, C, D, ψ_i, etc. = constants.

A. Linear Objective Function without Constraint Has No Optimum

Consider the linear objective function with one control variable:

$$F(Y, X) = A + BX + CY \tag{7.5-1}$$

where the linear physical process model is

$$Y = D + EX \tag{7.5-2}$$

Substituting the linear physical process model into the objective function F, the G form of the objective function is

$$G(X) = A + BX + CD + CEX \tag{7.5-3}$$

The general form is

$$G(X) = \psi_0 + \psi_1 X \tag{7.5-4}$$

By examining Equation 7.5-4 and the plot of $G(X)$ in Figure 7.5-1, it is obvious that the optimum is

$$\text{maximum at } X = +\infty$$

$$\text{minimum at } X = -\infty$$

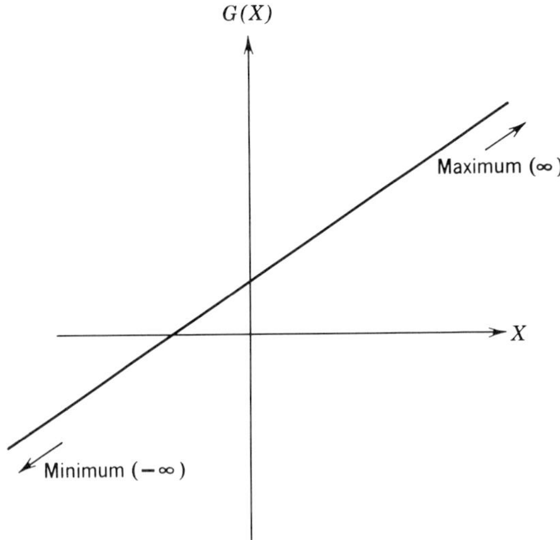

Fig. 7.5-1   Linear objective function $G(X)$ without constraints has minimum at $(-\infty)$ and maximum at $(+\infty)$.

Thus linear objective functions with one control variable and without constraints have no finite optimum. By extension it can be shown that objective functions with n control variables and without constraints also do not have a finite optimum.

B. The Linear Programming Problem in General Form

The optimal control of a linear objective function subject to linear constraints can be stated as

(1) optimize the objective function

$$G = \psi_0 + \sum_{i=1}^{n} \psi_i X_i \tag{7.5-5}$$

(2) subject to the linear constraints

$$R_{iL} \leq X_i \leq R_{iU} \tag{7.5-6}$$

$$Q_{kL} \leq H_k(X_i) \leq Q_{kU} \tag{7.5-7}$$

For example, in a blending process, the physical meaning of X_i is the relative amount of the ith material blended into the process, and ψ_i is the unit cost of the ith material. Equation 7.5-6 is the constraint on the limits of control variables, and Equation 7.5-7 is the constraint which expresses the relationships among the control variables.

C. Constraints

The first type of constraint is the constraint on the limits of control variables. These constraints restrict the allowable range of each variable to an upper and lower limit:

$$R_{iL} \leq X_i \leq R_{iU} \tag{7.5-8}$$

where R_{iL} is the lower limit and R_{iU} is the upper limit. Generally, the lower limit will be equal to zero, and the upper limit is the physical limit of the control variable, e.g., valve full open.

Constraints on the limit of control variables may be expressed as equalities by the introduction of positive slack variables Z_{iL} and Z_{iU}. Thus

$$X_i + Z_{iU} = R_{iU} \tag{7.5-9}$$

$$X_i - Z_{iL} = R_{iL} \tag{7.5-10}$$

Notice that X_i is at the constraint boundary when $Z_{iU} = 0$ or $Z_{iL} = 0$.

The second type of constraint is the constraint which expresses the relationships among the control variables. These constraints express the desired characteristics of the state variables, but are expressed as functions of the control variables. The general form for constraints on the control variables is

$$Q_{kL} \leq H_k(X_1, X_2, \ldots, X_n) \leq Q_{kU} \tag{7.5-11}$$

where H_k is the linear constraint function:

$$H_k = \sum_{i=1}^{n} K_{ki} X_i \tag{7.5-12}$$

and Q_{kU} and Q_{kL} are the high and low limits of H_k. The inequality in Equation 7.5-11 may be expressed as an equality by the introduction of positive slack variables W_{kU} and W_{kL}:

$$\sum_{i=1}^{n} K_{ki} X_i + W_{kU} = Q_{kU} \tag{7.5-13}$$

$$\sum_{i=1}^{n} K_{ki} X_i - W_{kL} = Q_{kL} \tag{7.5-14}$$

For an objective function with two control variables, the single constraint on one of the control variables may be stated as

$$K_1 X_1 + K_2 X_2 + W_{1U} = Q_{1U} \tag{7.5-15}$$

$$K_1 X_1 + K_2 X_2 - W_{1L} = Q_{1L} \tag{7.5-16}$$

By setting $W_{1U} = 0$ first and then $W_{1L} = 0$, it is possible to obtain the boundaries of the constraints:

$$K_1 X_1 + K_2 X_2 = Q_{1U} \quad \text{for} \quad W_{1U} = 0 \quad \text{and operation on the}$$
$$\text{upper boundary} \qquad (7.5\text{-}17)$$

$$K_1 X_1 + K_2 X_2 = Q_{1L} \quad \text{for} \quad W_{1L} = 0 \quad \text{and operation on the}$$
$$\text{lower boundary} \qquad (7.5\text{-}18)$$

Figure 7.5-2 shows Equations 7.5-17 and 7.5-18 plotted as straight lines with slope $-K_1/K_2$. The boundary of the upper constraint intersects the X_2 axis at Q_{1U}/K_2, and the boundary of the lower constraint intersects the X_2 axis at Q_{1L}/K_2.

Figure 7.5-3 shows three constraints which express the relationships among the control variables plotted on the (X_1, X_2) plane of an objective function with two control variables. Only the area $(ABCDEF)$ is *feasible for solution* because it satisfies all three constraints. The constraint on the limits of control variables prevents X_1 and X_2 from being negative.

Figure 7.5-4 shows how the constraints which express the relationships among the control variables may sometimes prevent any feasible solution from taking place. There are no common feasible areas between the two constraints in the upper right-hand quadrant. In other words, for $X_1 \geq 0$ and $X_2 \geq 0$, there are no values of X_1 and X_2 that will satisfy both constraints on the control variables.

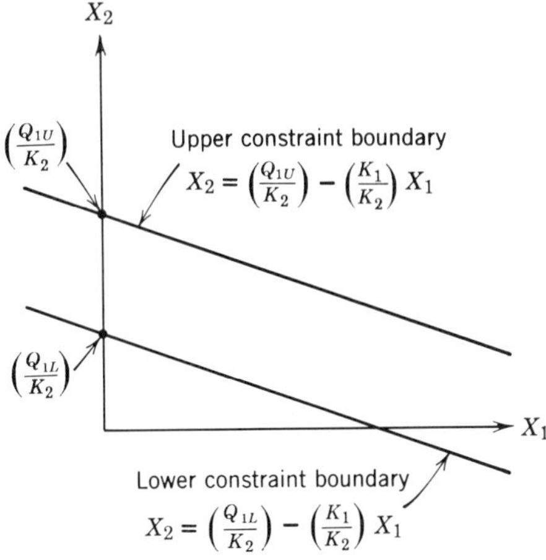

Fig. 7.5-2 Constraint on control variables for objective function with two control variables.

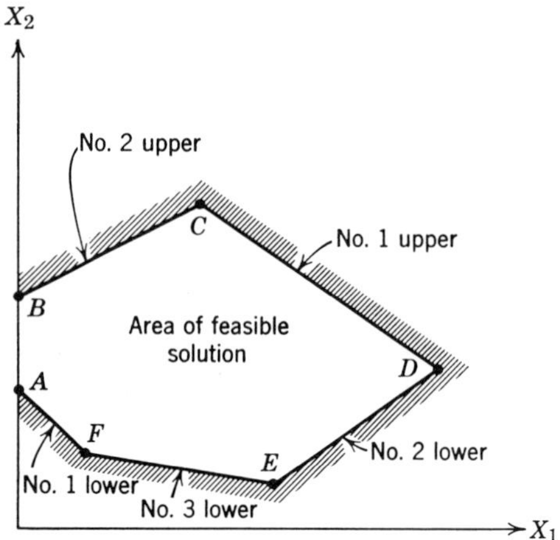

Fig. 7.5-3 Area of feasible solution of objective function with constraints on control variables.

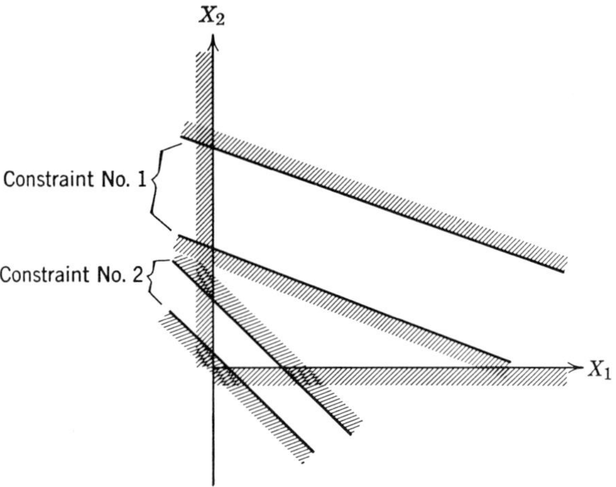

Fig. 7.5-4 Unfeasible solution due to constraint on control variables.

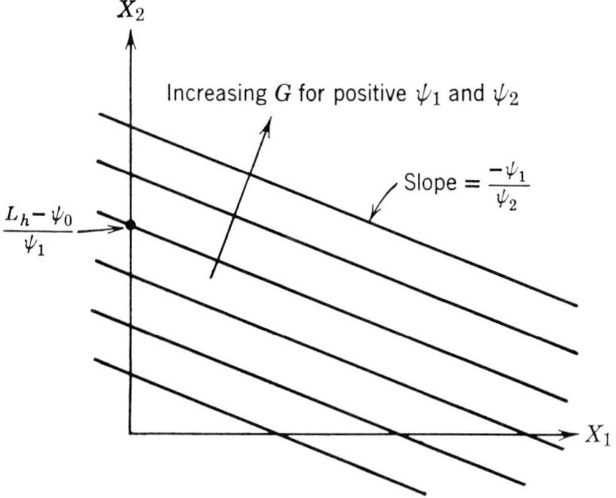

Fig. 7.5-5   Contours of objective function plotted on the (X_1, X_2) plane.

D. Contours of Objective Function

Contours of objective functions are lines which join together points where the values of the objective function are equal. These are similar to contours used in topology and for mapping purposes. For a linear objective function with two control variables:

$$G = \psi_0 + \psi_1 X_1 + \psi_2 X_2 \tag{7.5-19}$$

The contours plotted on the (X_1, X_2) plane are straight lines for the objective function with the values $G = L_1$, $G = L_2, \ldots, G = L_h$:

$$L_h = \psi_0 + \psi_1 X_1 + \psi_2 X_2 \tag{7.5-20}$$

or

$$X_2 = \frac{(L_h - \psi_0)}{\psi_2} - \left(\frac{\psi_1}{\psi_2}\right) X_1 \tag{7.5-21}$$

From Equation 7.5-21, the X_2 axis intercept is $(L_h - \psi_0)/\psi_2$, and the slope of the straight line is $-\psi_1/\psi_2$. Figure 7.5-5 shows the contours of the objective function plotted on the (X_1, X_2) plane. Note that these are parallel straight lines.

E. Pictorial Solution and Example

From examining the plot of constraints and the contours of the objective function, it is possible to conclude that the maximum (or minimum) of the objective function is on one of the intersections of the constraint boundaries. Figure 7.5-6 shows that the area of feasible solution lies between the area

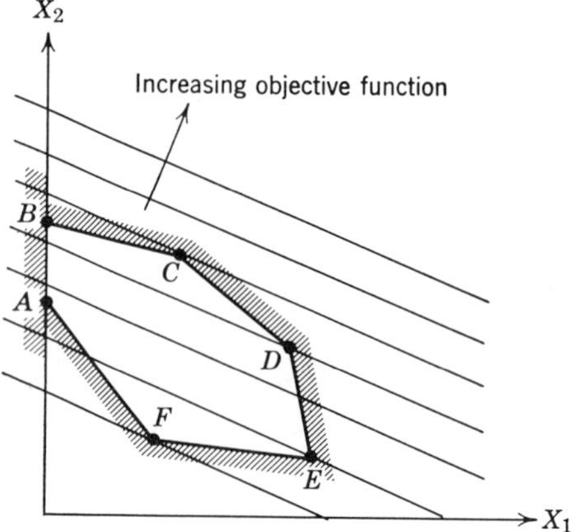

Fig. 7.5-6   Optimum solution of linear objective function is at one of the intersections of the constraint boundaries.

bounded by ($ABCDEF$), and the maximum (or minimum) is on one of these points. If the direction of increasing objective function is as shown, then the maximum is at the point C, and minimum is the point F. *Solutions for the optimum may be obtained by evaluating the objective function at the various intersections of the constraint boundaries.*

As an example, consider the simple example of maximizing

$$G = 25X_1 + 50X_2 \tag{7.5-22}$$

subject to the constraints

$$3X_1 + X_2 \geq 8 \tag{7.5-23}$$

$$4X_1 + 3X_2 \geq 19 \tag{7.5-24}$$

$$X_1 + 3X_2 \geq 7 \tag{7.5-25}$$

$$0 \leq X_1 \leq 10 \tag{7.5-26}$$

$$0 \leq X_2 \leq 9 \tag{7.5-27}$$

Figure 7.5-7 shows the area of feasible solution and the contours of the objective function. The coordinates of the intersections of the constraint boundaries can be solved for by algebra or by graphic means.

$$
\begin{aligned}
A&: \quad (0, 8)\\
B&: \quad (1, 5)\\
C&: \quad (4, 1)\\
D&: \quad (7, 0)\\
E&: \quad (10, 0)\\
F&: \quad (10, 9)\\
G&: \quad (0, 9)
\end{aligned} \tag{7.5-28}
$$

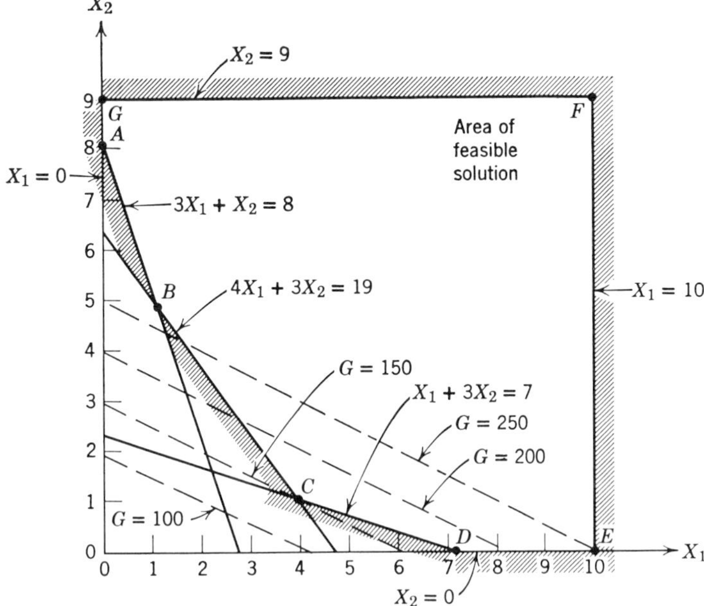

Fig. 7.5-7   Example of linear programming solution by graphic means.

The minimum is at C, and the maximum is at F. Thus

$$G_{\min} = 150 \tag{7.5-29}$$

$$G_{\max} = 700 \tag{7.5-30}$$

F. Standard and Canonical Forms

Beyond two or three control variables, graphical solutions of linear programming problems are difficult to implement. By transforming the linear programming problem to the standard and canonical forms iterative solutions based on the Simplex method can be used. Let the following symbols be assigned (bold face is used for matrix):

G = objective function.
ψ_0 = constant term of objective function.
ψ = other objective function coefficients, (1) by $(n + p)$ row vector.
X_c = control variables, (n) by (1) column vector.
X_s = slack variables, (p) by (1) column vector.
$\mathbf{K}$ = coefficient matrix, (p) by $(n + p)$.
$\mathbf{X}$ = control and slack variables, $(n + p)$ by (1) column vector.
$\mathbf{R}$ = constraint coefficients, (p) by (1) column vector.
k = index for constraints, $(k = 1)$ to $(k = p)$.

i = index for control variables, $(i = 1)$ to $(i = n)$; also index for slack variables, $(i = n + 1)$ to $(i = n + p)$.

q = pivot column.

s = pivot row.

$\mathbf{0}$ = matrix with zero elements.

The *standard form of the linear programming problem* in matrix notation is

$$\text{minimize:} \; G = \psi_0 + \psi X \tag{7.5-31}$$

subject to the constraints

$$\mathbf{KX = R} \tag{7.5-32}$$

$$\mathbf{X \geq 0} \tag{7.5-33}$$

The *linear programming problem is said to be in canonical* form if the following conditions of the standard form are satisfied:*

(1) $\psi_i = 0$ for $(i = n + 1)$ to $(i = n + p)$ $\qquad$ (7.5-34)

$$
(2) \quad
\begin{aligned}
K_{ki} &= 0 \;\; \text{for } (i = n + 1) \;\; \text{to} \;\; (i = n + p), \;\; \text{and} \;\; (i \neq k + n) \\
K_{ki} &= 1 \;\; \text{for } (i = n + 1) \;\; \text{to} \;\; (i = n + p), \;\; \text{and} \;\; (i = k + n)
\end{aligned}
\tag{7.5-35}
$$

(3) $R_k \geq 0$ $\qquad$ (7.5-36)

Since this is an "overdetermined system," as shown in Equation 7.3-10, not all the control variables are needed to satisfy these equations. The variables X which satisfy the conditions of (1) and (2) above are called *basic variables*. The others are called *nonbasic variables*. The *basic variables* may be solved for in terms of the *nonbasic variables*.

As an example, consider the problem of minimizing

$$G = 50 + 3X_1 - 2X_2 - 4X_3 + 7X_4 \tag{7.5-37}$$

subject to the constraints

$$X_1 - 2X_2 + X_3 + 2X_4 + X_5 = 3 \tag{7.5-38}$$

$$2X_1 + X_2 - 2X_3 + X_4 + X_6 = 7 \tag{7.5-39}$$

$$-3X_1 + 2X_2 + X_3 + 3X_4 + X_7 = 2 \tag{7.5-40}$$

and

$$X_1 \geq 0, \;\; X_2 \geq 0, \;\; X_3 \geq 0, \;\; X_4 \geq 0, \;\; X_5 \geq 0, \;\; X_6 \geq 0, \;\; X_7 \geq 0 \tag{7.5-41}$$

Stating the problem in standard form, the problem is to minimize

$$\mathbf{G} = \psi_0 + \psi \mathbf{X} \tag{7.5-42}$$

---

* Canonical form merely guarantees that a feasible solution always exists.

subject to

$$\mathbf{KX} = \mathbf{R} \tag{7.5-43}$$

and

$$\mathbf{X} \geq \mathbf{0} \tag{7.5-44}$$

where

$$\mathbf{K} = \begin{pmatrix} 1 & -2 & 1 & 2 & 1 & 0 & 0 \\ 2 & 1 & -2 & 1 & 0 & 1 & 0 \\ -3 & 2 & 1 & 3 & 0 & 0 & 1 \end{pmatrix} \tag{7.5-45}$$

$$\mathbf{R} = \begin{pmatrix} 3 \\ 7 \\ 2 \end{pmatrix} \tag{7.5-46}$$

$$\mathbf{\psi} = (3 \quad -2 \quad -4 \quad 7 \quad 0 \quad 0 \quad 0) \tag{7.5-47}$$

$$\psi_0 = 50 \tag{7.5-48}$$

The example is in standard form because it satisfies the form of Equations 7.5-31 through 7.5-33. The example is also in *canonical form* because it satisfies the conditions for canonical form:

(1) $\quad \psi_5 = \psi_6 = \psi_7 = 0$ $\hspace{3cm}$ (7.5-49)

$$
(2) \quad \begin{aligned} K_{16} = K_{17} = K_{25} = K_{27} = K_{35} = K_{36} = 0 \\ K_{15} = K_{26} = K_{37} = 1 \end{aligned} \tag{7.5-50}
$$

(3) $\quad R_1 \geq 0, R_2 \geq 0, R_3 \geq 0$ $\hspace{3cm}$ (7.5-51)

Note that letting $X_1 = X_2 = X_3 = X_4 = 0$, $X_5 = 3$, $X_6 = 7$, and $X_7 = 2$, the constraint equations are satisfied, and this is a solution (though not optimum). Therefore, the basic variables are X_5, X_6, and X_7, and the nonbasic variables are X_1, X_2, X_3, and X_4.

G. Optimal Solution by the Simplex Method

The Simplex method is an iterative procedure that solves for the optimal solution to linear programming problems stated in canonical form. Each iteration will improve on the previous one so that the Simplex method will not recycle or close on itself. An optimal solution is always assured in a finite number of steps. This discussion will only present the general outline of the Simplex method with an example. Special conditions such as degeneracy, multiple optimums, and so forth, will not be discussed. The procedure presented is designed for clarity and understanding rather than for efficiency in computer solution.

(1) The linear programming problem in canonical form is to minimize

$$G = \psi_0 + \psi X \tag{7.5-52}$$

subject to the constraints

$$KX = R \tag{7.5-53}$$

and

$$X \geq 0 \tag{7.5-54}$$

where

$$\psi_i = 0 \quad \text{for} \quad (i = n + 1) \quad \text{to} \quad (i = n + p) \tag{7.5-55}$$

$$\begin{aligned}
K_{ki} &= 0 \quad \text{for} \quad (i = n + 1) \quad \text{to} \quad (i = n + p), \quad \text{and} \quad (i \neq k + n) \\
K_{ki} &= 1 \quad \text{for} \quad (i = n + 1) \quad \text{to} \quad (i = n + p), \quad \text{and} \quad (i = k + n)
\end{aligned} \tag{7.5-56}$$

and

$$R_k \geq 0 \tag{7.5-57}$$

(2) The example to be solved is the same as that given in Equations 7.5-37 through 7.5-41. The problem is to minimize

$$G = 50 + 3X_1 - 2X_2 - 4X_3 + 7X_4 \tag{7.5-58}$$

subject to the contraints

$$X_1 - 2X_2 + X_3 + 2X_4 + X_5 = 3 \tag{7.5-59}$$

$$2X_1 + X_2 - 2X_3 + X_4 + X_6 = 7 \tag{7.5-60}$$

$$-3X_1 + 2X_2 + X_3 + 3X_4 + X_7 = 2 \tag{7.5-61}$$

and

$$X_1 \geq 0, \quad X_2 \geq 0, \quad X_3 \geq 0, \quad X_4 \geq 0, \quad X_5 \geq 0, \quad X_6 \geq 0, \quad X_7 \geq 0 \tag{7.5-62}$$

Figure 7.5-8 shows the problem stated in matrix notation. It can be seen that the problem is in canonical form by applying the tests given in Equations 7.5-34 through 7.5-36.

(3) The general flow chart for the Simplex method is given in Figure 7.5-9. The procedure assumes an initial feasible solution for all the nonbasic variables to be equal to zero. Then tests are made to determine if the solution can be improved. If it can be improved then a nonbasic variable is selected to become basic, and a basic variable is selected to become nonbasic. The interchange of basic and nonbasic variables is governed

(1) Minimize

$$G = 50 + (3 \quad -2 \quad -4 \quad 7 \quad 0 \quad 0 \quad 0) \cdot \begin{pmatrix} X_1 \\ X_2 \\ X_3 \\ X_4 \\ X_5 \\ X_6 \\ X_7 \end{pmatrix}$$

(2) subject to the contraints

$$\overset{\text{pivot column}}{\downarrow}$$

$$\begin{pmatrix} 1 & -2 & 1 & 2 & 1 & 0 & 0 \\ 2 & 1 & -2 & 1 & 0 & 1 & 0 \\ -3 & 2 & 1 & 3 & 0 & 0 & 1 \end{pmatrix} \cdot \begin{pmatrix} X_1 \\ X_2 \\ X_3 \\ X_4 \\ X_5 \\ X_6 \\ X_7 \end{pmatrix} = \begin{pmatrix} 3 \\ 7 \\ 2 \end{pmatrix}$$

and

$$X_i \geq 0 \quad \text{for} \quad (i = 1) \text{ to } (i = 7)$$

(3) A feasible solution is to set the nonbasic variables X_1, X_2, X_3, X_4 all to zero:

$$X_1 = 0, \qquad X_2 = 0, \qquad X_3 = 0, \qquad X_4 = 0$$

Then the basic variables may be obtained by solving Equations 7.5-59 through 7.5-61:

$$X_5 = 3, \qquad X_6 = 7, \qquad X_7 = 2$$

Fig. 7.5-8   Example to be optimized by the Simplex method.

by the conditions that the canonical form has to be preserved. The algorithm for the interchange is similar to that used in Gaussian elimination for the solution of simultaneous linear algebraic equations.

(4) *Step 1 of the Simplex Method.* Three possibilities exist when any feasible solution is examined. The solution is the true optimum, a relative optimum, or has no optimum.

(a) *The condition for the true optimum.* If all the terms in the objective function are equal to or greater than zero, then any change of the basic variables can only increase the objective function. Thus the condition for true optimum is

$$\psi_i \geq 0 \tag{7.5-63}$$

In the example, the initial solution of

$$X_1 = 0, \quad X_2 = 0, \quad X_3 = 0, \quad X_4 = 0 \tag{7.5-64}$$

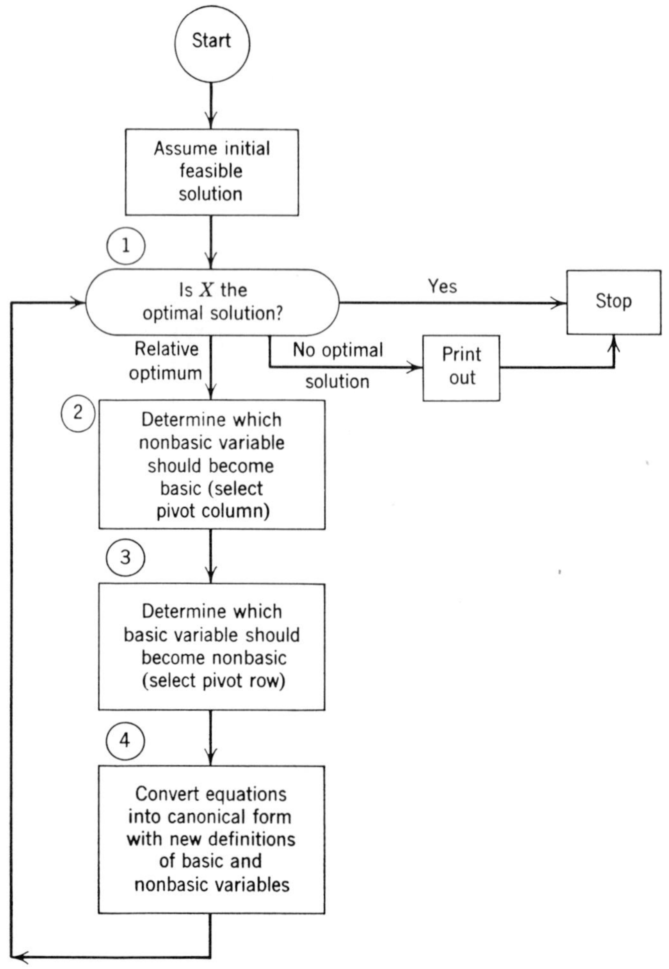

Fig. 7.5-9   Flow chart of the Simplex method.

does not satisfy this condition. By examining Equation 7.5-58, it can be seen that the objective function can be further reduced by introducing nonnegative values of X_2 or X_3.

(b) *The condition for a relative optimum.* In contrast to (a) above, if one of the terms in the objective function is negative, then further improvement is possible. The condition for a relative optimum is

$$\psi_i < 0 \text{ for any } i, \text{ but there is at least one } K_{ki} \text{ which}$$
$$\text{satisfies } K_{ki} > 0 \text{ for the same } i. \tag{7.5-65}$$

In the example of Figure 7.5-8

$$\psi_2 < 0 \quad \text{and} \quad \psi_3 < 0, \quad \text{but} \quad K_{22} > 0, K_{32} > 0$$

$$\text{and} \quad K_{13} > 0, K_{33} > 0 \qquad (7.5\text{-}66)$$

so the initial solution is a relative optimum.

(c) *The conditions for not having any optimum.* The following are the conditions for not having any optimum:

$$\psi_i < 0 \quad \text{for any } i \quad \text{and} \quad K_{ki} \leq 0 \quad \text{for that same } i \qquad (7.5\text{-}67)$$

In order to satisfy this condition, all the elements of the entire column of the K_{ki} must be equal to or less than zero for any coefficient ψ_i which is negative in the objective function.

(5) *Step 2 of the Simplex Method.* This step follows the tests made in Step 1 when the initial feasible solution is found to be a relative optimum. *Step 2 determines which nonbasic variable should become basic, i.e., selects the pivot column.* Any column which has a negative ψ_i in the objective function may be selected to be the pivot column. In other words, any nonbasic variable which can improve the solution can be selected to be a new basic variable. In the example, either X_2 or X_3 may be selected to be the new basic variable because they both may take on nonnegative values to improve the solution. However, generally the term with the largest coefficient is selected since this offers the greatest improvement. Suppose the pivot column is selected to be $i = 2$ in the example, so X_2 is the nonbasic variable selected to become the new basic variable.

(6) *Step 3 of the Simplex Method.* Once the incoming basic variable has been selected, *Step 3 determines the outgoing basic variable by determining which basic variable should become nonbasic, i.e., selects the pivot row.* The selection of the pivot row must satisfy the constraints:

$$\mathbf{K}\,\mathbf{X} = \mathbf{R}, \qquad \mathbf{X} \geq \mathbf{0} \qquad (7.5\text{-}68)$$

The largest value that the incoming basic variable can take and satisfy the constraints can be determined by selecting the row k which has the smallest factor P_{kq}, defined as

$$P_{kq} = \frac{R_k}{K_{kq}} \quad \text{for} \quad K_{kq} > 0 \qquad (7.5\text{-}69)$$

where q designates the pivot column. For the example, with $q = 2$ as the pivot column,

$$P_{22} = \frac{R_2}{K_{22}} = \frac{7}{1} \tag{7.5-70}$$

$$P_{32} = \frac{R_3}{K_{32}} = \frac{2}{2} \tag{7.5-71}$$

A nonbasic variable can be substituted for a basic variable by solving the simultaneous equations in such a manner that the original basic variable becomes zero, and the new basic variable assumes a positive value. A simple way to picture this is to visualize the pivot column of the constraint matrix in Figure 7.5-8. The plan is to reduce this column so that one coefficient is equal to one and the other two are zero. This can be done by multiplying the selected pivot row by the appropriate constant so the element of the pivot row in the pivot column is equal to one, and by subtracting the pivot row from the other rows after the pivot row is multiplied by another constant. Now the $\mathbf{R}$ vector must remain positive during this operation or the incoming basic variable would not be positive. Therefore the selection of the pivot row and the choice of the constants used must meet the criteria of Equation 7.5-69. Thus $k = 3$ is selected to be the pivot row, i.e., X_7 is the outgoing basic variable. The largest value that can be assumed by the incoming basic variable is $X_2 = 1$. The improvement in the objective function is equal to

$$\Delta G = \psi \mathbf{P} \tag{7.5-72}$$

In the example, the improvement in the objective function through the introduction of X_2 as the incoming basic variable to replace X_7 results in the decrease of the objective function by

$$\Delta G = (-2)(1) = -2 \tag{7.5-73}$$

An added sophistication in the Simplex method is to select the pivot column which will bring the greatest improvement where there are more than one available. Suppose $i = 3$ is selected as the pivot column, then

$$P_{13} = \frac{R_1}{K_{13}} = 3 \tag{7.5-74}$$

$$P_{33} = \frac{R_3}{K_{33}} = 2 \tag{7.5-75}$$

Then the pivot row should be $k = 3$, and the outgoing basic variable should be X_7. The decrease of the objective function is

$$\Delta G = (-4)(2) = -8 \qquad (7.5\text{-}76)$$

Selecting X_3 as the incoming basic variable will result in a greater improvement than X_2. Thus X_3 will be the incoming basic variable and X_7 will be the outgoing basic variable.

(7) *Step 4 of the Simplex Method.* As a result of Steps 2 and 3 of the Simplex method, it is now possible to proceed to Step 4 to convert Equations 7.5-52 and 7.5-53 into canonical form with the new definitions of basic and nonbasic variables. In the example the transformation is

	Old	New
Basic:	X_5, X_6, X_7	X_3, X_5, X_6
Nonbasic:	X_1, X_2, X_3, X_4	X_1, X_2, X_4, X_7
Solution:	$X_1 = X_2 = X_3 = X_4 = 0$	$X_1 = X_2 = X_4 = X_7 = 0, \ X_3 = 2$

The operations to be performed are governed by the condition that the equations must be in canonical form, i.e., the conditions stated in Equations 7.5-55 through 7.5-57 must be satisfied. In other words, the coefficient K_{ki} of the incoming basic variable must be (1) in the pivot row and (0) in all the nonpivot rows. This can be accomplished by performing the following operations on Equation 7.5-53:

(a) Divide all the coefficients of the pivot row by K_{sq} to obtain the new coefficients

$$K'_{si} = \frac{K_{si}}{K_{sq}} \qquad (7.5\text{-}77)$$

$$R'_s = \frac{R_s}{K_{sq}} \qquad (7.5\text{-}78)$$

where s designates the pivot row.

(b) Compute the new coefficients of any nonpivot row to eliminate the incoming basic variable X_q. Thus

$$K'_{ki} = K_{ki} - \left(\frac{K_{kq}}{K_{sq}}\right)(K_{si}) \qquad (7.5\text{-}79)$$

$$R'_k = R_q - \left(\frac{K_{kq}}{K_{sq}}\right)(R_s) \qquad (7.5\text{-}80)$$

where s designates the pivot row and q designates the pivot column.

$$\begin{pmatrix} 4 & -4 & 0 & -1 & 1 & 0 & -1 \\ -4 & 5 & 0 & 7 & 0 & 1 & 2 \\ -3 & 2 & 1 & 3 & 0 & 0 & 1 \end{pmatrix} \cdot \begin{pmatrix} X_1 \\ X_2 \\ X_3 \\ X_4 \\ X_5 \\ X_6 \\ X_7 \end{pmatrix} = \begin{pmatrix} 1 \\ 11 \\ 2 \end{pmatrix}$$

Fig. 7.5-10   Constraint equations after X_3 is introduced as basic variable to replace X_7.

Apply the operation defined by Equations 7.5-79 and 7.5-80 to the first row (nonpivot):

$$\left[1 - \left(\frac{1}{1}\right)(-3)\right]X_1 + \left[-2 - \left(\frac{1}{1}\right)(2)\right]X_2 + \left[1 - \left(\frac{1}{1}\right)(1)\right]X_3$$

$$+ \left[2 - \left(\frac{1}{1}\right)(3)\right]X_4 + X_5 - X_7 = \left[3 - \left(\frac{1}{1}\right)(2)\right]$$

$$\therefore \quad 4X_1 - 4X_2 - X_4 + X_5 - X_7 = 1 \tag{7.5-81}$$

Apply the operation defined by Equations 7.5-79 and 7.5-80 to the second row (nonpivot):

$$\left[2 - \left(\frac{-2}{1}\right)(-3)\right]X_1 + \left[1 - \left(\frac{-2}{1}\right)(2)\right]X_2 + \left[-2 - \left(\frac{-2}{1}\right)(1)\right]X_3$$

$$+ \left[1 - \left(\frac{-2}{1}\right)(3)\right]X_4 + X_6 + 2X_7 = \left[7 - \left(\frac{-2}{1}\right)(2)\right]$$

$$\therefore \quad -4X_1 + 5X_2 + 7X_4 + X_6 + 2X_7 = 11 \tag{7.5-82}$$

Apply the operation defined by Equations 7.5-77 and 7.5-78 to the third row (pivot):

$$-3X_1 + 2X_2 + X_3 + 3X_4 + X_7 = 2 \tag{7.5-83}$$

Figure 7.5-10 shows the constraint equation in matrix notation after this transformation. The same equation may be written as shown in Figure 7.5-11 by interchanging the positions of X_3 and X_7. To reduce

$$\begin{pmatrix} 4 & -4 & -1 & -1 & 1 & 0 & 0 \\ -4 & 5 & 2 & 7 & 0 & 1 & 0 \\ -3 & 2 & 1 & 3 & 0 & 0 & 1 \end{pmatrix} \cdot \begin{pmatrix} X_1 \\ X_2 \\ X_7 \\ X_4 \\ X_5 \\ X_6 \\ X_3 \end{pmatrix} = \begin{pmatrix} 1 \\ 11 \\ 2 \end{pmatrix}$$

Fig. 7.5-11   Constraint equation in canonical form after X_3 is introduced as a basic variable to replace X_7.

the objective function to canonical form, it is necessary to reduce the coefficient of X_3 to zero in the objective function. This may be done by substituting the constraint equation which only contains X_3 into the objective function, i.e., solve the pivot row for the incoming basic variable and substitute into the objective function. The new coefficients of the objective function may be computed by

$$\psi_i' = \psi_i - \psi_q\left(\frac{K_{si}}{K_{sq}}\right) \tag{7.5-84}$$

$$\psi_0' = \psi_0 + \psi_q\left(\frac{R_s}{K_{sq}}\right) \tag{7.5-85}$$

and

$$\psi_q = 0 \tag{7.5-86}$$

where s designates the pivot row and q designates the pivot column. The new objective function is

$$G = \left[50 + (-4)\left(\frac{2}{1}\right)\right] + \left[3 - (-4)\left(\frac{-3}{1}\right)\right]X_1 + \left[-2 - (-4)\left(\frac{2}{1}\right)\right]X_2$$

$$+ \left[7 - (-4)\left(\frac{3}{1}\right)\right]X_4 + \left[0 - (-4)\left(\frac{1}{1}\right)\right]X_7$$

$$G = 42 - 9X_1 + 6X_2 + 19X_4 + 4X_7 \tag{7.5-87}$$

(8) *Second Iteration of the Simplex Method Example.* Figure 7.5-12 shows the result of the first iteration. By applying the tests of Equations 7.5-63 through 7.5-67, it can be seen that the feasible solution is only a relative optimum. The incoming basic variable is selected to be X_1 because it has a negative coefficient in the objective function. The outgoing basic variable is selected to be X_5 by using the test given in Equation 7.5-69. Thus the first column and first row will be the pivot in the second iteration. The objective function will be improved by

$$\Delta G = (-9)(.25) = -2.25 \tag{7.5-88}$$

The result of the second iteration is shown in Figure 7.5-13.

(9) *Third Iteration of the Simplex Method.* Figure 7.5-13, the result of the second iteration, shows that the solution is only a relative optimum. The objective function can be further minimized by introducing X_2 as the incoming basic variable to replace X_6. Thus the second column

(1) Minimize

$$G = 42 - 9X_1 + 6X_2 + 19X_4 + 4X_7$$

(2) subject to the constraints

$$\begin{pmatrix} 4 & -4 & -1 & -1 & 1 & 0 & 0 \\ -4 & 5 & 2 & 7 & 0 & 1 & 0 \\ -3 & 2 & 1 & 3 & 0 & 0 & 1 \end{pmatrix} \cdot \begin{pmatrix} X_1 \\ X_2 \\ X_7 \\ X_4 \\ X_5 \\ X_6 \\ X_3 \end{pmatrix} = \begin{pmatrix} 1 \\ 11 \\ 2 \end{pmatrix}$$

and

$$X_i \geq 0 \quad \text{for} \quad (i = 1) \text{ to } (i = 7)$$

(3) A feasible solution is

$$X_1 = 0, \quad X_2 = 0, \quad X_7 = 0, \quad X_4 = 0, \quad X_5 = 1, \quad X_6 = 11, \quad X_3 = 2$$

(4) This is a relative optimum.

Fig. 7.5-12 Result of first iteration of the Simplex method.

(1) Minimize

$$G = 39.75 + 2.25X_5 - 3X_2 + 1.75X_7 + 16.75X_4$$

(2) subject to the constraints

$$\begin{pmatrix} 0.25 & -1 & -0.25 & -0.25 & 1 & 0 & 0 \\ 1 & 1 & 1 & 6 & 0 & 1 & 0 \\ 0.75 & -1 & 0.25 & 2.25 & 0 & 0 & 1 \end{pmatrix} \cdot \begin{pmatrix} X_5 \\ X_2 \\ X_7 \\ X_4 \\ X_1 \\ X_6 \\ X_3 \end{pmatrix} = \begin{pmatrix} 0.25 \\ 12 \\ 2.75 \end{pmatrix}$$

and

$$X_i \geq 0 \quad \text{for} \quad (i = 1) \text{ to } (i = 7)$$

(3) A feasible solution is

$$X_5 = 0, \quad X_2 = 0, \quad X_7 = 0 \quad X_4 = 0, \quad X_1 = 0.25, \quad X_6 = 12.0, \quad X_3 = 2.75$$

(4) This is a relative optimum.

Fig. 7.5-13 Result of second iteration of the Simplex method.

(1) Minimize

$$G = 3.75 + 5.25X_5 + 3X_6 + 4.75X_7 + 34.75X_4$$

(2) subject to the constraints

$$\begin{pmatrix} 1.25 & 1 & 0.75 & 5.75 & 1 & 0 & 0 \\ 1 & 1 & 1 & 6 & 0 & 1 & 0 \\ 1.75 & 1 & 1.25 & 8.25 & 0 & 0 & 1 \end{pmatrix} \cdot \begin{pmatrix} X_5 \\ X_6 \\ X_7 \\ X_4 \\ X_1 \\ X_2 \\ X_3 \end{pmatrix} = \begin{pmatrix} 12.25 \\ 12 \\ 14.75 \end{pmatrix}$$

and

$$X_i \geq 0 \quad \text{for} \quad (i = 1) \text{ to } (i = 7)$$

(3) A feasible solution is

$$X_5 = 0, \quad X_6 = 0, \quad X_7 = 0, \quad X_4 = 0, \quad X_1 = 12.25, \quad X_2 = 12, \quad X_3 = 14.75$$

(4) This is the true optimum.

Fig. 7.5-14 Result of third iteration of the Simplex method.

and the second row will be selected as pivot. The objective function will be improved by

$$\Delta G = (-3)\left(\frac{12}{1}\right) = -36 \tag{7.5-89}$$

The result of the third iteration is shown in Figure 7.5-14. The result of the third iteration indicates that the objective function now has all positive coefficients, i.e., it is now at the optimum by the tests given in Equation 7.5-63. The optimum solution is

$$\begin{aligned} X_1 &= 12.25 \\ X_2 &= 12 \\ X_3 &= 14.75 \\ X_4 &= 0 \end{aligned} \tag{7.5-90}$$

and the minimum value of the objective function is

$$G = 3.75 \tag{7.5-91}$$

7.6 SUMMARY

This chapter has presented the basic principles of optimal solutions of well-defined objective functions and processes. While the cases examined are relatively simple, they illustrate the *basic concepts and approaches to optimal control.* The same approaches may be applied to more complex systems.

At this point, the reader should have clearly in mind the basic approaches to optimal control of well-defined nonlinear objective functions subject to constraints, and optimal control of well-defined linear objective functions subject to linear inequalities. Chapter 8 will present approaches to " adapting " these techniques to practical situations where the physical process model and parameters are changing. Chapter 9 will present the evolutionary optimization (EVOP) approach for poorly defined processes, where the physical process models and the objective function may not even be defined. Chapter 9 will also present a combined feedforward and feedback approach for optimal control of well-defined processes, utilizing both the linear programming and EVOP techniques.

8

Adaptive Control of Moderately Well-Defined Processes— Steady-State

8.0 INTRODUCTION

Adaptive control is defined as the optimal control of processes in the presence of definable changes, i.e., *optimal control of moderately well-defined processes.* For moderately well-defined processes, the nature of the changes should be definable even though the exact magnitude of the changes may not be predictable. The optimal control techniques presented in Chapter 7 assumed that the physical models are perfect and would never be changed. This is unrealistic in all but theoretical situations. Some of the realities of change to which adaptive control techniques can be applied are the following:

Changes in the coefficients of the physical process model.
Changes in the structure of the physical process model caused by constraints.
Changes in the environment.
Noise in signal measurements.

Changes in the coefficients of the physical process model take place because their values, which are usually averages, may deviate from the actual instantaneous values. For example, the instantaneous material compositions may deviate from the average material compositions assumed in the linear programming coefficients. *Imperfections in the physical process model* may be caused by the inherent errors in using regression techniques to derive the coefficients from experimental data, and by physical wear and tear of the plant and equipment. *The structure of the physical process model* may be

245

changed by the introduction of constraints on control variables and state variables. As a given control variable exceeds the constraint limit, the control variable is replaced by a constant equal to the constraint limit, thus eliminating that variable from the objective function and the physical process model. Also, constraints on state and control variables reduce the number of independent control variables. *Changes in the environment* refer to uncontrolled variables which act on the process as disturbances, e.g., ambient temperature, humidity, and cooling water temperature. *Adapting to noise in signal measurement* is outside the scope of this book. Using analog resistance-capacitance filters for each pair of signal wires is one of the most practical techniques for reducing noise. Other techniques include digital filtering by averaging a number of successive measurements of the same signal, and digital smoothing of these successive measurements. Section 3.5C discussed the function of digital accumulation and averaging for noise reduction.

This chapter will continue the discussion of feedforward optimal control techniques developed in Chapter 7, except that the physical process model will be updated to adapt to changes in the process coefficients. Next, feedback optimal control using an incremental physical process model to adapt to changes in the process coefficients will be introduced. Then, the problems of adapting to changes in the structure of the physical process model caused by constraints on control variables will be discussed. Finally, the chapter will conclude with a discussion on how to adapt to changes in the environment, i.e., to the effects of the uncontrolled variables.

8.1 ADAPTING TO CHANGES IN THE COEFFICIENTS OF THE PHYSICAL PROCESS MODEL

Adaptive control in its simplest form can be applied when only the coefficients of the physical process model change and other factors such as the structure of the model, etc., are either perfect or unchanged. This section will first show why the coefficients of the physical process change, then indicate the effect of such changes, and finally develop an on-line experimental technique to update the coefficients. An example from Section 7.2 will be used to illustrate the updating technique.

A. Why Physical Process Model Coefficients Change

Section 7.3 developed the general form of the physical process model with n control variables and m state variables. From Equation 7.3-6, the general form for a linear physical process model is

$$Y_j = \sum_{i=1}^{n} A_{ji} X_i \qquad (8.1\text{-}1)$$

Changes in physical process model coefficients occur when the operating point exceeds the assumed linear region, e.g., while the physical process model is linear at full load, it may not be so at half-load. Thus a different set of coefficients may have to be developed for the operation at half-load. Changes in product mix and hysteresis effects after an upset in the process may also cause the coefficients to change. Other factors affecting the physical process model coefficients include the effects of a new batch of raw materials with different compositions and changes in process equipment characteristics due to temperature, wear, and other effects.

B. Need to Adapt Feedforward Optimal Control to Changing Coefficients

Consider the steady-state physical process model with one control variable and one state variable, as shown in Figure 8.1-1. Let the following symbols be assigned:

Y = state variable.
X = control variable.
A_0, A_1 = *actual* values of the coefficients measured in the plant.
A'_0, A'_1 = *average* values of the coefficients used in the model.
$\overline{Y}$ = desired operating point of the state variable.

From Section 7.1, the physical process model using the actual values of the coefficients is

$$Y = A_0 + A_1 X \tag{8.1-2}$$

The objective function to be minimized is the square of the difference between $\overline{Y}$ and Y:

$$F = (\overline{Y} - Y)^2 \tag{8.1-3}$$

Substituting Equation 8.1-2 into Equation 8.1-3, the objective function in G form is

$$G = (\overline{Y} - A_0 - A_1 X)^2 \tag{8.1-4}$$

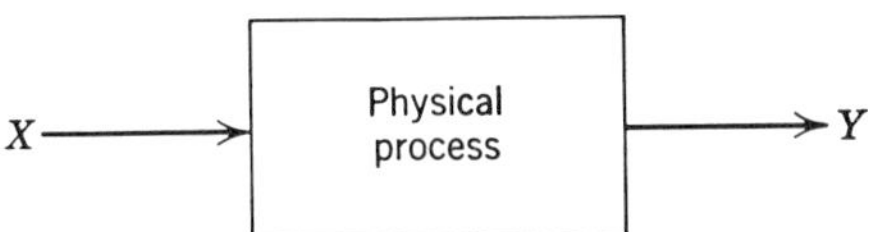

Physical process model $Y = A_0 + A_1 X$

Fig. 8.1-1   Physical process model with one control variable and one state variable.

Also, from Section 7.1, the condition of optimum is

$$\frac{dG}{dX} = 0 \qquad (8.1\text{-}5)$$

Solving Equation 8.1-5, the value of X at the optimum is

$$X = \frac{(\bar{Y} - A_0)}{A_1} \qquad (8.1\text{-}6)$$

Equation 8.1-6 is the optimal control equation for minimizing G. *The key to success in applying the optimal control equation is the accuracy of the coefficients A_0 and A_1.* If A_0 and A_1 can be determined accurately, or if the variations of A_0 and A_1 are small, then Equation 8.1-6 can be successfully applied. Otherwise Equation 8.1-6 will not yield a good solution for optimal control.

If, however, the *average* values of the coefficients A_0' and A_1' are used in the physical process model, the optimal solution is

$$X = \frac{(\bar{Y} - A_0')}{A_1'} \qquad (8.1\text{-}7)$$

By setting the control variable X according to Equation 8.1-7, the resulting state variable of the process is

$$Y = A_0 + A_1\left(\frac{\bar{Y} - A_0'}{A_1'}\right) \qquad (8.1\text{-}8)$$

$$Y = \bar{Y}\left(\frac{A_1}{A_1'}\right) + A_0\left[1 - \left(\frac{A_1}{A_1'}\right)\left(\frac{A_0'}{A_0}\right)\right] \qquad (8.1\text{-}9)$$

From Equation 8.1-9 it is obvious that generally

$$Y \neq \bar{Y} \qquad (8.1\text{-}10)$$

The only time $Y = \bar{Y}$ is when the *actual* and *average* values of the coefficients are equal:

$$A_0 = A_0' \quad \text{and} \quad A_1 = A_1' \qquad (8.1\text{-}11)$$

The error introduced into Y by the physical process model depends on the ratios of A_0 to A_0' and A_1 to A_1'.

Let the following values be assigned to the coefficients in the example:

$$\begin{aligned} A_0 &= 5 \\ A_0' &= 5.5 \\ A_1 &= 50 \\ A_1' &= 45 \end{aligned} \qquad (8.1\text{-}12)$$

and the desired operating point is

$$\overline{Y} = 50 \tag{8.1-13}$$

From Equation 8.1-7 the optimum value of X is

$$X = \frac{(\overline{Y} - A_0')}{A_1'} = 0.988 \tag{8.1-14}$$

By use of this value of X, the actual value of Y obtained from Equation 8.1-9 is

$$Y = 1.11(\overline{Y}) - 1.12 = 54.34 \tag{8.1-15}$$

Thus a 10% variation in the value of the coefficients resulted in an 8.7% error from the optimal value of the state variable Y.

From the example it can be seen that open-loop feedforward optimal control will have inherent errors if the *actual* values of the coefficients in the plant differ from the *average* values used. A practical compromise is to specify a range of allowable error for the state variable, e.g.,

$$Y = 50 \pm 2 \tag{8.1-16}$$

Then the values of the coefficients used in the physical process model must be analyzed to determine if the variations from the actual values will produce results within the range of the allowable error.

C. Adapting Feedforward Optimal Control to Changing Coefficients

Section 8.1B showed that the *ratios of the actual and average values of the physical process model coefficients* provide a measure of the errors introduced in the optimal control equation. The problem of improving the accuracy of feedforward optimal control reduces to the problem of improving the accuracy of the physical model coefficients. The process computer control system can be used to improve the accuracy by updating the physical process model coefficients based on experimental measurements. The regression analysis techniques discussed in Chapter 4 can be used to develop the coefficients in an on-line fashion.

Figure 8.1-2 shows that the control and state variables are measured to provide data for updating the physical process model coefficients. If the *actual* values are within a specified percentage of the *average* values of the coefficients used, then the errors introduced are within the specified accuracy and updating is not required. However, if the allowable error is exceeded, a new set of optimal control equations can be developed by using the *actual* values of the coefficients. For a linear physical process model with quadratic

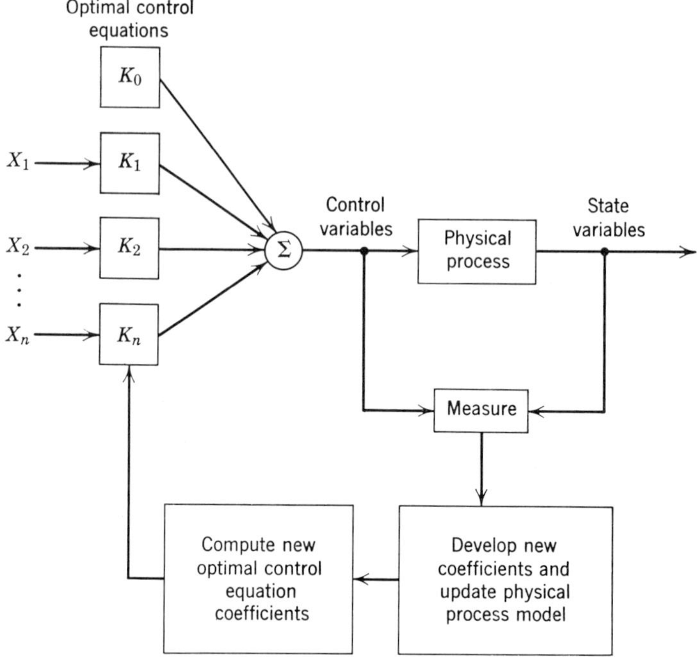

Fig. 8.1-2 Adapting feedforward optimal control to changing process model coefficients by updating physical model experimentally.

objective function the development of the new optimal control equation involves a matrix inversion.

As a precaution against noise or other uncontrolled disturbances causing unwarranted large changes, the *average* and *actual* values of the coefficients may be averaged to obtain the new optimal control coefficients. The time required to develop a new set of coefficients and compute the new optimal control equations must be matched to the rate of change of the process, otherwise the *updated model* will become an *outdated model* instead.

D. Example of Adapting Feedforward Optimal Control to Changing Physical Process Model

As an example of adapting feedforward optimal control to a changing physical process model, consider the two control variable case discussed in Section 7.2D. The physical process model defined in Equations 7.2-26 and 7.2-27 is

$$Y_1 = A_1 + B_1 X_1 + C_1 X_2 \qquad (8.1\text{-}17)$$

$$Y_2 = A_2 + B_2 X_1 + C_2 X_2 \qquad (8.1\text{-}18)$$

The objective function defined in Equation 7.2-28 is

$$F = \psi_1(\overline{Y}_1 - Y_1)^2 + \psi_2(\overline{Y}_2 - Y_2)^2 + \psi_3(\overline{X}_1 - X_1)^2 + \psi_4(\overline{X}_2 - X_2)^2$$

$$(8.1\text{-}19)$$

Let the values of (X_1, X_2, Y_1, Y_2) at time t_p be defined as $(X_1^p, X_2^p, Y_1^p, Y_2^p)$. Thus the objective function at time t_p is

$$F^p = \psi_1(\overline{Y}_1 - Y_1^p)^2 + \psi_2(\overline{Y}_2 - Y_2^p)^2 + \psi_3(\overline{X}_1 - X_1^p)^2 + \psi_4(\overline{X}_2 - X_2^p)^2$$

$$(8.1\text{-}20)$$

By sampling the values of (X_1, X_2, Y_1, Y_2) at intervals of time t_{p-q+1}, $t_{p-q+2}, \ldots, t_{p-1}, t_p$, a table of q sets of values may be stored in the memory of the computer process control system, as shown in Table 8.1-1. The q sets

Table 8.1-1 q Sets of Control and State Variables and Objective Function

Entry No.	X_1	X_2	Y_1	Y_2	F_{actual}
1	X_1^p	X_2^p	Y_1^p	Y_2^p	F^p
2	X_1^{p-1}	X_2^{p-1}	Y_1^{p-1}	Y_2^{p-1}	F^{p-1}
3	X_1^{p-2}	X_2^{p-2}	Y_1^{p-2}	Y_2^{p-2}	F^{p-2}
$\vdots$	$\vdots$	$\vdots$	$\vdots$	$\vdots$	$\vdots$
$q-1$	X_1^{p-q+2}	X_2^{p-q+2}	Y_1^{p-q+2}	Y_2^{p-q+2}	F^{p-q+2}
q	X_1^{p-q+1}	X_2^{p-q+1}	Y_1^{p-q+1}	Y_2^{p-q+1}	F^{p-q+1}

of values are always the most recent ones. In other words, the table is continually updated by deleting the last item and adding the most recent one at the top of the list. The rest of the entries are in effect "pushed down." Thus at the time t_{p+1}, the first entry is F_{actual}^{p+1}, the second entry is F_{actual}^p, and the last entry is $F_{\text{actual}}^{p-q+2}$.

By comparing the values of the actual F_{actual} against the predicted $F_{\text{predicted}}$ at each interval of time, a decision can be made as to whether the physical process model is valid. Given a positive Δ, if

$$\Delta \geq |F_{\text{predicted}}^p - F_{\text{actual}}^p| \qquad (8.1\text{-}21)$$

The process is considered to be *adequately controlled,* and there is no need to update the physical process model. If, however,

$$\Delta < |F_{\text{predicted}}^p - F_{\text{actual}}^p| \qquad (8.1\text{-}22)$$

the process is considered to be *poorly controlled.* The physical process model

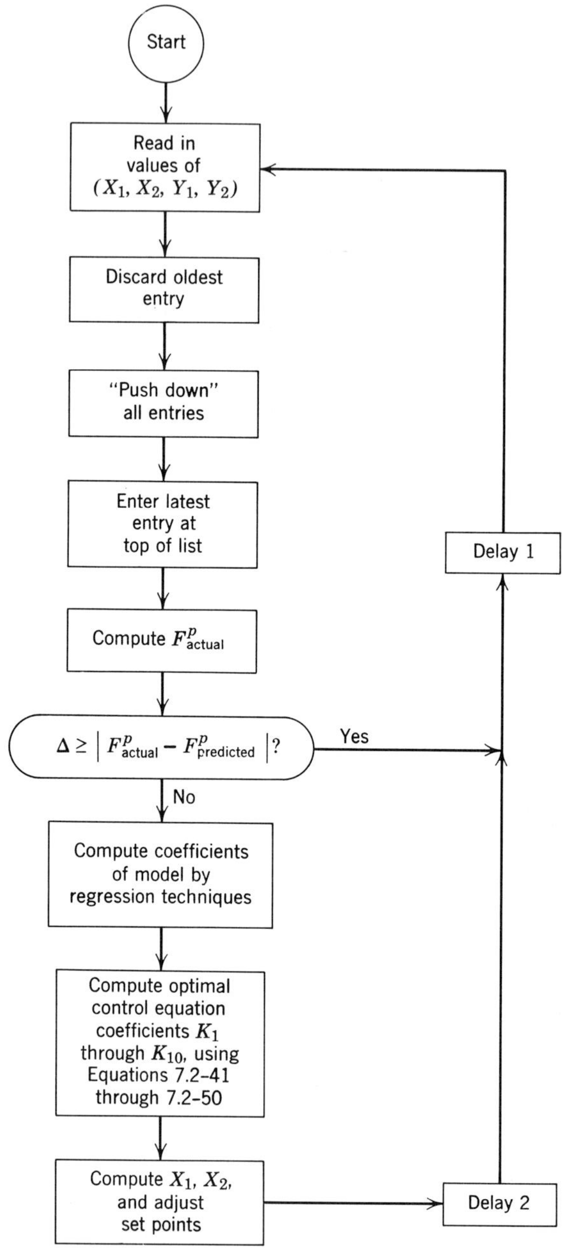

Fig. 8.1-3   Flow chart of adaptive feedforward optimal control.

252

coefficients must then be updated to improve the accuracy of the model, i.e., to further reduce the value F_{actual}^p.

Table 8.1-1 can be used to provide the data for a regression analysis to solve for the coefficients $(A_1, B_1, C_1, A_2, B_2, C_2)$. The use of q sets of data from successive intervals rather than a single set of data from the most recent interval will develop a smoother set of coefficients because the effect of noise is reduced.

The overall flow chart for the adaptive feedforward optimal control procedure is shown in Figure 8.1-3. The set (X_1, X_2, Y_1, Y_2) is read into the computer process control system and is entered into the top of the table after the oldest set of values is discarded and the remaining values are "pushed down." A test is made to determine if the coefficients of the physical process model are sufficiently accurate. If they are, the operation is delayed before it is repeated. If the test should indicate that the coefficients are not sufficiently accurate, a regression model is derived by using all the data stored in the table. The new coefficients will be used to derive the new optimal control equation, as shown in Equations 7.2-36 through 7.2-50. The control variables X_1 and X_2 are adjusted for optimal F. An additional delay is allowed for the process to settle before the procedure is repeated again.

Several refinements can be made to the procedure shown in Figure 8.1-3. Some form of signal filtering can be applied to F_{actual} before the comparison shown in Equation 8.1-22 is made. For example, a simple averaging of the several past values of F_{actual} may be made. Another alternate is to average the several past values of Δ computed by Equation 8.1-22. This will reduce the effect of random fluctuations of the process signal readings. However, if the errors of the coefficients of the physical process model result from a slow drift, a weighted regression analysis may be used. Weighting factors can be introduced to the expression of the error to be minimized so that the effect of older data is reduced. The result will be a set of coefficients which would depend more strongly on the more recent measurements.

8.2 FEEDBACK OPTIMAL CONTROL USING AN INCREMENTAL PHYSICAL PROCESS MODEL

Section 8.1 discussed how the physical process model coefficients may be updated to adapt to changes in the coefficients. This section will describe an alternative method of adapting to changes in the physical process model co-efficients. The technique is to *use a feedback approach together with an incremental physical process model*. The physical process models developed in Chapter 4 and used in Chapter 7 relate the control variables to the state variables. Incremental physical process models, however, *relate the changes in the control variables to the changes in the state variables.* The feedback approach used in this chapter takes into account the deviations of the actual values of the coefficients from the average values.

This section will first introduce the feedback optimal control technique for one control variable and one state variable, and then extend this to two control variables and two state variables. The general case which may be derived by extension will not be developed.

A. Feedback Optimal Control Using an Incremental Physical Process Model with One Control Variable and One State Variable

A different approach to optimal control is to apply *feedback by using the incremental physical process model.* In contrast to the open-loop feedforward approach, this approach does not require the coefficients of the physical process model to be accurate for successful optimal control. Let the following additional symbols be assigned:

t_p = an instant of time.

t_{p+1} = an instant of time sufficiently long after t_p so that the process will settle into steady-state after an adjustment.

Y^p, X^p, G^p = values as previously defined at t_p.

Y^{p+1}, X^{p+1}, G^{p+1} = values as previously defined at t_{p+1}.

The objective function to be minimized is

$$G^p = (\overline{Y} - Y^p)^2 \qquad (8.2\text{-}1)$$

and the physical process model is

$$Y = A_0 + A_1 X \qquad (8.2\text{-}2)$$

If $Y^p = \overline{Y}$, then the objective function will be minimum, and $G^p = 0$. However, if $G^p \neq 0$, then the control variable X^p can be adjusted to minimize G^{p+1}. The incremental change to be made to the control variable is

$$\Delta X = X^{p+1} - X^p \qquad (8.2\text{-}3)$$

This change in the control variable will result in a corresponding change in the state variable Y^{p+1}, provided sufficient time has elapsed for the process to settle into steady-state condition. Thus

$$\Delta Y = Y^{p+1} - Y^p \qquad (8.2\text{-}4)$$

At t_{p+1}, the objective function is

$$G^{p+1} = (\overline{Y} - Y^{p+1})^2 \qquad (8.2\text{-}5)$$

$$G^{p+1} = (\overline{Y} - Y^p - \Delta Y)^2 \qquad (8.2\text{-}6)$$

The objective function at t_{p+1} can be minimized to solve for the required change in the control variable ΔX. The *condition of optimum for* G^{p+1} *is*

$$\frac{\partial G^{p+1}}{\partial(\Delta X)} = 2(\overline{Y} - Y^p - \Delta Y)\left[\frac{-\partial(\Delta Y)}{\partial(\Delta X)}\right] = 0 \tag{8.2-7}$$

The *incremental physical process model* may be derived by subtracting Equation 8.2-9 from Equation 8.2-8:

$$Y^{p+1} = A_0' + A_1' X^{p+1} \tag{8.2-8}$$

$$Y^p = A_0' + A_1' X^p \tag{8.2-9}$$

$$\therefore \quad (\Delta Y) = A_1'(\Delta X) \tag{8.2-10}$$

From Equation 8.2-10

$$\frac{\partial(\Delta Y)}{\partial(\Delta X)} = A_1' \tag{8.2-11}$$

By substituting Equations 8.2-10 and 8.2-11 into Equation 8.2-7,

$$\frac{\partial G^{p+1}}{\partial(\Delta X)} = 2[\overline{Y} - Y^p - A_1'(\Delta X)][-A_1'] = 0 \tag{8.2-12}$$

By solving Equation 8.2-12, *the optimal control equation* is

$$\Delta X = \frac{(\overline{Y} - Y^p)}{A_1'} \tag{8.2-13}$$

Equation 8.2-13 may be applied iteratively to minimize the objective function G^{p+1}, G^{p+2}, etc., at t_{p+1}, t_{p+2}, etc. Table 8.2-1 shows the step-by-

Table 8.2-1   Iterative Solution for Feedback Optimal Control Using an Incremental Physical Process Model

Time	Y^p *	$(\overline{Y} - Y^p)$	$\Delta X = \dfrac{(\overline{Y} - Y^p)}{A_1'}$	X^p	$X^{p+1} = X^p + \Delta X$	G^p
t_p	54.4	-4.4	-0.0978	0.988	0.8902	19.3
t_{p+1}	49.5	0.5	0.011	0.8902	0.9012	0.25
t_{p+2}	50.06	-0.06	-0.00134	0.9012	0.90086	0.0036
$\vdots$	$\vdots$	$\vdots$	$\vdots$	$\vdots$	$\vdots$	$\vdots$
t_{p+q}	50.0	0	0	0.9	0.9	0

* Y^p should actually be measured from the process, i.e., through feedback. For this example, however, Y^p is simulated by using Equation 8.2-2, with $A_0 = 5$ and $A_1 = 50$.

step iterative solution for the optimal ΔX. Let the starting values at the pth instant of time for the example be

$$X^p = 0.988$$
$$Y^p = 54.4 \qquad (8.2\text{-}14)$$

and the desired operating point is

$$\overline{Y} = 50 \qquad (8.2\text{-}15)$$

The *average* values of the coefficients are

$$A_0' = 5$$
$$A_1' = 45 \qquad (8.2\text{-}16)$$

The values of Y^p, Y^{p+1}, etc., ..., should actually be measured from the process. For the purpose of this example, however, the actual values of Y^p, Y^{p+1}, ..., are simulated by using the *actual* values of the coefficients $A_0 = 5$ and $A_1 = 50$, and the physical process model given by Equation 8.2-1. Figure 8.2-1 shows the diagram for feedback optimal control to minimize the objective function

$$G^p = (\overline{Y} - Y^p)^2 \qquad (8.2\text{-}17)$$

The diagram shows that the measured Y is sampled and held for one interval, so Y^p is compared with $\overline{Y}$ at t_p. The difference is used to compute ΔX by using Equation 8.2-13. The computed ΔX is then added to the value of the control variable X^p to obtain X^{p+1}. The new control variable X^{p+1} is sampled and held for one interval of time to be used for the next period.

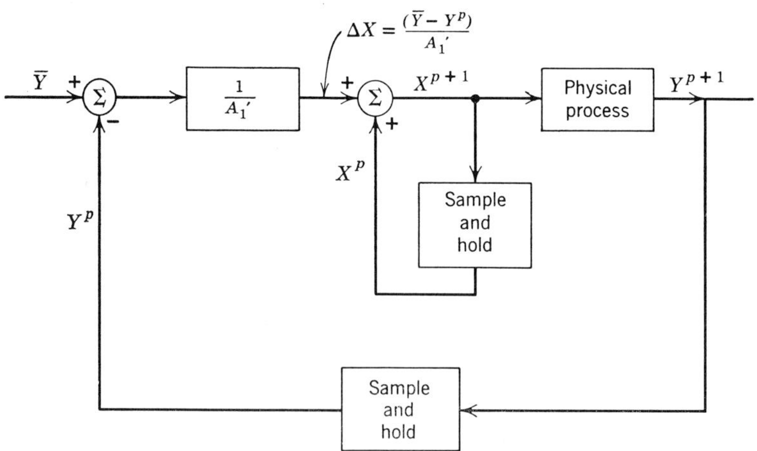

Fig. 8.2-1   Feedback optimal control using incremental physical process model to minimize $G^p = (\overline{Y} - Y^p)^2$.

In summary, the feedback optimal control approach using an incremental physical process model can be developed by using the following three-step procedure:

STEP 1. Measure the state variable and compare it with the desired value. The difference is used to compute the change in the control variable to minimize the objective function at the next time interval.

STEP 2. Use the optimal control equation to compute the change in the control variable ΔX, as shown in Equation 8.2-13.

STEP 3. Adjust the control variable X by ΔX. Repeat Steps 1, 2, 3 to minimize the objective function.

This procedure will produce a far superior result compared to the feed-forward optimal control approach even when the *actual* coefficients of the process may be quite different from the *average* values used in the incremental physical process model.

B. Feedback Optimal Control Using Incremental Physical Process Model with Two Control Variables and Two State Variables

The previous derivation can be extended to two control variables and two state variables. Let the following symbols be assigned:

Y_1^p, Y_2^p = state variables at the pth instant of time.
X_1^p, X_2^p = control variables at the pth instant of time.
A_{11}, A_{12}, A_{21}, A_{22} = coefficients of the physical process model.
$\overline{Y}_1$, $\overline{Y}_2$ = desired operating points for Y_1 and Y_2.
ψ_1, ψ_2 = coefficients of the objective function.
t_p, t_{p+1}, ... = instants of time spaced sufficiently apart so the process has time to settle after an adjustment.

Let

$$\Delta X_1 = X_1^{p+1} - X_1^p \tag{8.2-18}$$

$$\Delta X_2 = X_2^{p+1} - X_2^p \tag{8.2-19}$$

$$\Delta Y_1 = Y_1^{p+1} - Y_1^p \tag{8.2-20}$$

$$\Delta Y_2 = Y_2^{p+1} - Y_2^p \tag{8.2-21}$$

Let the physical process model be given by

$$Y_1^p = A_{11}X_1^p + A_{12}X_2^p \tag{8.2-22}$$

$$Y_1^{p+1} = A_{11}X_1^{p+1} + A_{12}X_2^{p+1} \tag{8.2-23}$$

$$Y_2^p = A_{21}X_1^p + A_{22}X_2^p \tag{8.2-24}$$

$$Y_2^{p+1} = A_{21}X_1^{p+1} + A_{22}X_2^{p+1} \tag{8.2-25}$$

The incremental physical process model may be derived by subtracting Equation 8.2-22 from 8.2-23 and Equation 8.2-24 from 8.2-25:

$$\therefore \quad (\Delta Y_1) = A_{11}(\Delta X_1) + A_{12}(\Delta X_2) \tag{8.2-26}$$

and

$$(\Delta Y_2) = A_{21}(\Delta X_1) + A_{22}(\Delta X_2) \tag{8.2-27}$$

The objective function to be minimized is

$$G^{p+1} = \psi_1(\overline{Y}_1 - Y_1^{p+1})^2 + \psi_2(\overline{Y}_2 - Y_2^{p+1})^2 \tag{8.2-28}$$

By substituting Equations 8.2-23 and 8.2-25 into Equation 8.2-28:

$$\begin{aligned}
G^{p+1} = {}&\psi_1(\overline{Y}_1 - A_{11}X_1^{p+1} - A_{12}X_2^{p+1})^2 \\
&+ \psi_2(\overline{Y}_2 - A_{21}X_1^{p+1} - A_{22}X_2^{p+1})^2
\end{aligned} \tag{8.2-29}$$

By substituting Equations 8.2-18 and 8.2-19 into Equation 8.2-29:

$$\begin{aligned}
G^{p+1} = {}&\psi_1(\overline{Y}_1 - A_{11}X_1^p - A_{12}X_2^p - A_{11}\Delta X_1^p - A_{12}\,\Delta X_2^p)^2 \\
&+ \psi_2(\overline{Y}_2 - A_{21}X_1^p - A_{22}X_2^p - A_{21}\Delta X_1^p - A_{22}\,\Delta X_2^p)^2
\end{aligned} \tag{8.2-30}$$

By substituting Equations 8.2-22 and 8.2-24 into Equation 8.2-30:

$$\begin{aligned}
G^{p+1} = {}&\psi_1(\overline{Y}_1 - Y_1^p - A_{11}\,\Delta X_1^p - A_{12}\,\Delta X_2^p)^2 \\
&+ \psi_2(\overline{Y}_2 - Y_2^p - A_{21}\,\Delta X_1^p - A_{22}\,\Delta X_2^p)^2
\end{aligned} \tag{8.2-31}$$

By substituting Equations 8.2-26 and 8.2-27 into Equation 8.2-31:

$$G^{p+1} = \psi_1(\overline{Y}_1 - Y_1^p - \Delta Y_1)^2 + \psi_2(\overline{Y}_2 - Y_2^p - \Delta Y_1)^2 \tag{8.2-32}$$

The conditions for optimum for minimizing the objective function G^{p+1} at t_{p+1} are

$$\begin{aligned}
\frac{\partial G^{p+1}}{\partial(\Delta X_1)} = {}&2\psi_1[\overline{Y}_1 - Y_1^p - A_{11}(\Delta X_1) - A_{12}(\Delta X_2)][-A_{11}] \\
&+ 2\psi_2[\overline{Y}_2 - Y_2^p - A_{21}(\Delta X_1) - A_{22}(\Delta X_2)][-A_{21}] = 0
\end{aligned} \tag{8.2-33}$$

$$\begin{aligned}
\frac{\partial G^{p+1}}{\partial(\Delta X_2)} = {}&2\psi_1[\overline{Y}_1 - Y_1^p - A_{11}(\Delta X_1) - A_{12}(\Delta X_2)][-A_{12}] \\
&+ 2\psi_2[\overline{Y}_2 - Y_2^p - A_{21}(\Delta X_1) - A_{22}(\Delta X_2)][-A_{22}] = 0
\end{aligned} \tag{8.2-34}$$

By dividing Equations 8.2-33 and 8.2-34 by $2\psi_1$ and collecting terms for ΔX_1 and ΔX_2,

$$\begin{aligned}
(\Delta X_1)&\left[A_{11}{}^2 + \left(\frac{\psi_2}{\psi_1}\right)A_{21}{}^2\right] + (\Delta X_2)\left[A_{11}A_{12} + \left(\frac{\psi_2}{\psi_1}\right)A_{21}A_{22}\right] \\
&= (\overline{Y}_1 - Y_1{}^p)A_{11} + \left(\frac{\psi_2}{\psi_1}\right)(\overline{Y}_2 - Y_2{}^p)A_{21}
\end{aligned} \tag{8.2-35}$$

$$(\Delta X_1)\left[A_{11}A_{12} + \left(\frac{\psi_2}{\psi_1}\right)A_{21}A_{22}\right] + (\Delta X_2)\left[A_{12}^2 + \left(\frac{\psi_2}{\psi_1}\right)A_{22}^2\right]$$

$$= (\overline{Y}_1 - Y_1^p)A_{12} + \left(\frac{\psi_2}{\psi_1}\right)(\overline{Y}_2 - Y_2^p)A_{22} \qquad (8.2\text{-}36)$$

Equations 8.2-35 and 8.2-36 are the optimal control equations which can be used to solve for the ΔX_1 and ΔX_2 that will minimize G^{p+1} at t_{p+1}. In matrix notation Equations 8.2-35 and 8.2-36 become

$$\begin{pmatrix} \alpha_{11} & \alpha_{12} \\ \alpha_{21} & \alpha_{22} \end{pmatrix} \cdot \begin{pmatrix} (\Delta X_1) \\ (\Delta X_2) \end{pmatrix} = \begin{pmatrix} \beta_1 \\ \beta_2 \end{pmatrix} \qquad (8.2\text{-}37)$$

where

$$\alpha_{11} = A_{11}^2 + \left(\frac{\psi_2}{\psi_1}\right)A_{21}^2 \qquad (8.2\text{-}38)$$

$$\alpha_{12} = A_{11}A_{12} + \left(\frac{\psi_2}{\psi_1}\right)A_{21}A_{22} \qquad (8.2\text{-}39)$$

$$\alpha_{21} = A_{11}A_{12} + \left(\frac{\psi_2}{\psi_1}\right)A_{21}A_{22} \qquad (8.2\text{-}40)$$

$$\alpha_{22} = A_{12}^2 + \left(\frac{\psi_2}{\psi_1}\right)A_{22}^2 \qquad (8.2\text{-}41)$$

$$\beta_1 = (\overline{Y}_1 - Y_1^p)A_{11} + \left(\frac{\psi_2}{\psi_1}\right)(\overline{Y}_2 - Y_2^p)A_{21} \qquad (8.2\text{-}42)$$

$$\beta_2 = (\overline{Y}_1 - Y_1^p)A_{12} + \left(\frac{\psi_2}{\psi_1}\right)(\overline{Y}_2 - Y_2^p)A_{22} \qquad (8.2\text{-}43)$$

The optimal changes to the control variables to minimize G^{p+1} can be obtained by solving Equation 8.2-37. Thus

$$(\Delta X_1) = \frac{(\alpha_{12}\beta_2 - \alpha_{22}\beta_1)}{(\alpha_{21}\alpha_{12} - \alpha_{11}\alpha_{22})} \qquad (8.2\text{-}44)$$

$$(\Delta X_2) = \frac{(\alpha_{21}\beta_1 - \alpha_{11}\beta_2)}{(\alpha_{21}\alpha_{12} - \alpha_{11}\alpha_{22})} \qquad (8.2\text{-}45)$$

The values of ΔX_1 and ΔX_2 may be applied to adjust the control variables X_1^p and X_2^p to minimize G^{p+1} at t_{p+1}. The procedure may be applied iteratively to minimize G^{p+2}, G^{p+3}, etc., at t_{p+2}, t_{p+3}, etc. The block diagram for feedback optimal control with two control variables and two state variables is shown in Figure 8.2-2.

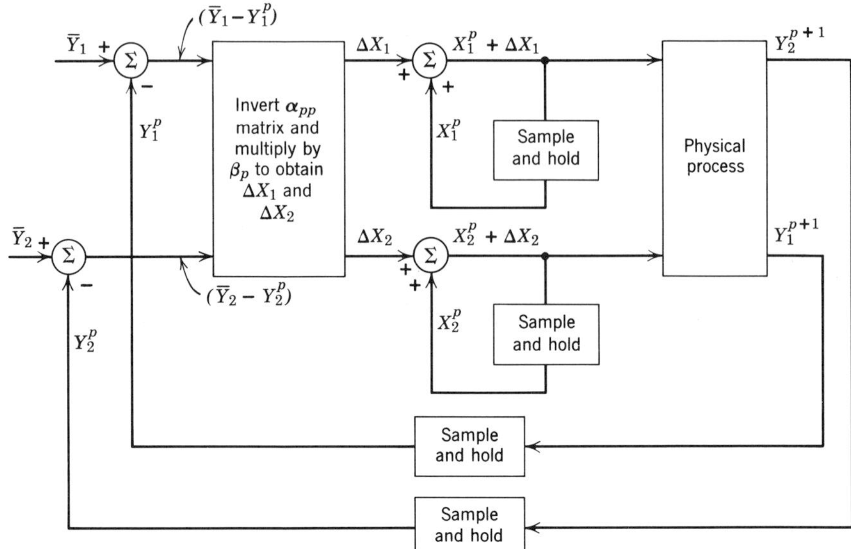

Fig. 8.2-2 Feedback optimal control using incremental physical process model with two control variables and two state variables.

An alternative method of deriving the optimal control equations is to state the conditions of optimum, Equations 8.2-33 and 8.2-34, in matrix notation:

$$\begin{pmatrix} 2\psi_1 A_{11} & 2\psi_2 A_{21} \\ 2\psi_1 A_{12} & 2\psi_2 A_{22} \end{pmatrix} \cdot \left[\begin{pmatrix} (\overline{Y}_1 - Y_1^p) \\ (\overline{Y}_2 - Y_2^p) \end{pmatrix} - \begin{pmatrix} A_{11} & A_{12} \\ A_{21} & A_{22} \end{pmatrix} \cdot \begin{pmatrix} \Delta X_1 \\ \Delta X_2 \end{pmatrix} \right] = 0 \quad (8.2\text{-}46)$$

Equation 8.2-46 may be multiplied out to verify that it indeed restates Equations 8.2-33 and 8.2-34. If both sides of Equation 8.2-46 are premultiplied by the inverse of

$$\begin{pmatrix} 2\psi_1 A_{11} & 2\psi_2 A_{21} \\ 2\psi_1 A_{12} & 2\psi_2 A_{22} \end{pmatrix} \quad (8.2\text{-}47)$$

The result is

$$\begin{pmatrix} 1 & 0 \\ 0 & 1 \end{pmatrix} \cdot \left[\begin{pmatrix} (\overline{Y}_1 - Y_1^p) \\ (\overline{Y}_2 - Y_2^p) \end{pmatrix} - \begin{pmatrix} A_{11} & A_{12} \\ A_{21} & A_{22} \end{pmatrix} \cdot \begin{pmatrix} \Delta X_1 \\ \Delta X_2 \end{pmatrix} \right] = 0 \quad (8.2\text{-}48)$$

$$\therefore \quad \begin{pmatrix} (\overline{Y}_1 - Y_1^p) \\ (\overline{Y}_2 - Y_2^p) \end{pmatrix} - \begin{pmatrix} A_{11} & A_{12} \\ A_{21} & A_{22} \end{pmatrix} \cdot \begin{pmatrix} \Delta X_1 \\ \Delta X_2 \end{pmatrix} = 0 \quad (8.2\text{-}49)$$

The optimal changes in the control variables are given by:

$$\begin{pmatrix} \Delta X_1 \\ \Delta X_2 \end{pmatrix} = \frac{1}{A_{11}A_{22} - A_{12}A_{21}} \cdot \begin{pmatrix} A_{11} & -A_{21} \\ -A_{12} & A_{22} \end{pmatrix} \cdot \begin{pmatrix} (\overline{Y}_1 - Y_1^p) \\ (\overline{Y}_2 - Y_2^p) \end{pmatrix} \quad (8.2\text{-}50)$$

Equation 8.2-50 is valid only for the special *determined system* where the number of control variables equals the number of state variables, and the objective function specifies the desired state variables, $\overline{Y}_1$ and $\overline{Y}_2$ only.

8.3 ADAPTING TO CHANGES IN THE STRUCTURE OF THE PHYSICAL PROCESS MODEL DUE TO CONSTRAINTS

Chapters 6 and 7 pointed out that constraints on the limits of control variables and constraints on state and control variables may alter the structure of the physical process model and the objective function. Constraints on the limits of control variables specify high and low limits which should not be exceeded. If the optimal control equations should result in an optimal control variable which exceeds the constraint limit, that particular control variable will be replaced by a constant equal to the constraint limit. The physical process model and the objective function will be altered because of the constant replacing the control variable. Constraints on state and control variables may be regarded as constraints on control variables by substituting the physical process model equations into the constraints to eliminate the state variables. Nonlinear constraints modify the objective function to be optimized through the Lagrange multiplier technique, which was discussed in Section 7.4. Thus, constraints on state and control variables also alter the structure of the physical process model and objective function.

This section will first discuss the techniques used to adapt to changes in the structure of the physical process model and the objective function caused by constraints on the limits of control variables.

A. Adapting to the Constraint on the Limits of One Control Variable

The constraint on the limits of the control variable X_i is expressed as

$$R_{iL} \leq X_i \leq R_{iU} \tag{8.3-1}$$

where R_{iL} is the lower limit and R_{iU} is the upper limit. This deceptively simple problem can become very complex for a system with a large number of control variables. For example, when one control variable exceeds the constraint limit and is replaced by a constant, the other control variables may in turn exceed the constraint limits. Each time a control variable exceeds the constraint limits, the structure of the physical process model and the objective function are changed, and the problem must be solved again.

A simple case involving one state variable and one control variable will be presented to develop an intuitive feeling for the problem. The physical process model is given by

$$Y = A_0 + A_1 X \tag{8.3-2}$$

And the objective function to be minimized is

$$F = (\bar{Y} - Y)^2 \qquad (8.3\text{-}3)$$

Substituting Equation 8.3-2 into Equation 8.3-3, the objective function in G form is

$$G = (\bar{Y} - A_0 - A_1 X)^2 \qquad (8.3\text{-}4)$$

From Equation 8.1-6, the feedforward optimal control equation is

$$X = \frac{(\bar{Y} - A_0)}{A_1} \qquad (8.3\text{-}5)$$

Let the following constraints be imposed on the limits of the control variable.

$$X \geq 0 \qquad (8.3\text{-}6)$$

$$X \leq R_U \qquad (8.3\text{-}7)$$

One method of adapting the optimal control equation to the constraints given by Equations 8.3-6 and 8.3-7 is to make the optimal control equation valid only for a range of allowable values of the control variable X as specified by Equations 8.3-6 and 8.3-7. The allowable limits of $\bar{Y}$ can be obtained by substituting Equations 8.3-6 and 8.3-7 as equalities into Equation 8.3-5:

$$\therefore \quad \bar{Y} = A_0 \qquad \text{for } X = 0 \qquad (8.3\text{-}8)$$

$$\bar{Y} = A_0 + A_1 R_U \quad \text{for } X = R_U \qquad (8.3\text{-}9)$$

Thus the optimal control equation is

$$X = \frac{(\bar{Y} - A_0)}{A_1} \qquad (8.3\text{-}10)$$

subject to the constraints:

$$\bar{Y} \geq A_0 \qquad (8.3\text{-}11)$$

$$\bar{Y} \leq (A_0 + A_1 R_U) \qquad (8.3\text{-}12)$$

The desired values of $\bar{Y}$ may be compared against the constraints given by Equations 8.3-11 and 8.3-12 before the optimal control equation is used to solve for X. If the desired $\bar{Y}$ is less than A_0, the optimal X is equal to (0). If the desired $\bar{Y}$ is greater than $(A_0 + A_1 R_U)$, the optimal X is equal to R_U. Otherwise, the optimal X is given by Equation 8.3-10. This approach is adequate for feedforward control.

An alternative method of adapting the optimal control equation to the constraints of Equations 8.3-6 and 8.3-7 is to use the optimal control equation to solve directly for the optimal control variable X. Then the optimal X can

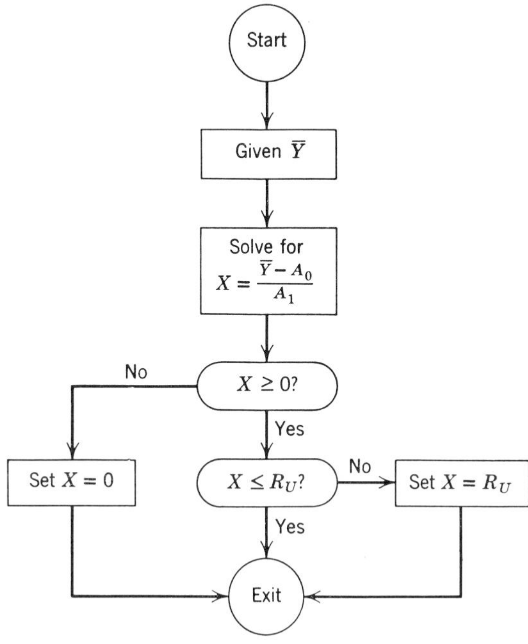

Fig. 8.3-1 Flow chart to minimize $(\overline{Y} - Y)^2$ subject to constraints on the limits of control variable X.

be compared against the limits set by the constraint. Any time the calculated optimal X exceeds the constraint limits, the control variable X is made equal to the value of the constraint limit. This approach is valid for both feedforward and feedback control. Figure 8.3-1 shows a flow chart for a procedure which assures that the optimal control variable X satisfies the constraints of Equations 8.3-6 and 8.3-7. Table 8.3-1 shows the table of possible conditions of the control variable X, the resulting state variable Y, and the objective function G.

Table 8.3-1 Possible Values of the State Variable Y and the Objective Function G Due to Constraints on the Limits of the Control Variable X

Conditions for X	Optimal X	Y	G
$X \leq 0$	0	A_0	$(\overline{Y} - A_0)^2$
$0 \leq X \leq R_U$	$\dfrac{(\overline{Y} - A_0)}{A_1}$	$\overline{Y}$	0
$X \geq R_U$	R_U	$(A_0 + A_1 R_U)$	$(\overline{Y} - A_0 - A_1 R_U)^2$

B. Adapting to Constraints on the Limits of Two Control Variables

Adapting optimal control to constraints on the limits of two control variables will use the feedback optimal control technique discussed in Section 8.2B. The symbols used will be the same as those in Section 8.2B unless otherwise specified. The objective function to be minimized is

$$G^p = \psi_1(\overline{Y}_1 - Y_1^p)^2 + \psi_2(\overline{Y}_2 - Y_2^p)^2 \tag{8.3-13}$$

and the incremental physical process model is

$$\Delta Y_1 = A_{11}(\Delta X_1) + A_{12}(\Delta X_2) \tag{8.3-14}$$

$$\Delta Y_2 = A_{21}(\Delta X_1) + A_{22}(\Delta X_2) \tag{8.3-15}$$

subject to the constraints on the limits of the control variables:

$$R_{1L} \leq X_1 \leq R_{1U} \tag{8.3-16}$$

$$R_{2L} \leq X_2 \leq R_{2U} \tag{8.3-17}$$

The optimal changes to the control variables ΔX_1 and ΔX_2 without constraints are given by Equation 8.2-50:

$$\begin{pmatrix} \Delta X_1 \\ \Delta X_2 \end{pmatrix} = \frac{1}{A_{11}A_{22} - A_{12}A_{21}} \cdot \begin{pmatrix} A_{11} & -A_{21} \\ -A_{12} & A_{22} \end{pmatrix} \cdot \begin{pmatrix} (\overline{Y}_1 - Y_1^p) \\ (\overline{Y}_2 - Y_2^p) \end{pmatrix} \tag{8.3-18}$$

The optimal changes to the control variables must be added to the values of the control variables at t_p to obtain the new values of the control variables at t_{p+1}:

$$X_1^{p+1} = X_1^p + (\Delta X_1) \tag{8.3-19}$$

$$X_2^{p+1} = X_2^p + (\Delta X_2) \tag{8.3-20}$$

With the introduction of the constraints given by Equations 8.3-16 and 8.3-17, several possibilities exist when the calculated X_1^{p+1} and X_2^{p+1} are compared against the limits, as shown in Figure 8.3-2. The flow chart for

X_2^{p+1}	X_1^{p+1}	
	X_1^{p+1} in limits	X_1^{p+1} out of limits
X_2^{p+1} in limits	Case 1	Case 2
X_2^{p+1} out of limits	Case 3	Case 4

Fig. 8.3-2   Decision table of calculated control variables compared with constraint limits.

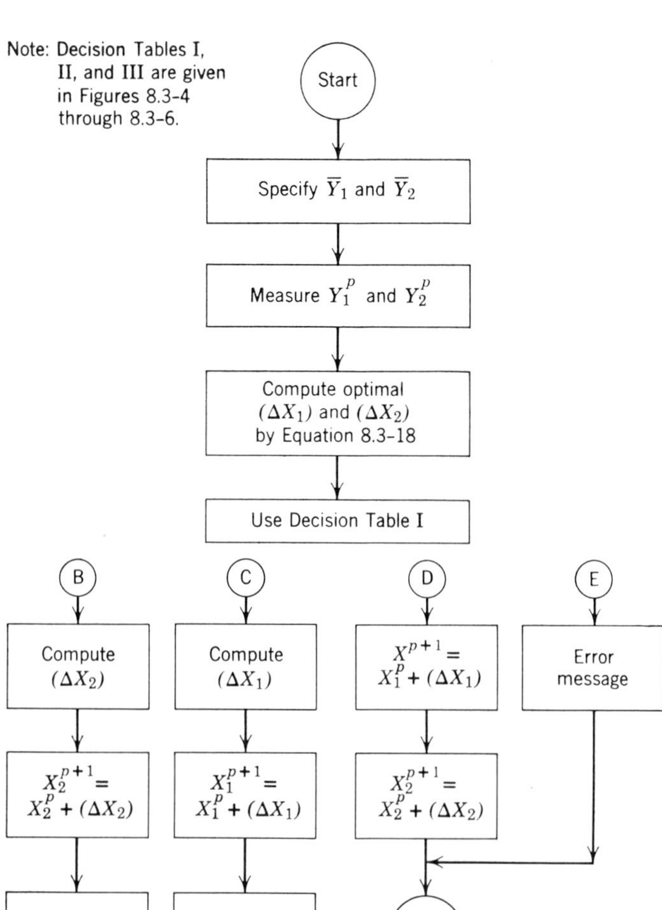

Fig. 8.3-3 Flow chart for adapting feedback optimal control to constraint on two control variables.

adapting feedback optimal control to constraints on two control variables is given in Figure 8.3-3. The truth table for the continuing actions in the flow chart are given in Figures 8.3-4 through 8.3-6.

From Figures 8.3-4 through 8.3-6, the following possibilities exist:

Case 1

Both X_1^{p+1} and X_2^{p+1} are within the constraint limits, and no action is required.

Note: $0 =$ false
$1 =$ true

Case	$X_1^{p+1} < R_{1L}$	$X_1^{p+1} > R_{1U}$	$X_2^{p+1} < R_{2L}$	$X_2^{p+1} > R_{2U}$	Action	Go To
1	0	0	0	0	None	D
3	0	0	0	1	$\Delta X_2 = R_{2U} - X_2^p$	C
3	0	0	1	0	$\Delta X_2 = R_{2L} - X_2^p$	C
3	0	0	1	1	Not possible	Error
2	0	1	0	0	$\Delta X_1 = R_{1U} - X_1^p$	B
4	0	1	0	1	$\Delta X_1 = R_{1U} - X_1^p$ $\Delta X_2 = R_{2U} - X_2^p$	D
4	0	1	1	0	$\Delta X_1 = R_{1U} - X_1^p$ $\Delta X_2 = R_{2L} - X_2^p$	D
4	0	1	1	1	Not possible	Error
2	1	0	0	0	$\Delta X_1 = R_{1L} - X_1^p$	B
4	1	0	0	1	$\Delta X_1 = R_{1L} - X_1^p$ $\Delta X_2 = R_{2U} - X_2^p$	D
4	1	0	1	0	$\Delta X_1 = R_{1L} - X_1^p$ $\Delta X_2 = R_{2L} - X_2^p$	D
4	1	0	1	1	Not possible	Error
2	1	1	0	0	Not possible	Error
4	1	1	0	1	Not possible	Error
4	1	1	1	0	Not possible	Error
4	1	1	1	1	Not possible	Error

Fig. 8.3-4 Decision Table I for Flow chart shown in Figure 8.3-3.

Note: 0 = false
1 = true

$X_2^{p+1} < R_{2L}$	$X_2^{p+1} > R_{2U}$	Action	Go To
0	0	None	D
0	1	$\Delta X_2 = R_{2U} - X_2^p$	D
1	0	$\Delta X_2 = R_{2L} - X_2^p$	D
1	1	Not possible	Error

Fig. 8.3-5 Decision Table II for flow chart shown in Figure 8.3-3.

Case 2

X_1^{p+1} is out of the constraint limits, but X_2^{p+1} is within the constraint limits. The constraint limit which was exceeded can be used to replace the new control variable setting X_1^{p+1}. ΔX_1 can be derived to satisfy this condition, i.e.,

$$\Delta X_1 = R_{1L} - X_1^p \tag{8.3-21}$$

or

$$\Delta X_1 = R_{1U} - X_1^p \tag{8.3-22}$$

Case 3

X_1^{p+1} is within the constraint limits, but X_2^{p+1} is out of the constraint limits. The constraint limit which was exceeded can be used to replace the new

Note: 0 = false
1 = true

$X_1^{p+1} < R_{1L}$	$X_1^{p+1} > R_{1U}$	Action	Go To
0	0	None	D
0	1	$\Delta X_1 = R_{1U} - X_1^p$	D
1	0	$\Delta X_1 = R_{1U} - X_1^p$	D
1	1	Not possible	Error

Fig. 8.3-6 Decision Table III for flow chart shown in Figure 8.3-3.

control variable setting X_2^{p+1}. ΔX_2 can be derived to satisfy this condition, i.e.:

$$\Delta X_2 = R_{2L} - X_2^p \tag{8.3-23}$$

or

$$\Delta X_2 = R_{2U} - X_2^p \tag{8.3-24}$$

Case 4

Both X_1^{p+1} and X_2^{p+1} are out of the constraint limits. The constraint limits which were exceeded can be used to replace the new control variable settings X_1^{p+1} and X_2^{p+1}; ΔX_1 and ΔX_2 can be derived as before to satisfy these conditions.

8.4   ADAPTING TO CHANGES IN THE ENVIRONMENT

Sections 8.1 through 8.3 describe how optimal control techniques may be adapted to changes in the physical process model coefficients and to changes in the structure of the model caused by the constraints on state and control variables. This section will discuss *adapting optimal control techniques to changes in the environment*, i.e., *the effects of uncontrolled variables.* Examples of uncontrolled variables are ambient temperature, barometric pressure, humidity, cooling water temperature, wind velocity, etc. In addition to the symbols previously assigned in this chapter, the following symbols will be assigned:

$$U = \text{uncontrolled variable.}$$

$$A_0, B_0 = \text{coefficients.}$$

The uncontrolled variable can be assumed to have its effect on the state variable added to the effect of the control variable, as shown in Figure

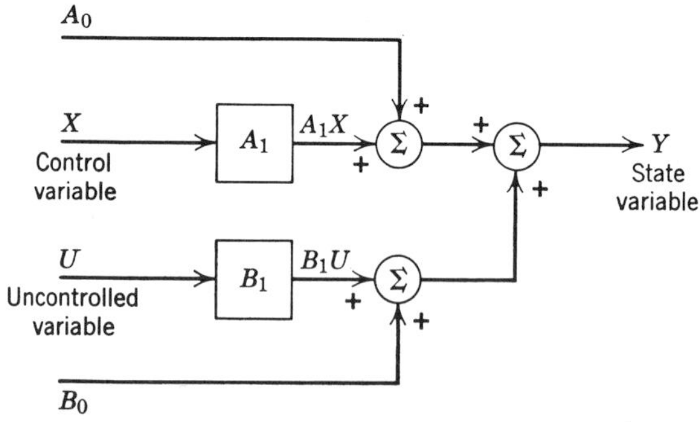

Fig. 8.4-1   Physical process model including the effect of an uncontrolled variable.

8.4-1. Thus the physical process model for one control variable and one state variable is

$$Y = A_0 + A_1 X + B_1 U + B_0 \qquad (8.4\text{-}1)$$

This section will use the simple one-variable case to show how feedforward optimal control can be adapted to the effect of the uncontrolled variable, then show how feedback optimal control can be adapted to the effect of the uncontrolled variable, and conclude with a discussion of the combined feedback and feedforward optimal control approach.

A. Adapting Feedforward Optimal Control to the Effects of Uncontrolled Variables

Feedforward optimal control techniques can be adapted to changes in the environment, i.e., the effects of uncontrolled variables; for example, the objective function to be minimized is

$$G = (\overline{Y} - Y)^2 \qquad (8.4\text{-}2)$$

where $\overline{Y}$ is the desired state variable.

Substituting the physical process model given by Equation 8.4-1 into the objective function:

$$G = (\overline{Y} - A_0 - A_1 X - B_0 - B_1 U)^2 \qquad (8.4\text{-}3)$$

The condition for optimum is

$$\frac{dG}{dX} = 2(\overline{Y} - A_0 - A_1 X - B_0 - B_1 U)(-A_1) = 0 \qquad (8.4\text{-}4)$$

Solving Equation 8.4-4, the optimal control variable is

$$X = \frac{(\overline{Y} - A_0 - B_0 - B_1 U)}{A_1} \qquad (8.4\text{-}5)$$

Thus the optimal control variable X is dependent on the uncontrolled variable U. One alternative is to estimate the uncontrolled variable. Let the *estimated* uncontrolled variable be

$$U = U_e \qquad (8.4\text{-}6)$$

Then U_e can be used to compute the optimal control variable X. Thus

$$X = \frac{(\overline{Y} - A_0 - B_0 - B_1 U_e)}{A_1} \qquad (8.4\text{-}7)$$

If the *actual* value of the uncontrolled variable is

$$U = U_T \tag{8.4-8}$$

The resulting state variable can be obtained by substituting Equations 8.4-7 and 8.4-8 into Equation 8.4-1:

$$Y = A_0 + A_1 \left[\frac{(\bar{Y} - A_0 - B_0 - B_1 U_e)}{(A_1)} \right] + B_0 + B_1 U_T \tag{8.4-9}$$

$$\therefore \quad Y = \bar{Y} + B_1(U_T - U_e) \tag{8.4-10}$$

and the objective function is equal to

$$G = [B_1(U_T - U_e)]^2 \tag{8.4-11}$$

From Equations 8.4-10 and 8.4-11, it is apparent that unless

$$U_T = U_e \tag{8.4-12}$$

Y will not equal $\bar{Y}$. The error is

$$\text{Error} = B_1(U_T - U_e) \tag{8.4-13}$$

The magnitude of the error will depend on the coefficient B_1 and the deviation of U_e from U_T. Thus feedforward optimal control can adapt satisfactorily to changes in the environment if the estimated uncontrolled variable U_e is within a reasonable range of the actual uncontrolled variable U_T.

B. Adapting Feedback Optimal Control to the Effects of the Uncontrolled Variables

In case feedforward optimal control cannot be adapted satisfactorily to changes in the environment, i.e., the estimated uncontrolled variable U_e deviates too much from the actual uncontrolled variable U_T, then feedback optimal control may be used. The effects of the uncontrolled variables need not be included in the incremental physical process model because these will be included in the measured state variable through feedback. Let the incremental physical process model be

$$\Delta Y = A_1(\Delta X) \tag{8.4-14}$$

Then the objective function to be minimized is

$$G^{p+1} = (\bar{Y} - Y^{p+1})^2 \tag{8.4-15}$$

Since

$$Y^{p+1} = Y^p + (\Delta Y) \tag{8.4-16}$$

$$G^{p+1} = [\bar{Y} - Y^p - A_1(\Delta X)]^2 \tag{8.4-17}$$

From Equation 8.2-12, the condition for minimum is

$$\frac{\partial G^{p+1}}{\partial(\Delta X)} = 2[\overline{Y} - Y^p - A_1(\Delta X)][-A_1] = 0 \qquad (8.4\text{-}18)$$

The optimal ΔX is

$$(\Delta X) = \frac{(\overline{Y} - Y^p)}{A_1} \qquad (8.4\text{-}19)$$

Thus the feedback optimal control technique described in Section 8.2 may be applied without any change to include the effects of the uncontrolled variable.

The block diagram for feedback optimal control adapting to the effect of controlled variables is given in Figure 8.4-2. Note that the only difference between Figure 8.4-2 and Figure 8.2-2 is the presence of the uncontrolled variable U.

C. Example of Adapting Feedback Optimal Control (with Perfect Physical Process Model) to the Effect of the Uncontrolled Variable

Let the coefficients of the perfect physical process model with one control variable and one state variable be

$$\begin{aligned} A_0 &= 5 \\ A_1 &= 50 \end{aligned} \qquad (8.4\text{-}20)$$

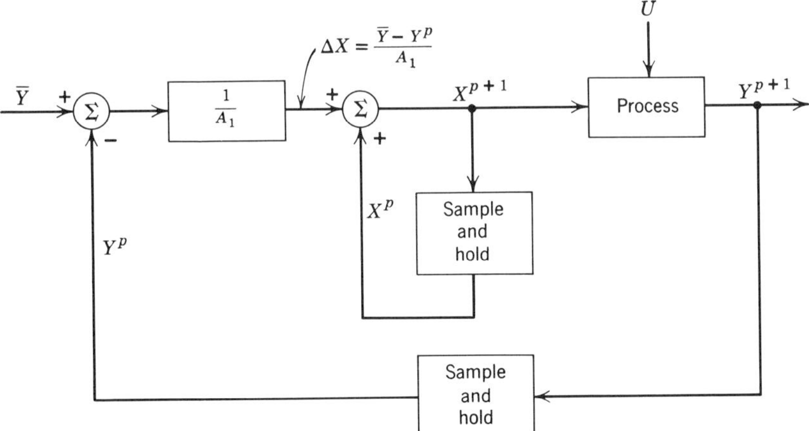

Fig. 8.4-2   Feedback optimal control adapting to effect of the uncontrolled variable.

And the problem is to attain the desired $\bar{Y} = 50$ with the initial $X = 0.9$. The uncontrolled variable is

$$U = 0.5 \qquad (8.4\text{-}21)$$

The coefficients of the uncontrolled variable are

$$\begin{aligned} B_0 &= -10 \\ B_1 &= 40 \end{aligned} \qquad (8.4\text{-}22)$$

For this example the measured state variable Y^p will be simulated by the perfect physical process model:

$$Y = A_0 + A_1 X + B_0 + B_1 U \qquad (8.4\text{-}23)$$

The optimal change in the control variable is

$$(\Delta X) = \frac{\bar{Y} - Y^p}{A_1} = \frac{50 - Y^p}{50} \qquad (8.4\text{-}24)$$

and the simulated physical process model is

$$Y = 5 + 50X - 10 + 40(0.5) \qquad (8.4\text{-}25)$$

$$Y = 15 + 50X \qquad (8.4\text{-}26)$$

Table 8.4-1 Adapting Feedback Optimal Control (with Perfect Physical Process Model) to Uncontrolled Variable

Time	Y^p	$(\bar{Y} - Y^p)$	$\Delta X = \dfrac{(\bar{Y} - Y^p)}{A_1}$	X^p	$X^{p+1} = X^p + \Delta X$	G^p
t_p	60	-10	-0.2	0.9	0.7	100
t_{p+1}	50	0	0	0.7	0.7	0
$\vdots$	$\vdots$	$\vdots$	$\vdots$	$\vdots$	$\vdots$	$\vdots$
t_{p+q}	50	0	0	0.7	0.7	0

The computations for the various intervals are given in Table 8.4-1. For this simple example with a perfect physical process model the use of feedback optimal control is very effective in adapting to the effects of the uncontrolled variable. If the uncontrolled variable is changing, the feedback optimal control technique will still be effective if the variations of U are slower than the rate of convergence used in the feedback calculations.

D. Example of Adapting Feedback Optimal Control (with Imperfect Physical Process Model) to the Effect of Uncontrolled Variable

Assume that the *imperfect* physical process model has the coefficients

$$A'_0 = 5$$
$$A'_1 = 45 \tag{8.4-27}$$

whereas the *perfect* physical process model should have

$$A_0 = 5$$
$$A_1 = 50 \tag{8.4-28}$$

The measured state variable is simulated by the *perfect* physical process model

$$Y = A_0 + A_1 X + B_0 + B_1 U \tag{8.4-29}$$
$$Y = 15 + 50X \tag{8.4-30}$$

and the *imperfect* incremental physical process model is

$$(\Delta Y) = 45(\Delta X) \tag{8.4-31}$$

The optimal change in the control variable derived by using the *imperfect* incremental physical process model is

$$(\Delta X) = \frac{(\bar{Y} - Y^p)}{45} \tag{8.4-32}$$

The computations for the various operations are given in Table 8.4-2. Note that two additional iterations are required because of the *imperfect* physical process model.

Table 8.4-2   Adapting Feedback Optimal Control (with Imperfect Physical Process Model) to Uncontrolled Variable

Time	Y^p	$(\bar{Y} - Y^p)$	$\Delta X = \dfrac{(\bar{Y} - Y^p)}{45}$	X^p	$X^{p+1} = X^p + \Delta X$	G^p
t_p	60.0	-10.0	-0.222	0.900	0.678	100.00
t_{p+1}	48.9	1.1	0.024	0.678	0.702	1.21
t_{p+2}	50.1	-0.1	-0.002	0.702	0.700	0.01
t_{p+3}	50.0	0	0	0.700	0.700	0
$\vdots$	$\vdots$	$\vdots$	$\vdots$	$\vdots$	$\vdots$	$\vdots$
t_{p+q}	50.0	0	0	0.700	0.700	0

E. Combined Feedback and Feedforward Approach to Adapt to the Uncontrolled Variable

If the physical process model relating the uncontrolled variable U and the state variable Y is available, it is possible to use a combined feedback and feedforward approach for optimal control of processes whose characteristics change, and are subject to uncontrolled variables. The feedback action will be used to adapt to the changes of the physical process characteristics, i.e., changes that make the *actual* physical process model coefficients deviate from the *average* values used. The feedforward action will be used to adapt to the effect of the uncontrolled variable.

Let the physical process model be

$$Y^{p+1} = A_0 + A_1 X^{p+1} + B_0 + B_1 U^p \tag{8.4-33}$$

Equation 8.4-33 embodies two concepts. The first concept is that the state variable Y can be computed as a function of the control variable X and the uncontrolled variable U. The second concept is how the state variable Y is computed at time t_{p+1}. Y^{p+1} is computed by using the projected value of the control variable X^{p+1} at time t_{p+1}, and the best information of the uncontrolled variable U^p that is available, i.e., at time t_p. Thus Equation 8.4-33 uses the combined effect of a projected variation of the control variable X^{p+1} with the known past value of the uncontrolled variable U^p to estimate the projected value of the state variable Y^{p+1}. Similarly, the state variable one time interval earlier is

$$Y^p = A_0 + A_1 X^p + B_0 + B_1 U^{p-1} \tag{8.4-34}$$

The incremental physical process model can be obtained by subtracting Equation 8.4-34 from Equation 8.4-33:

$$(\Delta Y) = Y^{p+1} - Y^p \tag{8.4-35}$$

$$(\Delta Y) = A_1(\Delta X) + B_1(\Delta U) \tag{8.4-36}$$

$$\therefore \quad Y^{p+1} = Y^p + A_1(\Delta X) + B_1(\Delta U) \tag{8.4-37}$$

where

$$\Delta X = X^{p+1} - X^p \tag{8.4-38}$$

$$\Delta U = U^p - U^{p-1} \tag{8.4-39}$$

The objective function to be minimized is

$$G^{p+1} = (\bar{Y} - Y^{p+1})^2 \tag{8.4-40}$$

By substituting Equation 8.4-37 into Equation 8.4-40,

$$G^{p+1} = [\bar{Y} - Y^p - A_1(\Delta X) - B_1(\Delta U)]^2 \tag{8.4-41}$$

The condition for optimum with respect to ΔX is

$$\frac{\partial G^{p+1}}{\partial(\Delta X)} = 0 \tag{8.4-42}$$

$$(-2A_1)[\overline{Y} - Y^p - A_1(\Delta X) - B_1(\Delta U)] = 0 \tag{8.4-43}$$

By solving for ΔX the optimal control equation is

$$\Delta X = \frac{1}{A_1}[\overline{Y} - Y^p] - \frac{B_1}{A_1}[(\Delta U)] \tag{8.4-44}$$

Substituting Equation 8.4-39 into Equation 8.4-44,

$$\Delta X = \frac{1}{A_1}[\overline{Y} - Y^p] - \frac{B_1}{A_1}[U^p - U^{p-1}] \tag{8.4-45}$$

If U^p and U^{p-1} can be estimated, then the optimal change of the control variable can be adjusted to minimize the objective function given by Equation 8.4-40. Equation 8.4-45 is the optimal control equation using a combined feedback and feedforward approach. Figure 8.4-3 shows how the optimal control equation can be implemented. Note that the feedback portion of the control scheme is identical to the diagram given in Figure 8.2-2. The feedforward portion of the control scheme uses the estimated values of the uncontrolled variable for the past two intervals, U^p and U^{p-1}.

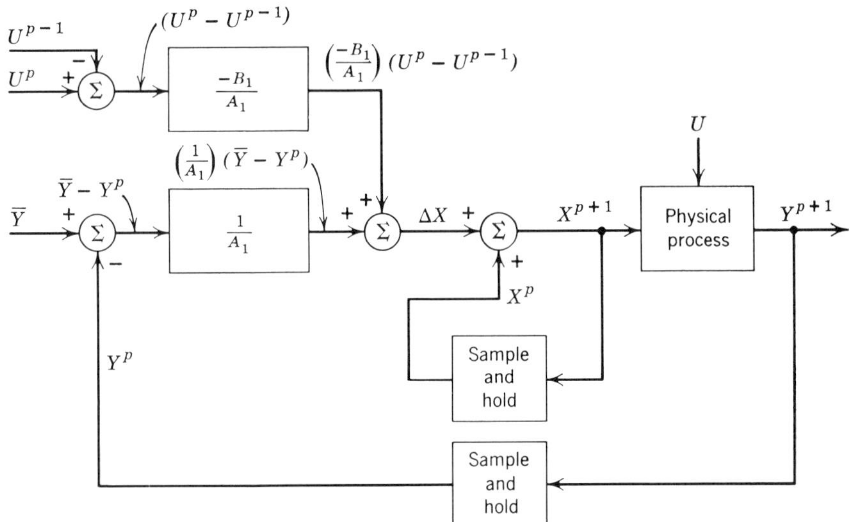

Fig. 8.4-3 Combined feedback and feedforward optimal control to adapt to the uncontrolled variable.

8.5 SUMMARY

This chapter discussed the adaptation of optimal control techniques to moderately well-defined steady-state processes. The topics discussed included adaptive control techniques to cope with changes in the physical process model, and changes in the environment.

Since all real physical processes have these characteristics of change, the techniques presented in this chapter are perhaps the most important among all the optimization techniques presented in Chapters 7 through 10. Without these adaptive control techniques, most optimization problems will stay in the realm of theoretical exercises and off-line research. These adaptive control techniques make it practical to implement optimal control.

Chapter 9 will present the evolutionary optimization (EVOP) approach for poorly defined steady-state processes, and the combined feedforward and feedback optimal control for well-defined processes, using linear programming and EVOP techniques. Chapter 10 will present the dynamic optimization techniques for time-dependent processes.

9

Evolutionary Optimization (EVOP) of Poorly Defined and Well-Defined Processes—Steady-State

9.0 INTRODUCTION

This chapter will present an approach to optimal control which does not require a mathematical model of the process. The process itself is used instead of the physical process model. The *evolutionary optimization (EVOP) approach is particularly applicable to steady-state optimization of poorly defined processes.* EVOP uses an iterative procedure which adjusts the control variables in successive moves to arrive at the optimum of the objective function. The objective function and process may be nonlinear and even undefined. The procedure is similar to the Simplex method of solving linear programming problems, as discussed in Section 7.5, in the sense that each iteration is designed to bring the solution closer to the optimum of the objective function.

Figure 9.0-1 shows the evolutionary optimization (EVOP) approach for poorly defined processes, which uses the measured state and control variables

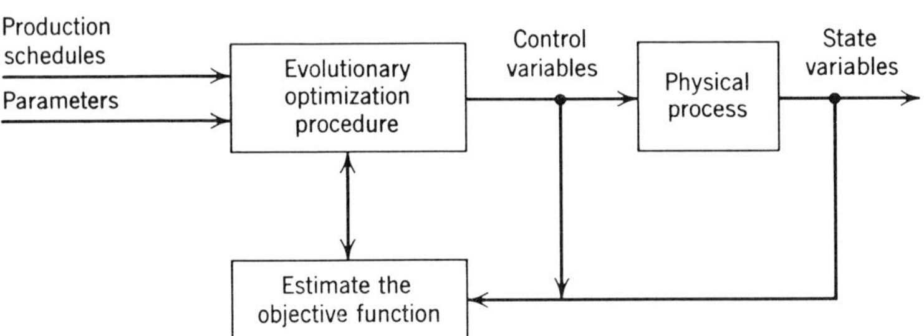

Fig. 9.0-1 Evolutionary optimization (EVOP) procedure for poorly-defined processes.

277

to solve for the value of the objective function. It is a feedback approach because the evolutionary optimization procedure uses the estimated objective function values to determine the next move of the control variables.

In general the EVOP approach for steady-state optimization may be used advantageously where one or more of the following conditions of the physical process model exist:

1. It is technically unfeasible to develop.
2. It is uneconomical to develop.
3. It is impractical to update.
4. It is unreliable (or inaccurate) over the range of operating conditions contemplated.
5. The goal is "relative process improvement" at a lower investment cost, rather than "perfect optimum" at a higher investment cost.

Figure 9.0-2 shows a combined feedforward and feedback approach for optimal control of well-defined processes. The objective function and the physical process model may be nonlinear. The physical process model is simulated by the on-line process computer. Feedback measurements of the state and control variables will be used to determine the operating point. The EVOP approach will be applied to the objective function and the simulated physical process model to estimate the linear coefficients of the objective function. Then the linear programming technique discussed in Section 7.5 will be used to solve for the optimal objective function. In order to make the linearization assumption valid, the solution should be limited by constraints to the neighborhood of the operating point. The results of the linear programming solution will be applied in a feedforward approach to adjust the control variables. This process is repeated over and over again to maintain or improve on the optimum of the objective function.

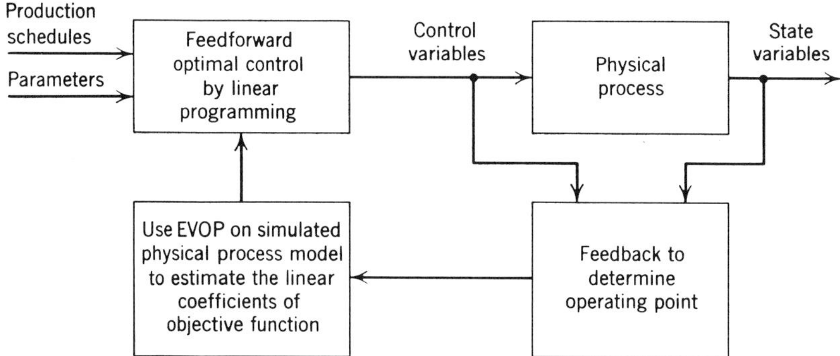

Fig. 9.0-2 Combined feedforward and feedback approach for optimal control of well-defined processes.

This chapter will first present the general approach used for EVOP for poorly defined processes, and introduce some commonly used strategies. Then the various gradient measurement techniques and the realities of constraints and stochastic noise will be discussed. Finally the combined feedforward and feedback approach for optimal control of well-defined processes using EVOP and linear programming will be presented with an example.

9.1 EVOP APPROACH FOR POORLY DEFINED PROCESSES

The basic EVOP approach for poorly defined processes is to develop an iterative procedure which adjusts the control variables so that the optimum of the objective function may be reached *effectively*. *An effective procedure is one which is fast and not easily confused by unusual characteristics of the objective function.* For example, local optimums and ridges in the objective function may mislead the procedure to settle for a nonoptimum as the final condition or to oscillate. The general EVOP approach for poorly defined processes consists of a four-step procedure being applied repetitively, as shown in Figure 9.1-1. The *first step* derives the required changes to be made

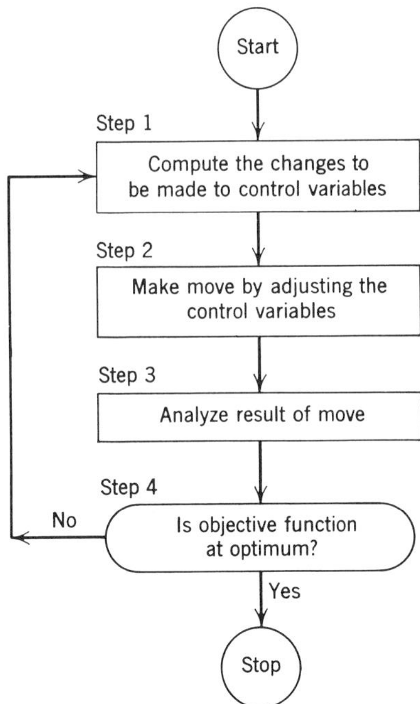

Fig. 9.1-1 Four step iterative procedure for EVOP.

in the control variable settings for the next move toward the optimum of the objective function; the *second step* actually makes the move by adjusting the control variables; the *third step* analyzes the result of the move just made; the *fourth step* determines if the objective function is at the optimum. If the objective function has not attained an optimum value, the procedure is repeated.

This section will present some elementary definitions and nomenclature used for the EVOP approach which should be useful for following the material in the rest of the chapter.

A. Response Surface

The value of the objective function plotted as a function of the control variables is called the response surface. For one control variable, the response surface is simply a line. For two control variables, the response surface is a three-dimensional surface, as shown in Figure 9.1-2. *The lines which join equal response surface values together are called contour lines*, which are similar to those used for mapping and topological purposes. Note that in Figure 9.1-2 the optimum of the objective function is $Q = 10.0$, and the three contours shown have the values of $c = 8$, $b = 5$, and $a = 3$.

Let the following symbols be assigned:

p = index of points along the EVOP trajectory.
X_i^p = control variables at the pth point, $(i = 1)$ to $(i = n)$.
F = objective function or response surface.
G_i^p = component of the gradient along the axis of the ith control variable, evaluated at the operating point p.

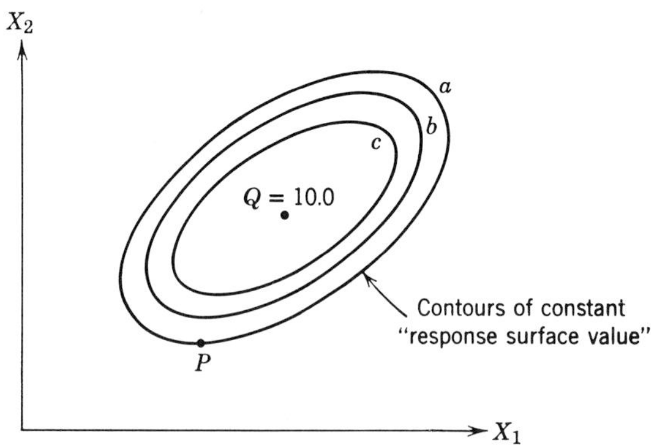

Fig. 9.1-2 Response surface with two control variables.

M^p = magnitude of the gradient at the operating point p.

$\mathbf{D}^p$ = gradient directional vector, which is a unit vector defining the direction of the gradient at the operating point p.

In general, the objective function or response surface can be expressed as

$$F = F(X_1, X_2, \ldots, X_n) \tag{9.1-1}$$

The objective function F for n control variables will be a surface in $(n + 1)$ dimensional space, and the contour lines are lines which have equal objective function values in the $(n + 1)$ dimensional space. The shape of the response surface is referred to as the characteristics of the response surface. Some typical response surface characteristics which present difficulties to the EVOP approach include flat surfaces, local optimums, and long narrow ridges. Response surfaces with more than two control variables are impractical to present pictorially; thus response surfaces will generally be presented mathematically after the introduction.

B. Strategy

Strategy is the logic or computations used to determine the changes in the control variables for the next move in the EVOP approach for poorly defined processes. Strategy corresponds to Step 1 in Figure 9.1-1. Several different strategies may be used in a given EVOP approach. For example, the strategy may be changed when the operating conditions approach the optimum of the response surface.

The effectiveness of a strategy is measured by the following:

1. The speed of arriving at the response surface optimum.
2. The ability to cope with unusual process or response surface characteristics.
3. The assurance that the optimum selected is indeed the true optimum.
4. The simplicity of the strategy.

The *speed of arriving at the response surface optimum* is important because of the long time constants involved with some processes, e.g., 4–6 hours. A strategy involving too many steps for a slow process may make it impractical to ever arrive at the optimum because of the difficulty of holding everything else in the process constant, e.g., material composition and catalyst condition. The *ability to cope with unusual process or response surface characteristics* will enable the strategy to overcome or by-pass these difficulties instead of being stalled or confused, e.g., hunting around an optimum near two narrow ridges. Usually, a number of strategies may be used to allow for the flexibility required. The *assurance of a true optimum* corresponds to the work involved in Step 4 of Figure 9.1-1. Examples of false optimums are *local optimums* and *wandering optimums*. Local optimums are suboptimums, as shown in Figure 9.1-3. Wandering optimums are those that are sensitive to slight changes in uncontrolled variables, e.g., humidity and ambient temperature. Naturally, the speed of the strategy must be sufficient to keep up with the speed at which the true optimum wanders. The *simplicity or complexity of a strategy* will determine the computer resources* and instrumentation required to implement the strategy. Once again, common sense and judgment should prevail in selecting a compromise between *speed* and *simplicity*.

C. Gradient

The gradient is a directional quantity having components along the axes of the control variables. Each component is equal to the partial derivative of the objective function with respect to the corresponding control variable.

The components of the gradient are defined as

$$G_1^p = \frac{\partial F}{\partial X_1}\bigg|_p$$

$$G_2^p = \frac{\partial F}{\partial X_2}\bigg|_p$$

$$\vdots$$

$$G_n^p = \frac{\partial F}{\partial X_n}\bigg|_p$$

(9.1-2)

The components of the gradient indicate the direction along a path to the optimum point. If a particular component of the gradient is positive, movement along the axis of the corresponding control variable in a positive direction should lead toward a maximum terminal point. Movement along the axis in a negative direction should lead to a minimum terminal point. If a particular component of the gradient is negative, movement along the axis in the positive direction leads to a minimum terminal point. Thus the gradient points towards a maximum and away from a minimum.

The magnitude of the gradient is given by

$$M^p = \left[\left(\frac{\partial F}{\partial X_1}\bigg|_p\right)^2 + \left(\frac{\partial F}{\partial X_2}\bigg|_p\right)^2 + \cdots + \left(\frac{\partial F}{\partial X_n}\bigg|_p\right)^2\right]^{\frac{1}{2}}$$

(9.1-3)

* In Section 3.2, computer resources were defined as *storage, execution time,* and *input-output devices.*

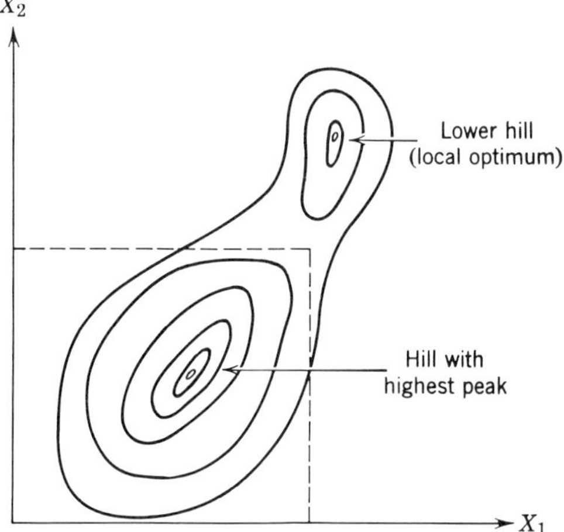

Fig. 9.1-3   Local optimum of response surface.

and the direction of the gradient is given by

$$\mathbf{D}^p = \frac{1}{M^p} \cdot \mathbf{G}^p$$

$$= \frac{1}{M^p} \cdot \begin{pmatrix} \left.\dfrac{\partial F}{\partial X_1}\right|_p \\[2ex] \left.\dfrac{\partial F}{\partial X_2}\right|_p \\[1ex] \vdots \\[1ex] \left.\dfrac{\partial F}{\partial X_n}\right|_p \end{pmatrix} \tag{9.1-4}$$

D. Trajectory

The trajectory is a series of steps followed by a strategy to arrive at the optimum of the response surface. Each step on the response surface is the difference between the operating points at the beginning and end of the move. Since EVOP for poorly defined processes is a steady-state optimization technique, the operating points of the response surface are independent of time. Thus the steps and trajectory in a given EVOP attempt are also independent of time. In an actual application, sufficient time must be allowed after a move for the process to "settle-out" to steady-state conditions.

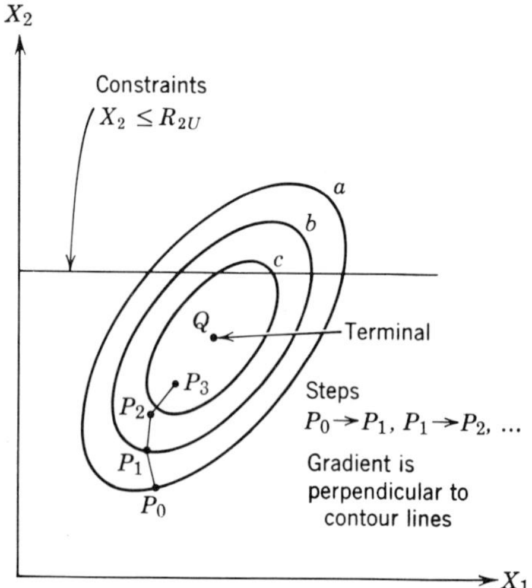

Fig. 9.1-4 Terminal, steps, gradient, and constraint of a response surface.

Each step along the response surface can be characterized by two components: *step size* or *magnitude* and *direction of the step*. Figure 9.1-4 shows the trajectory being made up of the following steps:

$$\text{Step 1: } P_0 \rightarrow P_1$$
$$\text{Step 2: } P_1 \rightarrow P_2$$
$$\text{Step 3: } P_2 \rightarrow P_3$$

E. Constraints

Constraints are physical or economic restrictions which limit the operating portion of the response surface to specific regions. In Figure 9.1-4, the constraint on the control variable X_2 is shown as

$$X_2 \leq R_{2U} \tag{9.1-5}$$

where R_{2U} is the upper limit of X_2. This constraint restricts the response surface to operate below and up to the line $X_2 = R_{2U}$. Constraints may be applied to other variables beside the control variables, i.e., state variables and economic variables. Constraints are also classified into *hard* and *soft* categories, which was discussed in Section 6.3.

F. Terminal

Terminal is the final result of the EVOP strategies. The terminal point should correspond to the response surface optimum. Special conditions exist when the operating point is at or near the terminal, e.g., the slope is zero at the terminal.

9.2   GRADIENT STRATEGIES FOR EVOP

As defined in Section 9.1, the gradient indicates the direction toward the optimum of the response surface. *Gradient strategies are those which base the EVOP approach on making the moves along the gradient.* This section will present the recurrence equation of a step, and discuss the following gradient strategies:

Steepest ascent (or descent).
Optimal gradient.
Gradient prediction.

A. Recurrence Equation of a Step

Let the following additional symbols be assigned:

X_i^p = value of ith control variable at the pth point.
X_i^{p+1} = value of ith control variable at the $(p + 1)$th point.
$\Delta X_i^p = X_i^{p+1} - X_i^p$.
$\mathbf{H}^p$ = scale factor operating on the component of the gradient at the pth step.

The recurrence equations for each control variable at the pth step are:*

$$X_i^{p+1} = X_i^p + \Delta X_i^p$$
$$= X_i^p + (H_i^p)(D_i^p) \tag{9.2-1}$$

where $(H_i^p)(D_i^p)$ is the adjustment to be made to the ith control variable at the pth point.

Equation 9.2-1 is valid for all other steps. Let $\mathbf{X}^{p+1}$, $\mathbf{X}^p$, and $\mathbf{D}^p$ be column vectors, and $\mathbf{H}^p$ be a diagonal matrix where the scale factor appears only on the main diagonal. Thus Equation 9.2-1 in matrix notation becomes

$$\mathbf{X}^{p+1} = \mathbf{X}^p + \Delta\mathbf{X}^p = \mathbf{X}^p + \mathbf{H}^p\mathbf{D}^p \tag{9.2-2}$$

For two control variables, Equation 9.2-2 becomes

$$\begin{pmatrix} X_1^{p+1} \\ X_2^{p+1} \end{pmatrix} = \begin{pmatrix} X_1^p \\ X_2^p \end{pmatrix} + \begin{pmatrix} H_1^p & 0 \\ 0 & H_2^p \end{pmatrix} \cdot \begin{pmatrix} D_1^p \\ D_2^p \end{pmatrix} \tag{9.2-3}$$

---

* Assuming that the response surface is to be maximized.

The various gradient strategies presented in this section offer different methods of computing the adjustments to the control variables necessary to arrive at the optimum.

B. Steepest Ascent (or Descent)

The strategy of steepest ascent (or descent) was developed by Cauchy in 1847. The principle is to *use the direction of the gradient as the directional vector for the next step*. The scale factor is a constant A, and is made equal for all the control variables. Thus:

$$\Delta \mathbf{X}^p = \mathbf{H}^p \mathbf{D}^p = A\mathbf{D}^p \tag{9.2-4}$$

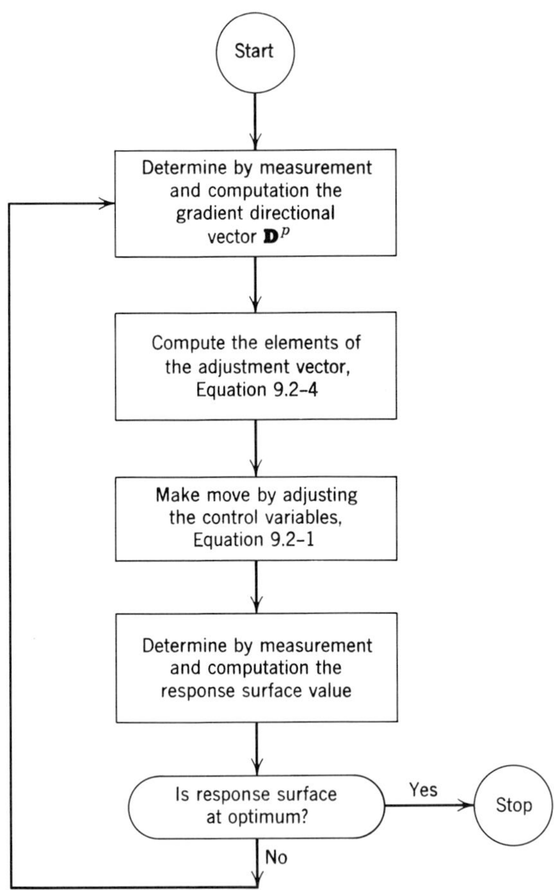

Fig. 9.2-1   Flow chart for the strategy of steepest ascent.

The result is a movement of distance A along a step defined by the direction of the gradient. A is a parameter to be chosen by the designer.

The flow chart for this strategy is shown in Figure 9.2-1. It is seen that each iteration requires that measurements and computations be made to determine the gradient at the operating point.

The shortcomings of the strategy of steepest ascent are

Slow rate of convergence.
Difficulties in the neighborhood of a long ridge in the response surface.
Difficulty in approaching the terminal.

The rate of convergence depends on the selection of the scale factor A. Although the selection of A can be guided by taking higher order partial derivatives, to do so experimentally is not an easy matter. In the neighborhood of a long narrow ridge, as shown in Figure 9.2-2, the strategy of steepest ascent may cause hunting back and forth across the ridge; e.g.,

$$P_0 \rightarrow P_1$$
$$P_1 \rightarrow P_2$$
$$P_2 \rightarrow P_3$$

Eventually, the trajectory will reach the terminal, but at the cost of many ineffectual moves. Also, as the trajectory approaches the terminal, the elements of the directional vector become smaller, until they approach zero at the terminal. This problem becomes more acute in the presence of noise and random fluctuations, as will be discussed in Section 9.4. The strategy of steepest ascent is generally not practical in the neighborhood of the terminal.

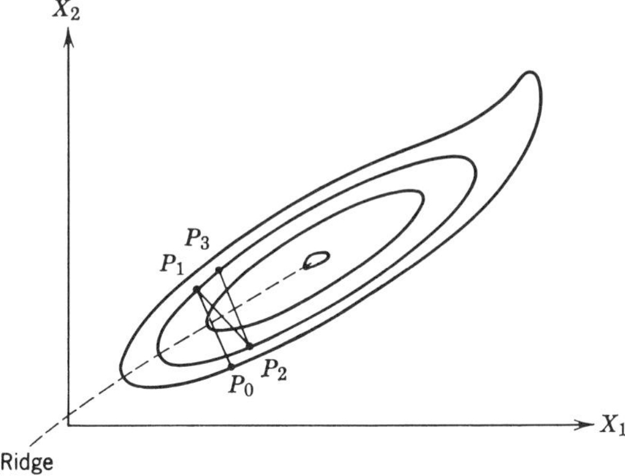

Fig. 9.2-2 Hunting across a long narrow ridge by using the strategy of steepest ascent.

C. Optimal Gradient

The optimal gradient strategy uses the same gradient directional vector and scale factor until the response surface value decreases. The flow chart in Figure 9.2-3 shows the logic for this strategy. As long as the response surface value is increasing, the same gradient will be used for successive steps. At the first sign of a decrease in the response surface value, the last step is retraced by a full or half step and the new gradient computed. The advantage of this strategy is the relative ease in measuring the response surface value compared to computing all the elements of the gradient vector.

Figure 9.2-4 shows an example of applying the optimal gradient strategy to a response surface with two control variables. Note that after 4 steps from P_0, the response surface value decreased. Thus the last step was retraced a full step, and a new gradient was used to reach P_1 in 6 additional steps. The next step from P_1 detected a decrease in the response surface value, so it was

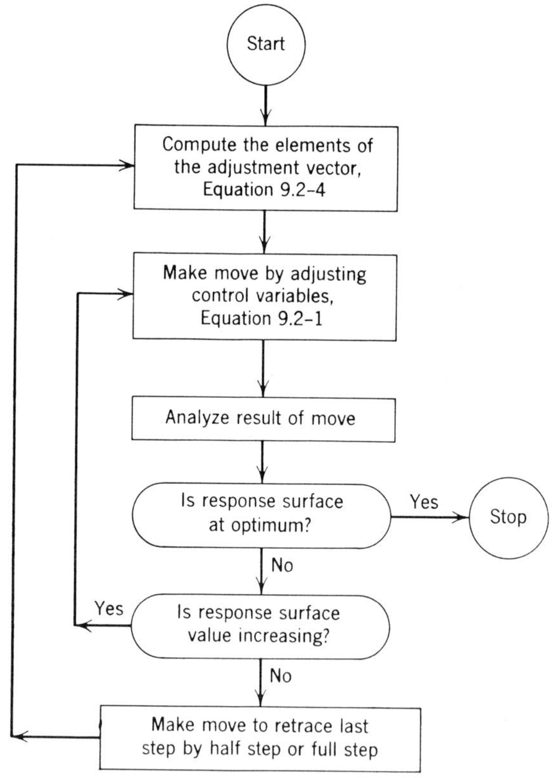

Fig. 9.2-3 Flow diagram of the optimal gradient strategy.

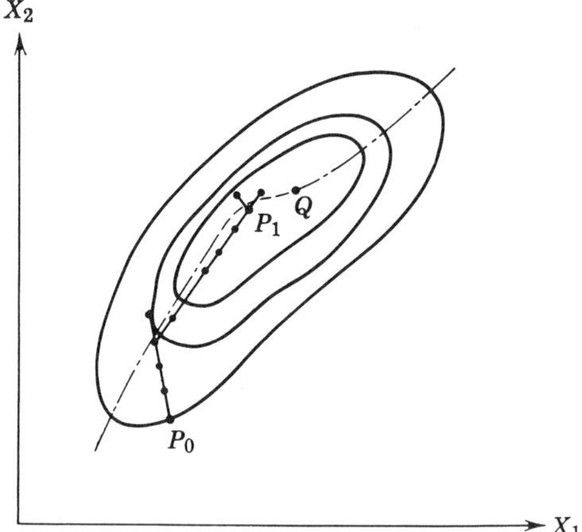

Fig. 9.2-4 Example of the optimal gradient strategy.

retraced a full step. The new gradient computed takes the step across the ridge instead of towards the terminal Q. It is possible for the optimal gradient strategy to hunt around a point near the terminal without ever reaching the terminal. Thus, although the optimal gradient strategy converges much faster than the strategy of steepest ascent, it shares a common shortcoming, i.e., difficulty in approaching the terminal. Modifying the step size can alleviate some of the difficulties around the terminal and speed up the convergence in other regions.

A useful technique for modifying the step size is based on comparing the ratio of the differences of the response surface values for two successive moves. If the response surface is changing more gradually, then the step size should be decreased. On the other hand, if the response surface is changing rapidly, then the step size should be increased. The reasoning is based on the fact that the response surface is usually changing more gradually near the terminal point. Let the following additional symbols be assigned:

R^p = ratio of the change in distance between the response surface values at the pth point.

K_1 = lower limit of R^p.

K_2 = upper limit of R^p.

By definition

$$R^p = \frac{\left[\sum_i (X_i^{p+1} - X_i^p)^2\right]^{1/2}}{\left[\sum_i (X_i^p - X_i^{p-1})^2\right]^{1/2}} \tag{9.2-5}$$

The step size *should not be changed if*

$$K_1 < |R^p| < K_2 \tag{9.2-6}$$

The step size *should be decreased* if

$$|R^p| \le K_1 \tag{9.2-7}$$

The step size *should be increased* if

$$R^p| \ge K_2 \tag{9.2-8}$$

D. Gradient Prediction

The gradient prediction strategy uses the same gradient direction and scale factor for successive steps until the response surface value falls outside the range of predicted values. Then the last step is retraced by a full or half step and a new gradient is computed. This is similar to the optimum gradient strategy. The strategy is also useful when it is necessary to predict the response surface values in the immediate neighborhood of an operating point. Let the following additional symbols be assigned:

$\mathbf{F}^p$ $\quad$ = actual response surface value for operating point p.

$\overline{\mathbf{F}^p}$ $\quad$ = predicted response surface value for operating point p.

$\overline{\Delta\mathbf{F}^p}$ $\;$ = $\overline{\mathbf{F}^{p+1}} - \mathbf{F}^p$ = predicted change in reponse surface value at step p.

$|\overline{\Delta F^p}|$ = predicted change in distance at step p.

K_3 $\quad$ = constant.

In the simple case of one control variable, the response surface value can be predicted by using linear extrapolation, i.e., the predicted change is equal to the change in the control variable times the slope (gradient) at the operating point. In the general case, the predicted change of the response surface value can be computed by using linear extrapolation in $(n + 1)$ dimensional space. The predicted value at operating point $(p + 1)$ is

$$\overline{\mathbf{F}^{p+1}} = \mathbf{F}^p + \mathbf{G}^p[\mathbf{H}^p\mathbf{D}^p] \tag{9.2-9}$$

and the predicted change in distance is

$$|\overline{\Delta F^p}| = \left[\sum_i (X_i^{p+1} - X_i^p)^2 (G_i^p)^2\right]^{1/2} \tag{9.2-10}$$

Comparing the change in value $|\overline{\Delta F^p}|$ against a known constant K_3 will be the criterion of deciding whether the gradient direction should be changed. The gradient direction *should not be changed* for the next step if

$$|\overline{\Delta F^p}| < K_3 \tag{9.2-11}$$

The gradient direction *should be changed* for the next step if

$$\overline{|\Delta F^p|} \geq K_3 \qquad\qquad (9.2\text{-}12)$$

9.3 MEASURING THE GRADIENT EXPERIMENTALLY

The purpose of this section is *to present some practical experimental techniques to measure the gradient for undefined response surfaces.* Section 9.2 showed that the strategy of steepest ascent moves the trajectory in the direction of the gradient. The other strategies also require the gradient to be calculated when the direction is to be changed. The general approach of measuring the gradient will first be presented. Then the factorial design of experiments for two levels and two variables will be introduced. Finally, the more general case of using the factorial design of experiments to approximate the gradient will be presented.

A. General Approaches

Case 1 Well-Defined Response Surface

For the case of the well-defined response surface, the gradient can be computed by evaluating the partial derivatives of the response surface F with respect to the control variables at point p. From Equation 9.2-4, the changes to be made to the control variables are given by the column vector $\mathbf{H}^p \mathbf{D}^p$. Thus

$$\Delta \mathbf{X} = \mathbf{H}^p \mathbf{D}^p$$

$$= \left. \left(\mathbf{H}^p \, \frac{\partial \mathbf{F}}{\partial \mathbf{X}} \right) \right|_p \qquad\qquad (9.3\text{-}1)$$

Case 2 Response Surface Approximated by Regression Models

For the case where the response surface is not known, it can be determined by using regression model approximations. In general, regression models have to be developed around each operating point where the gradient is needed. Linear regression models will be the simplest to develop because they require the least number of experimental measurements.

Linear models will also be the least accurate for predicting the gradient because the higher order or interaction effects of other control variables are ignored. However, second order polynomial regression models require more experimental samples and computational work.

Let the additional symbol be assigned:

$\theta_i =$ coefficients of response surface regression model.

From Equation 4.5-2 the linear regression model of n control variables around an operating point is

$$F = \theta_0 + \theta_1 X_1 + \theta_2 X_2 + \cdots + \theta_n X_n \qquad (9.3\text{-}2)$$

where the coefficients $\theta_0, \theta_1, \ldots, \theta_n$ are obtained by the techniques discussed in Section 4.5.

The gradient can be approximated by operating on the objective function given by Equation 9.3-2. The column vector for the gradient is

$$\frac{\partial \mathbf{F}}{\partial \mathbf{X}_i} = \boldsymbol{\theta}_i \qquad (9.3\text{-}3)$$

Thus if the linear regression model of the response surface with n control variables is available around an operating point p, the elements of the gradient vector can be approximated by the coefficients θ_i of the linear regression model. Higher order regression models of the response surface may also be used.

Case 3   Gradient Approximated by Difference Equation

A third alternative in measuring the gradient is by using a difference equation to approximate the partial derivative of F with respect to the control variables X_i. The difference equation can then be evaluated around the operating point p. Let the following additional symbols be assigned:

δX_i = small positive adjustment of the control variable X_i.

δF_i = resulting change in the response surface F as a result of adjusting only one of the control variables X_i.

ΔX_i = changes to be made to the control variables X_i for the next step.

A simple difference equation approximating the partial derivative of F with respect to X_i is

$$\left(\frac{\partial F}{\partial X_i}\right) \approx \frac{(\delta F_i)}{(\delta X_i)}\bigg|_p \qquad (9.3\text{-}4)$$

where

$$\delta F_i = F[X_1, X_2, \ldots, (X_i + \delta X_i), \ldots, X_n]$$
$$- F[X_1, X_2, \ldots, X_i, \ldots, X_n] \qquad (9.3\text{-}5)$$

Another method of approximating the partial derivative is to use the operating point p as the central point and make small forward and backward adjustments of $\pm(\delta X_i/2)$ from it. Thus:

$$\left(\frac{\partial F}{\partial X_i}\right) \approx \frac{(\delta F_i)}{(\delta X_i)}\bigg|_p \qquad (9.3\text{-}6)$$

where

$$\delta F_i = F\left[X_1, X_2, \ldots, \left(X_i + \frac{\delta X_i}{2}\right) \cdots X_n\right]$$

$$- F\left[X_1, X_2, \ldots, \left(X_i - \frac{\partial X_i}{2}\right) \cdots X_n\right] \qquad (9.3\text{-}7)$$

Figure 9.3-1 shows how the gradient may be approximated by using difference equations for a two control variable response surface. If the strategy of steepest ascent is used, then ΔX_1 will be proportional to δF_1 divided by δX_1, and ΔX_2 will be proportional to δF_2 divided by δX_2.

$$\therefore \quad \Delta X_1 \sim \left.\frac{(\delta F_1)}{(\delta X_1)}\right|_p \qquad (9.3\text{-}8)$$

and

$$\Delta X_2 \sim \left.\frac{(\delta F_2)}{(\delta X_2)}\right|_p \qquad (9.3\text{-}9)$$

The shortcoming of using simple difference equations to approximate the gradient is that there is no averaging, and therefore it is subject to effects of noise and other process disturbances.

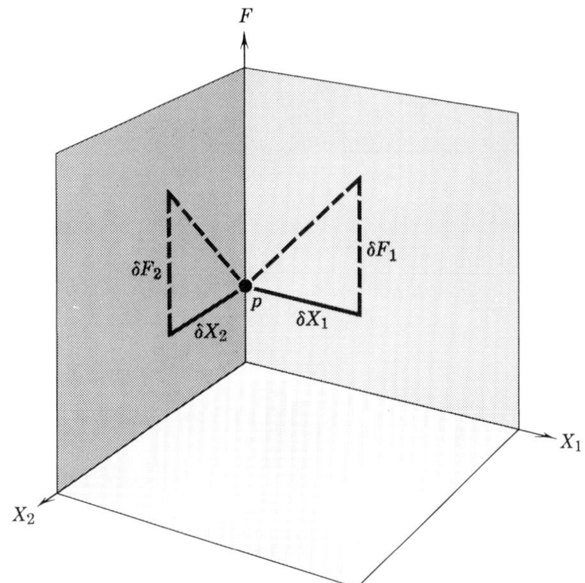

Fig. 9.3-1 Gradient approximated by difference equation.

B. Design of Experiments for Two Levels and Two Control Variables

An experiment involving two levels (values) of two control variables will be used to illustrate the *basic concept of design of experiments*. The presentation will be largely pictorial. Figure 9.3-2 shows a factorial design where small changes are made to the control variables around the operating point p. The coordinates of the point p and the three experimental points are shown in Figure 9.3-2.

An estimate of the average effect of X_1 alone on the response surface can be obtained by taking the difference between "the response surface values involving the change δX_1" and "the response surface values not involving the change δX_1." Similarly, an estimate of the average effect of X_2 alone on the response surface can be obtained by taking the difference between "the response surface values involving the change δX_2" and "the response surface values not involving the change δX_2." Thus an estimate of the average effect of X_1 on F is

$$\frac{(F^1 + F^2) - (F^3 + F^p)}{2} \tag{9.3-10}$$

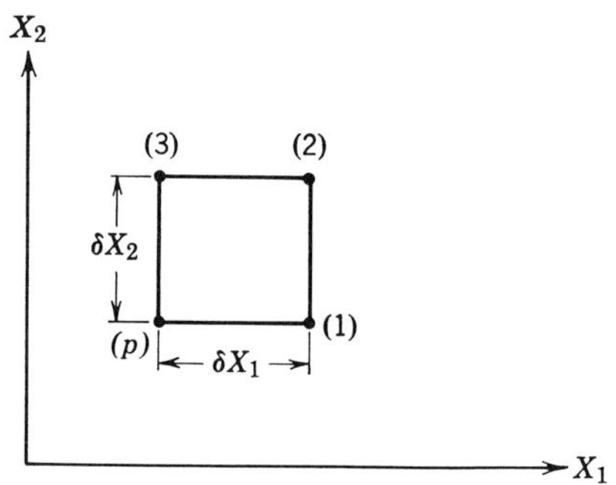

Coordinates for the points are

(p): (X_1^p, X_2^p)

(1): $(X_1^p + \delta X_1, X_2^p)$

(2): $(X_1^p + \delta X_1, X_2^p + \delta X_2)$

(3): $(X_1^p, X_2^p + \delta X_2)$

Fig. 9.3-2 2^2 factorial design of experiment.

and the estimate of the average effect of X_2 on F is

$$\frac{(F^2 + F^3) - (F^1 + F^p)}{2} \tag{9.3-11}$$

A simple first order approximation to the gradient is

$$\frac{\partial F}{\partial X_1} \approx \frac{(F^1 + F^2 - F^3 - F^p)}{(2\delta X_1)} \tag{9.3-12}$$

$$\frac{\partial F}{\partial X_2} \approx \frac{(F^2 + F^3 - F^1 - F^p)}{(2\delta X_2)} \tag{9.3-13}$$

Another technique for determining the gradient experimentally is to locate the experimental points centrally, as shown in Figure 9.3-3. The difference from Figure 9.3-2 is that the operating point p is centrally located between the four experimental points. The coordinates of the point p and the four experimental points are shown in Figure 9.3-3.

Using the same reasoning as before, the estimated average effect of X_1 on F is

$$\frac{(F^3 + F^2 - F^1 - F^4)}{2} \tag{9.3-14}$$

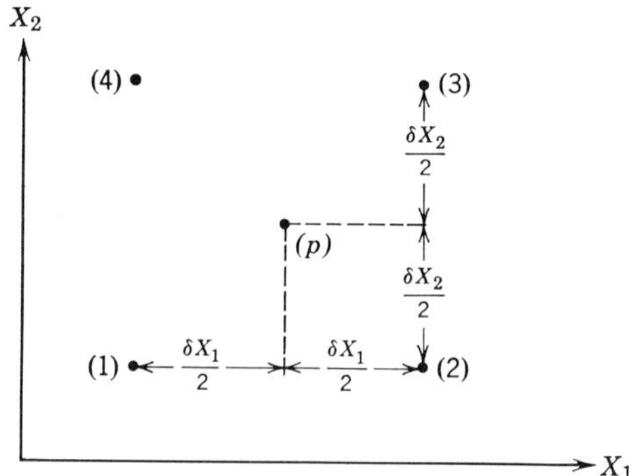

Coordinates for the points are

(p): (X_1^p, X_2^p)

$$(1): \left(X_1^p - \frac{\delta X_1}{2},\ X_2^p - \frac{\delta X_2}{2}\right)$$

$$(2): \left(X_1^p + \frac{\delta X_1}{2},\ X_2^p - \frac{\delta X_2}{2}\right)$$

$$(3): \left(X_1^p + \frac{\delta X_1}{2},\ X_2^p + \frac{\delta X_2}{2}\right)$$

$$(4): \left(X_1^p - \frac{\delta X_1}{2},\ X_2^p + \frac{\delta X_2}{2}\right)$$

Fig. 9.3-3   Centrally located 2^2 factorial design of experiment.

And the estimated average effect of X_2 on F is

$$\frac{(F^4 + F^3 - F^1 - F^2)}{2} \tag{9.3-15}$$

The first-order approximation to the gradient is

$$\frac{\partial F}{\partial X_1} \approx \frac{(F^3 + F^2 - F^1 - F^4)}{2(\delta X_1)} \tag{9.3-16}$$

$$\frac{\partial F}{\partial X_2} \approx \frac{(F^4 + F^3 - F^1 - F^2)}{2(\delta X_2)} \tag{9.3-17}$$

C. Estimating the Gradient by Regression Models

A general method of estimating the gradient is to use the regression model of the response surface in the neighborhood of an operating point. This model in turn may be obtained by using data derived from a centrally located design of experiments with the center at the origin; this would be the case if the point p in Figure 9.3-3 were located at the origin. The linear regression model corresponds to the multiple linear regression model which "goes through the mean," developed in Section 4.5. The regression model was given by Equations 4.5-8 through 4.5-10. Note, however, that the regression model is for the response surface in the neighborhood of the point p and there are $(k = 1)$ to $(k = q)$ experimental readings for each of the n control variables. With adjustments made for the differences in notation, the multiple linear regression model developed by a centrally located design of experiments with the center at the origin is

$$F^p = \sum_{i=1}^{n} \theta_i \, X_i \big|_p \tag{9.3-18}$$

and the coefficients θ_i may be solved for with the following equation:

$$\begin{pmatrix} \sum_{k=1}^{q}(X_{1k})^2 & \sum_{k=1}^{q}(X_{1k})(X_{2k}) \cdots & \sum_{k=1}^{q}(X_{1k})(X_{nk}) \\ \sum_{k=1}^{q}(X_{2k})(X_{1k}) & \sum_{k=1}^{q}(X_{2k})^2 \quad \cdots & \sum_{k=1}^{q}(X_{2k})(X_{nk}) \\ \vdots & \vdots & \vdots \\ \sum_{k=1}^{q}(X_{nk})(X_{1k}) & \sum_{k=1}^{q}(X_{nk})(X_{2k}) \cdots & \sum_{k=1}^{q}(X_{nk})^2 \end{pmatrix} \cdot \begin{pmatrix} \theta_1 \\ \theta_2 \\ \vdots \\ \theta_n \end{pmatrix} = \begin{pmatrix} \sum_{k=1}^{q}(X_{1k}F_k) \\ \sum_{k=1}^{q}(X_{2k}F_k) \\ \vdots \\ \sum_{k=1}^{q}(X_{nk}F_k) \end{pmatrix}$$

$$\tag{9.3-19}$$

It can be shown for a centrally designed experiment with the center at the origin that all terms in the matrix of Equation 9.3-19 except those on the main diagonal will be equal to zero. This is because

$$\sum_{k=1}^{q} (X_{ik})(X_{\gamma k}) = 0 \quad \text{for} \quad \gamma \neq i \tag{9.3-20}$$

Equation 9.3-20 may be checked by substituting the values of the experimental points of Figure 9.3-3 (with point p at the origin) into the equations.

Substituting Equation 9.3-20 into Equation 9.3-19,

$$\begin{pmatrix} \sum_{k=1}^{q}(X_{1k})^2 & 0 & \cdots & 0 \\ 0 & \sum_{k=1}^{q}(X_{2k})^2 & \cdots & 0 \\ \vdots & \vdots & & \vdots \\ 0 & 0 & \cdots & \sum_{k=1}^{q}(X_{nk})^2 \end{pmatrix} \cdot \begin{pmatrix} \theta_1 \\ \theta_2 \\ \vdots \\ \theta_n \end{pmatrix} = \begin{pmatrix} \sum_{k=1}^{q}(X_{1k}F_k) \\ \sum_{k=1}^{q}(X_{2k}F_k) \\ \vdots \\ \sum_{k=1}^{q}(X_{nk}F_k) \end{pmatrix}$$

$$\tag{9.3-21}$$

From Equation 9.3-21 the solution of the regression coefficients at operating point p yields

$$\theta_i|_p = \left. \frac{\sum_{k=1}^{q}(X_{ik}F_k)}{\sum_{k=1}^{q}(X_{ik})^2} \right|_p \tag{9.3-22}$$

The component of the gradient may now be estimated by

$$\left. \frac{\partial F}{\partial X_i} \right|_p = \theta_i|_p \tag{9.3-23}$$

9.4 CONSTRAINTS AND STOCHASTIC NOISE

Constraints on control variables limit the region of allowable operation of the response surface. Section 6.3 showed that all physical systems and processes inherently have constraints. This section will discuss the implications of constraints on the gradient strategies, then discuss the effect of stochastic noise on the EVOP technique, and show how the effect can be reduced by replicating the experimental design.

A. Constraints on the Limit of Control Variables

The EVOP procedure can be modified to restrict the control variables to within specified ranges. Before a move is made for a given step, the new values of the control variables must be checked to determine if they will be within the constraint limits. If one or more of the control variables will be out of the constraint limits as a result of the next move, the corresponding components of the adjustment vector must be modified. The modification of the control variables for *hard constraints* can be made so as to bring the control variables to the constraint limits, but not to exceed it. The procedure for adapting to hard constraints on control variables discussed in Section 8.3 can be directly applied to the EVOP technique. Upon reaching the hard constraint limit, the trajectory normally proceeds along the limit until it either finds the terminal on the boundary or leaves the boundary for a higher response surface value within the constraint limit. Figure 9.4-1 shows an example where the trajectory proceeds until it reaches the boundary of the hard constraint. Then it proceeds along the boundary until it settles on the optimum Q_1 on the boundary. The true optimum Q could not be reached because of the constraint.

Soft constraints are best implemented by using a modified response surface which is heavily penalized if the constraint limits are exceeded by a significant extent. The technique of modifying the objective function to take soft constraints into account was discussed in Section 6.3.

Figure 9.4-2 shows the profile of the actual and modified response surface along a plane perpendicular to the soft constraint limit. The modified response

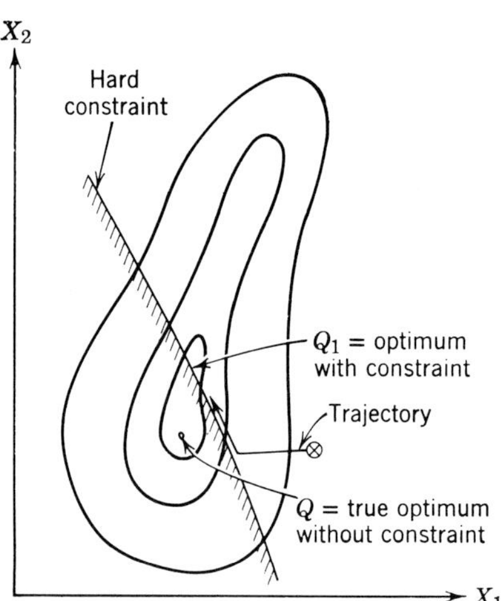

Fig. 9.4-1   EVOP modified for hard constraint.

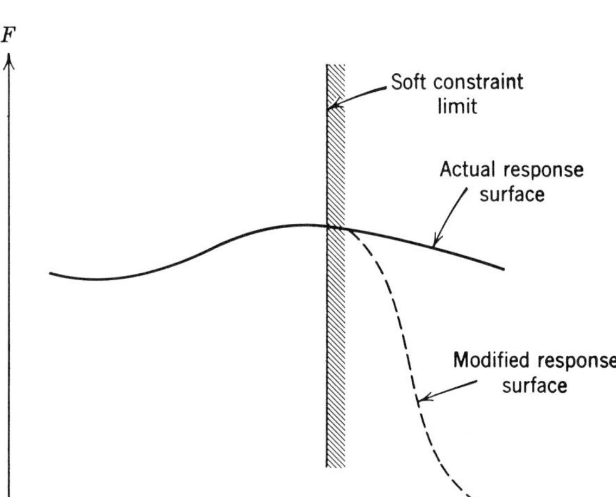

Fig. 9.4-2   Profile of actual and modified response surface in a plane perpendicular to the soft constraint limit.

surface is shown taking a sharp drop *beyond* the soft constraint limit. If there are any abrupt changes in the values of the objective function *near* the limit of the soft constraint, the modified response surface may result in a ridge. As discussed in Section 9.2, the EVOP trajectory may tend to oscillate across the ridge created by the soft constraint boundary, as shown in Figure 9.4-3.

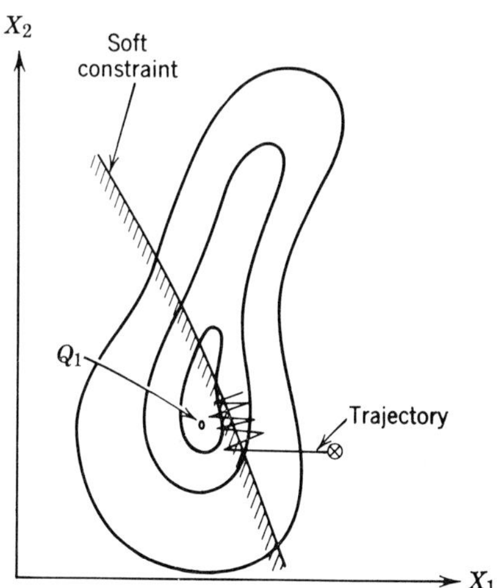

Fig. 9.4-3   EVOP modified for soft constraint.

B. Statistical Significance of the Gradient

Techniques discussed in Section 4.4 can be applied to determine whether the variations in the response surface values used to estimate the gradient represent

(1) a significant change due to changes in the control variables or
(2) random variations due to stochastic noise.

Let the following additional symbols be assigned:

F_k = actual response surface measurement at the kth sample.
$\overline{F}$ = average value of the response surface.
$\hat{F}_k$ = predicted value of the response surface.
θ_i = effect on the response surface due to the ith control variable in the design of experiments.
X_{ik} = setting for the ith control variable at the kth experimental point.
X_{i0} = 1, by definition.
$\dfrac{\delta X_i}{2}$ = small change made to the ith control variable.

σ^2 = variance of the response surface due to noise or measurement error.
SS_i = sum of squares for the ith variable component.
R_i = ratio of $\dfrac{\mathrm{av}(SS_i)}{\sigma^2}$.
q = number of experimental points.
r = number of times experiment is replicated.
n = number of control variables.
$rq - n - 1$ = degree of freedom for σ^2.

Since the gradient may be estimated from the regression coefficients, the problem of determining the statistical significance of the gradient may be redefined as the problem of determining the statistical significance of the regression coefficients θ_i.

The factorial designs of experiments, as discussed in Section 9.3, allowed independent estimates of the regression coefficients and an independent check of the statistical significance of the regression coefficients. The statistical significance of the regression coefficients provides clues to the questions of whether (1) the derived regression coefficients provide a good fit to the response surface or (2) whether the derived regression coefficients are largely caused by stochastic noise. For regression models the regression coefficients can be assumed to have a normal frequency distribution.

The variations of the response surface due to stochastic noise are given by the variance σ^2, which is estimated from the variance of the measurements

F_k from the predicted value $\hat{F}_k$:

$$\sigma^2 = \frac{1}{(rq - n - 1)} \sum_{k=1}^{q} (F_k - \hat{F}_k)^2 \qquad (9.4\text{-}1)$$

where the value $\hat{F}_k$ is

$$\hat{F}_k = \sum_{i=1}^{n} \theta_i X_{ik} \qquad (9.4\text{-}2)$$

The remainder of this section will treat the orthogonal experimental design only, i.e., where the coefficients are determined independently. From Equation 9.3-23 the regression coefficients of the linear model developed by a centrally located design of experiments with the center at the origin is

$$\theta_i = \left. \frac{\sum_{k=1}^{q} (X_{ik} F_k)}{\sum_{k=1}^{q} (X_{ik})^2} \right|_p \qquad (9.4\text{-}3)$$

The variations of the response surface due to changes in the control variables may be used to estimate the variance of the regression coefficients. Using the experimental design which provided orthogonal estimates of the effects of the control variables on the response surface, the variance of the regression coefficients may be estimated independently and each of the coefficients checked for statistical significance individually.

The main effect of the control variables on the response surface is

$$\theta_i = \left. \frac{\sum_{k=1}^{q} (X_{ik} F_k)}{\sum_{k=1}^{q} (X_{ik})^2} \right|_p \qquad (9.4\text{-}4)$$

An estimate of the significance of the ith regression coefficient may be made from the *average sum of the squares of the variations from the mean* caused by the ith control variable. The sum of the squares SS_i is defined by

$$SS_i = (\theta_i)^2 \sum_{k=1}^{q} (X_{ik})^2 \Big|_p \qquad (9.4\text{-}5)$$

By substituting Equation 9.4-4 into Equation 9.4-5,

$$SS_i = \left. \frac{\left(\sum_{k=1}^{q} X_{ik} F_k \right)^2}{\sum_{k=1}^{q} (X_{ik})^2} \right|_p \qquad (9.4\text{-}6)$$

The mean square, or the average of the SS_i is equal to

$$\text{av } SS_i = \frac{\sum\limits_i SS_i}{\text{degrees of freedom}} \tag{9.4-7}$$

$$\text{av}\left[\theta_i^2 \sum_{k=1}^{q} (X_{ik})^2\right]\Bigg|_p = \left[\sigma^2 + \theta_1^2 \sum_{k=1}^{q} (X_{ik})^2\right]\Big|_p \tag{9.4-8}$$

Thus the variance of the ith regression coefficient is equal to the variance of the regression surface σ^2 plus the term $\theta_i^2 \sum_{k=1}^{q} (X_{ik})^2$, all evaluated at operating point p. Table 9.4-1 shows the analysis of variance for a linear regression.

Table 9.4-1   Table of Analysis of Variance for Linear Regression Model.

Source of Variation	Sum of Squares	Degrees of Freedom	Average Value or Mean Square
X_i	$\dfrac{\left(\sum\limits_{k=1}^{q} X_{ik} F_k\right)^2}{\sum\limits_{k=1}^{q} (X_{ik})^2}$	1	$\sigma^2 + \theta_i^2 \sum\limits_{k=1}^{q} X_{ik}^2$
Residual	$\sum (F_k - \bar{F}_k)^2$	$rq - n - 1$	σ^2
Total	$\sum (F_k - \bar{F})^2$	$rq - 1$	$\dfrac{\Sigma (F_k - \bar{F})^2}{rq - 1}$

Note that $\bar{F}_k$ is assumed to be the predicted response surface. $\bar{F}$ is the average value of the response surface.

C. Testing for Statistical Significance

The statistical significance of the ith regression coefficient is indicated by the ratio

$$R_i = \frac{\text{av}(SS_i)}{\sigma^2} \tag{9.4-9}$$

The variance of the regression coefficient divided by the variance of the regression model can be interpreted as the "mean square of variations due to X_i," divided by the "mean square of variations due to stochastic noise." The ratio R_i is

$$R_i = 1 + \frac{\theta_i^2 \sum\limits_{k=1}^{q} (X_{ik})^2}{\sigma^2} \tag{9.4-10}$$

If R_i is nearly equal to 1, the effects of X_i on F are small compared with the effects of stochastic noise and the statistical significance is considered poor. If R_i is much larger than 1, the effects of X_i on F are much larger than the effects of stochastic noise and the statistical significance is considered good.

The intuitive reasoning given above for measuring the statistical significance may be assigned a *quantitative confidence level by using the Fisher F-test*, which was discussed in Section 4.4. Let the following additional symbols be assigned:

$F_{1,rq-n-1,\alpha} = $ F-distribution value at (1) and $(rq - n - 1)$ degrees of freedom, with certain chance of error α.

where $q = $ number of experimental measurements.

$r = $ number of times experiment is replicated.

$n = $ number of control variables.

$\alpha = $ certain chance of error; confidence level $= (1 - \alpha)$.

The Fisher F-test is used to test the hypothesis that the regression coefficient $\theta_i = 0$ by making the comparison

$$R_i > F_{1,(rq-n-1),\alpha} \tag{9.4-11}$$

If the test in Equation 9.4-11 is satisfied, the hypothesis that $\theta_i = 0$ can be rejected, with a confidence level of $(1 - \alpha)$. In other words, the F-test provides a comparison of the variance of the θ_i distribution against the variance of the response surface F distribution, all with a confidence level of $(1 - \alpha)$.

If the Fisher F-test should indicate that the regression coefficients have insufficient statistical significance, the *experiment may be replicated to improve the statistical significance.* Since R_i is

$$R_i = 1 + \left[\frac{\theta_i^2 \sum\limits_{k=1}^{q} (X_{ik})^2}{\sigma^2} \right] \tag{9.4-12}$$

the ratio R_i will be increased by increasing the term $\sum_{k=1}^{q} (X_{ik})^2$; for example, replicating the experiment a second time doubles the term $\sum_{k=1}^{q} (X_{ik})^2$. Thus, when there are r replicates,

$$R_i = 1 + r \left[\frac{\theta_i^2 \sum\limits_{k=1}^{q} (X_{ik})^2}{\sigma^2} \right] \tag{9.4-13}$$

The effect of stochastic noise can be reduced by taking more measurements of data for the same point and averaging the values.

9.5   COMBINED FEEDFORWARD AND FEEDBACK OPTIMAL CONTROL OF WELL-DEFINED PROCESSES

This section will present the combined feedforward and feedback approach by using EVOP and linear programming techniques for optimal control of well-defined steady-state processes. The physical process model and objective function may both be *nonlinear. At each measured operating point (feedback), EVOP techniques will be applied to the simulated physical process model to estimate the linear coefficients of the objective function. Then linear programming techniques will be applied to solve for the optimal control variables (feedforward).* The objective function and the physical process model are assumed to be linear in the neighborhood of the operating point. This section will begin with the general approach, then discuss the special case where the physical process model can be substituted into the objective function to eliminate the state variables, and finally, a two control variable example will be developed to illustrate the EVOP part of the approach. The linear programming part of the approach will not be discussed as it was already developed in Section 7.5.

A. The General Approach

There are many nonlinear physical processes which may be considered to be linear in the neighborhood of an operating point. If these nonlinear physical processes and objective functions can be linearized for the neighborhood of specific operating points, then the techniques of linear programming, as discussed in Section 7.5, can be used to solve for the optimal control variables in the neighborhood. When there are more than four or five control variables, it is not practical to store all the sets of linearized coefficients for the objective function and the physical process model in the process computer. A better approach is to use the EVOP technique with the simulated physical process model to estimate the components of the gradient of the objective function. The components of the gradient will be approximately equal to the linear coefficients of the objective function. Let the following symbols be assigned:

X_i = control variables $(i = 1)$ to $(i = n)$.
Y_j = state variables $(j = 1)$ to $(j = m)$.
p = operating point.

F = nonlinear objective function, as a function of X_i and Y_j.

ψ_i = coefficients of the nonlinear objective function.

G^p = linearized objective function, valid in the neighborhood of p, as a function of X_i only.

θ_i^p = coefficients of the linearized objective function G^p, valid in the neighborhood of p.

The nonlinear objective function is

$$F = F(X_i, Y_j) \tag{9.5-1}$$

The approach is to use the EVOP procedure developed in Section 9.3 to estimate the coefficients of the linearized objective function. The linearized objective function will be expressed as a function of the control variables X_i only, and will be valid only in the neighborhood of p. It will be derived from the measured control variables X_i^p and state variables Y_j^p. The linearized objective function is

$$G^p = \theta_0 + \sum_{i=1}^{n} \theta_i^p X_i \tag{9.5-2}$$

From Equation 9.3-3, the coefficients θ_i^p may be estimated by the components of the gradient, which in turn are equal to the partial derivatives of the objective function F with respect to the control variables X_i. θ_0 may be developed by comparing the constant terms of Equation 9.3-3 and Equation 9.5-2. Thus

$$\theta_i^p = \left. \frac{\partial F(X_i, Y_j)}{\partial X_i} \right|_p \tag{9.5-3}$$

Since F is a function of X_i and Y_j, θ_i^p will be functions of X_i^p, Y_j^p, and $\partial Y_i^p / \partial X_i$.

$$\therefore \quad \theta_1^p = \theta_1^p \left(X_i^p, Y_j^p, \frac{\partial Y_j^p}{\partial X_1} \right)$$

$$\theta_2^p = \theta_2^p \left(X_i^p, Y_j^p, \frac{\partial Y_j^p}{\partial X_2} \right)$$

$$\vdots$$

$$\theta_n^p = \theta_n^p \left(X_i^p, Y_j^p, \frac{\partial Y_j^p}{\partial X_n} \right)$$

(9.5-4)

$\partial Y_j^p / \partial X_1$, $\partial Y_j^p / \partial X_2, \ldots, \partial Y_j^p / \partial X_n$ are obtained by taking the partial derivatives of the state variables with respect to the control variables, i.e., the partial derivatives of the physical process model functions with respect to the control variable. Examining Equation 9.5-4 leads to the conclusion that the following variables are required to estimate the linear coefficients of the objective function:

(1) X_i^p = measured control variables at p.

(2) Y_j^p = measured state variables at p.

(3) $\dfrac{\partial Y_i^p}{\partial X_i}$ = partial derivatives of the state variables with respect to the control variables, evaluated at p.

Figure 9.5-1 shows the flow chart of the combined feedforward and feedback approach for optimal control of well-defined processes. Note that after the linear coefficients are estimated, it is necessary to add the *linearity constraints* which will restrict the linear programming solution to the neighborhood of p. The linearity constraints are added to the constraints on the limit

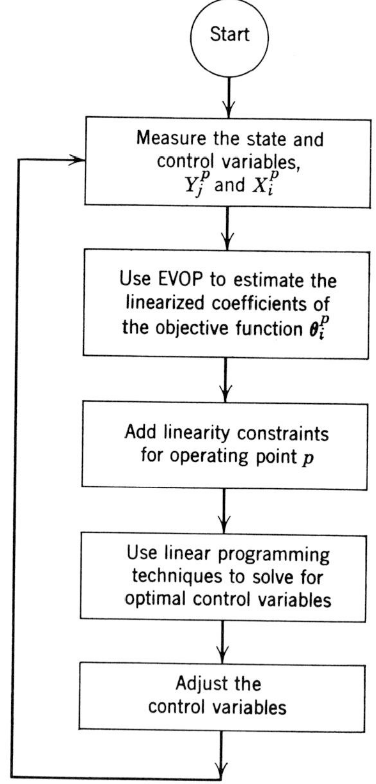

Fig. 9.5-1   Flow chart of combined feedforward and feedback approach for optimal control of well-defined processes.

of the control variables. Then the linear programming techniques can be directly applied to solve for the optimal control variables. The procedure is repeated so as to continually improve the optimal operating point. Logic may be used to stop the procedure if the absolute optimum should be reached. For example, the objective function at two successive points may be compared to see if the change is less than a prescribed value. If the change is less than a prescribed value, the objective function may be assumed to be at the optimum.

B. The Special Case

In the special case where the physical process model may be substituted into the objective function to eliminate the state variables, the EVOP procedure is somewhat simpler. It is also not necessary to evaluate the partial derivatives of the state variables with respect to the control variables, $\partial Y_j^p / \partial X_i$. The nonlinear objective function in G-form is

$$G = G(X_i) \tag{9.5-5}$$

Again, the linear coefficients of the objective function at point p can be estimated by

$$\theta_i^p = \left. \frac{\partial G(X_i)}{\partial X_i} \right|_p \tag{9.5-6}$$

$$\therefore \quad \theta_1^p = \theta_1^p(X_i^p)$$

$$\theta_2^p = \theta_2^p(X_i^p)$$
$$\vdots$$
$$\theta_n^p = \theta_n^p(X_i^p) \tag{9.5-7}$$

Comparing Equation 9.5-7 with Equation 9.5-4, it can be seen that only the measured control variables at point p are required to estimate the linear coefficients of the objective function for the special case.

C. A Two-Control Variable Example

A two-control variable example will be used to illustrate how the EVOP procedure can be applied to the simulated physical process model to estimate the linear coefficients of the objective function.

Let the nonlinear objective function be

$$F = \psi_0 + \psi_1 X_1 + \psi_2 X_2 + \psi_3 X_1 X_2 + \psi_4 Y_1 + \psi_5 Y_2 \tag{9.5-8}$$

The nonlinear physical process model is

$$Y_1 = A_{10} + A_{11}X_1 + A_{12}X_2 + A_{13}X_1X_2 \tag{9.5-9}$$

$$Y_2 = A_{20} + A_{21}X_1 + A_{22}X_2 + A_{23}X_1X_2 \tag{9.5-10}$$

Substituting the physical process model into the objective function to eliminate the state variables:

$$\begin{aligned}
G = \psi_0 &+ \psi_1 X_1 + \psi_2 X_2 + \psi_3 X_1 X_2 \\
&+ \psi_4(A_{10} + A_{11}X_1 + A_{12}X_2 + A_{13}X_1X_2) \\
&+ \psi_5(A_{20} + A_{21}X_1 + A_{22}X_2 + A_{23}X_1X_2)
\end{aligned} \tag{9.5-11}$$

$$\begin{aligned}
\therefore \quad G^p = (\psi_0 &+ \psi_4 A_{10} + \psi_5 A_{20}) \\
&+ X_1(\psi_1 + \psi_4 A_{11} + \psi_5 A_{21} \\
&+ X_2(\psi_2 + \psi_4 A_{12} + \psi_5 A_{22}) \\
&+ X_1 X_2(\psi_3 + \psi_4 A_{13} + \psi_5 A_{23})
\end{aligned} \tag{9.5-12}$$

The approximate linear objective function is

$$\hat{G}^p = \theta_0^p + \theta_1^p X_1 + \theta_2^p X_2 \tag{9.5-13}$$

By taking the partial derivatives of Equation 9.5-12 with respect to X_1 and X_2 and comparing the coefficients with Equation 9.5-13,

$$\theta_1^p \approx \left. \frac{\partial G}{\partial X_1} \right|_p = (\psi_1 + \psi_4 A_{11} + \psi_5 A_{21}) + X_2^p(\psi_3 + \psi_4 A_{13} + \psi_5 A_{23}) \tag{9.5-14}$$

$$\theta_2^p \approx \left. \frac{\partial G}{\partial X_2} \right|_p = (\psi_2 + \psi_4 A_{12} + \psi_5 A_{22}) + X_1^p(\psi_3 + \psi_4 A_{13} + \psi_5 A_{23}) \tag{9.5-15}$$

and

$$\theta_0 = \psi_0 + \psi_4 A_{10} + \psi_5 A_{20} \tag{9.6-16}$$

By substituting Equations 9.5-14 through 9.5-16 into Equation 9.5-13, the approximate linear objective function is

$$\begin{aligned}
\hat{G}^p = (\psi_0 &+ \psi_4 A_{10} + \psi_5 A_{20}) \\
&+ [(\psi_1 + \psi_4 A_{11} + \psi_5 A_{21}) + X_2^p(\psi_3 + \psi_4 A_{13} + \psi_5 A_{23})]X_1 \\
&+ [(\psi_2 + \psi_4 A_{12} + \psi_5 A_{22}) + X_1^p(\psi_3 + \psi_4 A_{13} + \psi_5 A_{23})]X_2
\end{aligned} \tag{9.5-17}$$

Equation 9.5-17 is the linear approximation of F in the neighborhood of the operating point p.

9.6 SUMMARY

This chapter concludes the series of three chapters on optimal control of steady-state processes. Chapter 7 presented the idealized optimal control techniques on well-behaved processes, and Chapter 8 presented the techniques to adapt the idealized optimal control techniques to moderately well-defined processes. This chapter presented the EVOP strategy for poorly defined processes and the combined feedforward and feedback approach for optimal control of well-defined processes.

The material in this chapter is also important to the study of computer process control systems because many basic concepts of the objective function are discussed directly or implied. For example, some of the topics on the objective function discussed in this chapter included:

Characteristics of the objective function at the optimum.
Slope and gradient of the objective function.
Effect of constraints on the objective function.
Modeling the objective function by experimentation.
Statistical significance of the objective function models.
Effect of stochastic noise on the objective function.
Simulation of the objective function.

Regardless of whether the reader plans to apply EVOP strategy, he will find the study of this chapter rewarding. Many insights to the problems and approaches of optimal control can be developed by this study. Chapter 10 will introduce the optimal control techniques of dynamic processes.

10

Dynamic Optimization of Well-Defined Processes

10.0 INTRODUCTION

Chapters 7 through 9 presented some of the practical optimization techniques for steady-state processes. Steady-state processes are approximations to the real-world *where time is assumed not to be a parameter*. Steady-state approximations are generally valid for cases where all the transients are allowed to decay before measurements of the state variables are made, or before adjustments are made to the control variables. Steady-state physical process models also assume that the processes are always in equilibrium.

This chapter will present several practical techniques for the optimization of processes which are time-dependent. The dynamic objective functions and the physical process models are no longer simple functions of the present values of the state variables, but become complex functions which depend on the past history and predicted future values of the state variables.

This chapter will first present the classical approach of applying *variational techniques* to the continuous-time objective function in integral equation form. This technique can transform the problem to one of solving the Euler-Lagrange equation in differential equation form for the optimal control variables. Or, the Euler-Lagrange equation may be directly implemented in a feedback control loop to adjust the control variables. This chapter will then present *dynamic programming concepts, techniques of optimizing discrete-time and discrete-value functions*, and various *feedforward and feedback approaches for implementing optimal control with dynamic programming techniques.*

310

10.1   POSING THE DYNAMIC OPTIMIZATION PROBLEM

This section will describe how to pose the following types of dynamic optimization problems:

Minimum time.
Minimum cost.
Optimal continuous operation.

A. Minimum-Time Problems

Minimum-time problems are those which have *the objective of proceeding between known initial conditions and pre-determined terminal conditions in minimum time, subject to constraints.* Typical applications of minimum-time problems are batch processes, where the material costs are relatively independent of the processing conditions, e.g., basic oxygen furnaces for steelmaking. The major costs which are dependent on the processing conditions are plant investment and labor. Plant investment is in the category of fixed expenses. However, if the processing time can be reduced, the same investment will be capable of producing more output. This will increase the productivity in output per dollar of invested capital, which was discussed in Chapter 1. Thus minimizing the time required for a batch process will reduce the investment cost allocation for each unit of product. The cost of labor and fuel is directly related to time, so minimizing the processing time will directly minimize the cost of labor and fuel for each unit of product.

Let the following symbols be assigned:

t = time.
T = elapsed time to be minimized.
t_0, t_f = initial and final times.
X_i = control variables, $(i = 1)$ to $(i = n)$.
Y_j = state variables, $(j = 1)$ to $(j = m)$.
H_j = physical process model functions, $(j = 1)$ to $(j = m)$.
W_q, Z_i = constraint functions for state and control variables, respectively,
$\qquad (q = 1)$ to $(q = r)$, and $(r \leq m)$.
L = transformed objective function.

From Section 4.7C, a general form of the dynamic physical process model with n control variables and m state variables is

$$\frac{dY_j}{dt} = H_j(Y_j, X_i, t) \qquad (10.1\text{-}1)$$

subject to the constraints (with slack variables introduced)

$$W_q(Y_j) = 0 \qquad\qquad (10.1\text{-}2)$$

$$Z_i(X_i) = 0 \qquad\qquad (10.1\text{-}3)$$

The objective function to be minimized in terms of the variable t is

$$T = \int_{t_0}^{t_f} dt \qquad\qquad (10.1\text{-}4)$$

Unfortunately, the form of objective function in Equation 10.1-4 is not very useful because it is not expressed in terms of the state and control variables. Let Y_α be the particular state variable whose initial and final values are specified:

$$Y_\alpha(t_0) = Y_{\alpha 0} \qquad\qquad (10.1\text{-}5)$$

$$Y_\alpha(t_f) = Y_{\alpha f} \qquad\qquad (10.1\text{-}6)$$

where α designates the specified state variables. The dynamic physical process model in Equation 10.1-1 can also be expressed in the form of state equations for the variable Y_α:

$$\frac{dY_\alpha}{dt} = G_\alpha(Y_1, Y_2, \ldots, Y_j, \ldots, Y_m; t) + J_\alpha(X_1, X_2, \ldots, X_i, \ldots, X_n; t)$$

$$(10.1\text{-}7)$$

where G_α represents the rate of change of the state variable Y_α as a function of all the state variables, and J_α represents the external influences on the state variable Y_α, e.g., changes due to the control variables.

In cases where the dynamic physical process model can be directly substituted into the objective function, the objective function can be transformed so that the variable of integration is the state variable Y_α. The initial and final conditions will also be transformed to $Y_{\alpha 0}$ and $Y_{\alpha f}$. The general form of the transformed objective function L to be minimized is then

$$T = \int_{Y_{\alpha 0}}^{Y_{\alpha f}} L(Y_j; X_i; t)\, d(Y_\alpha) \qquad\qquad (10.1\text{-}8)$$

subject to the constraints

$$W_q(Y_j) = 0 \qquad\qquad (10.1\text{-}9)$$

$$Z_i(X_i) = 0 \qquad\qquad (10.1\text{-}10)$$

In cases where it is not possible to substitute the dynamic physical process model into the objective function to replace t with Y_α as the variable of integration, the Lagrange multiplier approach may be used to produce a modified objective function by considering the state equations as constraints:

$$\frac{dY_j}{dt} - G_j - J_j = 0 \qquad (10.1\text{-}11)$$

The modified objective function to be minimized is

$$T = \int_{t_0}^{t_f} \left\{ 1 + \sum_{j=1}^{m} \lambda_j \left[\frac{dY_j}{dt} - G_j - J_j \right] \right\} dt \qquad (10.1\text{-}12)$$

Example 1

Let the dynamic physical process model be

$$\frac{dY}{dt} = C_1 Y + C_2 X(t) \qquad (10.1\text{-}13)$$

where C_1, C_2 are constants, Y is the state variable, and X is the control variable. From Equation 10.1-13,

$$dt = \frac{dY}{C_1 Y + C_2 X(t)} \qquad (10.1\text{-}14)$$

Substituting Equation 10.1-14 into the objective function in Equation 10.1-4, the objective function to be minimized is

$$T = \int_{Y_0}^{Y_f} \frac{dY}{C_1 Y + C_2 X(t)} \qquad (10.1\text{-}15)$$

Example 2

Let the dynamic physical process model be

$$\frac{dY_1}{dt} = C_1 Y_1 + C_2 Y_2 + C_3 X(t) \qquad (10.1\text{-}16)$$

$$\frac{dY_2}{dt} = C_4 Y_1 + C_5 Y_2 + C_6 X(t) \qquad (10.1\text{-}17)$$

Since the Equations 10.1-16 and 10.1-17 cannot be directly substituted into the objective function, the objective function will be modified by the Lagrange multiplier technique. Thus the objective function to be minimized is

$$T = \int_{t_0}^{t_f} \left\{ 1 + \lambda_1 \left[\frac{dY_1}{dt} - C_1 Y_1 - C_2 Y_2 - C_3 X(t) \right] \right.$$

$$\left. + \lambda_2 \left[\frac{dY_2}{dt} - C_4 Y_1 - C_5 Y_2 - C_6 X(t) \right] \right\} dt \qquad (10.1\text{-}18)$$

B. Minimum Cost Problems

Minimum cost dynamic optimization problems have the objective of minimizing the cost of operation for a period of time, or from a known initial condition to terminal condition. This type of problem is similar to the minimum time problem except the cost is minimized directly. This is also referred to as the optimal trajectory problem.

The operating economic models discussed in Chapter 5 are essentially steady-state models. To view the operating economic model as a time-dependent entity, it is necessary to use the *cost-rate function* which states the rate of cost expenditures, in dollars per hour. Thus the cost over a period of time will be the integral or summation of the cost-rate function over the period of time. Let the following additional symbols be assigned:

F = cost-rate function expressed in state and control variables, in dollars per hour.

ϕ = objective function to be minimized, in dollars.

The objective function to be minimized can be derived by integrating the cost-rate function from t_0 to t_f:

$$\phi = \int_{t_0}^{t_f} F(Y_j, X_i, t)\, dt \tag{10.1-19}$$

The objective function in discrete-time form may be derived by taking the summation of the cost-rate functions from the interval $(t = 0)$ to $(t = N\,\Delta t)$:

$$\phi = \sum_{p=0}^{N} F(Y_j, X_i, p\,\Delta t)\Delta t \tag{10.1-20}$$

where p is the index of time-intervals.

Example

The problem of combined hydro-thermal generation dispatching for electric utilities will be used as an example to illustrate how the minimum cost problem can be posed. The objective of operation is to minimize the thermal fuel cost required to meet a specified load demand over a twenty-four hour period, subject to the constraint that the amount of water equal to the river flow for that period be used for hydro-generation. Stated in another way, the constraint specifies that the reservoir storage at the beginning and end of the period be equal. Let the following symbols be defined:

F = thermal fuel cost, in dollars per hour.

T = thermal generation, in megawatts.

L = loss in transmission, in megawatts.
W = system connected load demand, in megawatts.
S = reservoir storage, in cubic feet.
H = hydraulic head, in feet.
M = forebay elevation, in feet.
N = tailrace elevation, in feet.
P = hydro-generation, in megawatts.
Q = river flow into the reservoir, in cubic feet per second.
U = nongenerating release from the reservoir, in cubic feet per second.
R = plant discharge, in cubic feet per second.
$C_0, C_1, C_2, \ldots, C_9$ = constants.

Figure 10.1-1 shows an elevation diagram for the reservoir. The following assumptions will be made:

1. The effective tailrace elevation N is constant.
2. The thermal system will not reach maximum capacity.
3. The coefficients of the cost-rate equation are fixed for all load conditions.
4. Scale factors are incorporated into the constants.

The state variables of the system are reservoir storage S, hydraulic head H, and hydro-generation P. The control variables are the plant discharge R and thermal generation T. The river flow is assumed to be a constant for the 24-hour period. The dynamic physical process model can be developed by expressing the state variables in terms of other state variables and the control variable.

The hydraulic head H is the difference between the forebay elevation M and tailrace elevation N:

$$H = M - N \tag{10.1-21}$$

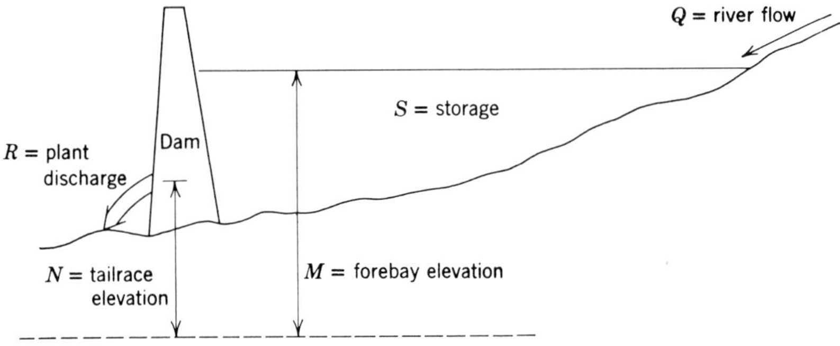

Fig. 10.1-1   A typical reservoir for hydro-generation.

The forebay elevation M can be empirically expressed as a function of reservoir storage S. The model depends on the shape of the reservoir. Regression techniques may be used to develop a model for the forebay elevation M:

$$M = C_0 + C_1 S + C_2 S^2 \tag{10.1-22}$$

The hydraulic head H may how be expressed as a function of the reservoir storage S by substituting Equation 10.1-22 into Equation 10.1-21.

$$H = (C_0 - N) + C_1 S + C_2 S^2 \tag{10.1-23}$$

The hydro-generation P is proportional to the hydraulic head H and the plant discharge R.

$$P = C_3(HR) \tag{10.1-24}$$

The system connected load demand may be assumed to vary according to a cosine wave within the 24-hours of the day. Thus:

$$W = \frac{C_4}{3}(2 + \cos 2\pi t) \tag{10.1-25}$$

where C_4 is the peak load of the 24-hour period, and the time t is given in units of day. Figure 10.1-2 shows the profile of the system connected load through a 24-hour day. Thus the peak is reached at $t = 0$ and $t = 1$, and the minimum is reached at $t = \frac{1}{2}$.

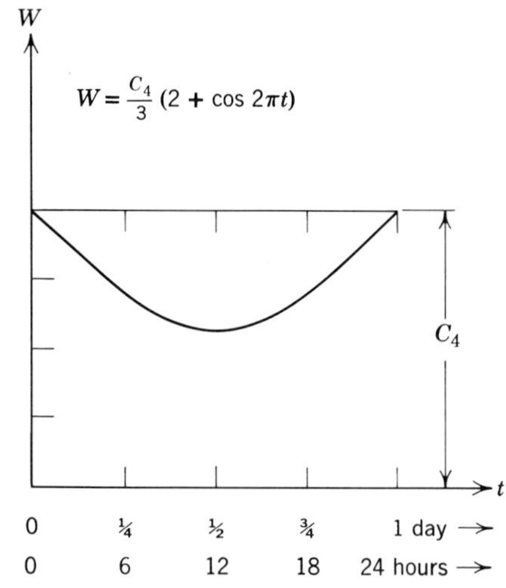

Fig. 10.1-2 Profile of system connected load demand.

The thermal generation T required is the difference between the system connected load W and the hydro-generation plus the transmission losses L. Thus

$$T = W - P + L \tag{10.1-26}$$

The rate of change of the reservoir storage is the difference between the river flow into the reservoir Q and the plant discharge R, less the non-generating release U:

$$\frac{dS}{dt} = Q - R - U \tag{10.1-27}$$

The cost-rate equation for operating the system is equal to the cost-rate of operating the thermal generating plant F, which may be approximated by

$$F = C_6 + C_7 T + C_8 T^2 + C_9 T^3 \tag{10.1-28}$$

where C_6, C_7, C_8, C_9 are constant coefficients.

The problem is to program the plant discharge rate $R(t)$ so as to minimize the total cost $\phi(R)$ over the 24-hour period, i.e., to minimize

$$\phi(R) = \int_0^1 F(T)\, dt \tag{10.1-29}$$

subject to the constraint

$$S(0) - S(1) = 0 \tag{10.1-30}$$

and subject to the constraints of the physical process model given by Equations 10.1-26 and 10.1-27, which may be rewritten as

$$L - T - P + W = 0 \tag{10.1-31}$$

$$\frac{dS}{dt} - Q + R + U = 0 \tag{10.1-32}$$

The Lagrange multiplier technique may be used to modify the dynamic objective function $\phi(R)$ to be subject to the physical process model constraints. The modified dynamic objective function is

$$\phi'(R) = \int_0^1 \left[F(T) + \lambda(L - P + W - T) + \gamma\left(\frac{dS}{dt} - Q + R + U\right) \right] dt$$

$$\tag{10.1-33}$$

where λ and γ are the Lagrange multipliers.

In discrete-time form the objective function to be minimized is

$$\phi = \sum_{p=0}^{N} F(T^p, S^p, p\,\Delta t)\Delta t \tag{10.1-34}$$

subject to the constraints of the boundary conditions and the physical process model:

$$L^p - P^p + W^p - T^p = 0 \qquad (10.1\text{-}35)$$

$$S^{p+1} - S^p = (Q^p - R^p - U^p)\Delta t \qquad (10.1\text{-}36)$$

where the superscript p denotes the time intervals.

The derivation given above is for one hydro-generating reservoir. The physical process model and the objective function for multiple hydro-generating reservoirs may be developed by extension.

C. Continuous Optimal Operation Problem

Continuous optimal operations have the objective of bringing a set of process state variables as economically as possible to their desired values. The optimizing procedure is repeated if the set of desired operating values are changed at a later time. For example, the continuous optimal operation of the chemical reactor discussed in Section 4.7 has the objective of bringing the concentrations of the reactants and products to their specified desired values *as rapidly as possible.*

The continuous optimal operation of a dynamic process has many facets which are similar to the steady-state optimization problems discussed in Sections 7.1, 7.2, and 7.3, and to the adaptive control problems discussed in Sections 8.1 and 8.2. The objective of dynamic continuous optimal operation differs, however, by the addition of the economics of the transient. Let the following additional symbols be assigned:

$\overline{Y}_j$ = desired values of the state variables, $(j = 1)$ to $(j = r)$.
ψ_j = constants, $(j = 1)$ to $(j = r)$, where $(r \leq n)$.
Δt = sampling period.
F = steady-state quadratic form objective function.
ϕ = dynamic objective function to be minimized.
α = a positive real constant.
p = index for time intervals.

Chapters 7 and 8 presented the quadratic form of the objective function for minimizing the differences between a set of state variables Y_j and their desired values $\overline{Y}_j$. The set of r state variables to be optimally controlled may not be the complete set of n state variables, thus $r \leq n$. The quadratic form of the steady-state objective function is

$$F = \sum_{j=1}^{r} \psi_j (\overline{Y}_j - Y_j)^2 \qquad (10.1\text{-}37)$$

The conditions for optimum may be imposed on the objective function in Equation 10.1-37 to derive a set of optimal control equations. The optimal control equations express the settings of the control variables X_i which will minimize the objective function F. This type of optimal control is called open-loop or feedforward optimal control for steady-state processes, i.e., processes which are assumed to be in equilibrium.

The major shortcoming of feedforward optimal control was discussed in Chapter 8, i.e., the errors introduced when the *actual values* of the process coefficients deviate from the *average values* of the coefficients of the physical process model. This shortcoming can be reduced by using the feedback optimal control technique which uses an incremental physical process model. However there still remain two major shortcomings of the steady-state feedforward optimal control technique: the problems of programming and stability. The problem of programming is the lack of information for the sequence of adjustments to the control variables to optimize the process. For example, should all the m control variables be adjusted in unison or should they be adjusted sequentially? Or, should the change to a specific control variable be made in a step change or a succession of smaller changes? The problems of stability for steady-state feedforward optimal control is raised because the optimal control variable settings for minimum F may lead to a region of unstable operations. Stability considerations, however, are outside the scope of this book.

The objective of continuous dynamic optimization is to find the set of control variables $X_i(t)$ which will minimize the objective function for a process which was at the initial state $Y_i(t_0)$ at $t = t_0$. The period of time involved is from $(t = t_0)$ to $(t = \infty)$. The objective function to be minimized depends on the initial values of the state variables. One form is

$$\phi[Y(t_0)] = \int_{t_0}^{\infty} (F)\exp\left[-\alpha(t - t_0)\right] dt \qquad (10.1\text{-}38)$$

The exponential term is an additional weighting constant which is time-dependent. Substituting the value of F from Equation 10.1-37 into Equation 10.1-38, the objective function to be minimized is

$$\phi[Y_j(t_0)] = \int_{t_0}^{\infty} \left\{ \sum_{j=1}^{r} \psi_j[\bar{Y}_j - Y_j(t)]^2 \right\}\exp[-\alpha(t - t_0)] dt \qquad (10.1\text{-}39)$$

Equation 10.1-39 may be simplified if the dynamic physical process model is expressed as a set of difference equations. Furthermore the assumption can be made that the values of the control variables X_i are constant over the sampling period Δt. Thus it is no longer necessary to solve for all possible functions of the control variables from $(t = t_0)$ to $(t = \infty)$ to optimize ϕ, i.e.,

$$X_i(t) \quad \text{from} \quad (t = t_0) \quad \text{to} \quad (t = \infty) \qquad (10.1\text{-}40)$$

It is necessary to solve for the following sequences of constant control variables to optimize ϕ:

$$X_i(t_0), X_i(t_0 + \Delta t), X_i(t_0 + 2\,\Delta t), \ldots, X_i(t_0 + p\,\Delta t), \ldots \qquad (10.1\text{-}41)$$

The objective function in Equation 10.1-39 may be restated in discrete-time form as

$$\phi[Y_j(t_0)] = \sum_{p=1}^{\infty} \left\{ \sum_{j=1}^{r} \psi_j [\overline{Y}_j - Y_j(t_0 + p\,\Delta t)]^2 \right\} \exp[-p(\Delta t)]\Delta t \qquad (10.1\text{-}42)$$

The optimization of $\phi[Y_j(t_0)]$ is a finite step-wise procedure. The choice of the set of control variables at $t = (t_0 + p\,\Delta t)$ generally depends on the preceding steps chosen and succeeding steps to be chosen.

10.2   VARIATIONAL TECHNIQUES FOR CONTINUOUS-TIME FUNCTIONS

The classical calculus of variations approach can be used to solve the simple problems of optimizing the integral involving two continuous-time variables. While this approach is generally not practical for real-life problems, the theoretical derivations can provide useful insights for the analytical approach to the dynamic optimization problem.

This section will first derive the Euler-Lagrange equation by using the classical variational approach. The Euler-Lagrange equation plays the role for dynamic optimization which the conditions of optimum played for steady-state optimization. Closed-form optimal solutions for certain simple objective functions and models can be developed from the Euler-Lagrange differential equations. A simple example will be used to illustrate the application of the Euler-Lagrange equation. Then the direct implementation of the Euler-Lagrange equation will be given.

A. Derivation of the Euler-Lagrange Equation

The classical calculus of variation problem consists of minimizing the integral

$$\phi(Y, X) = \int_{t_0}^{t_f} F(Y, X)\, dt \qquad (10.2\text{-}1)$$

where the symbols are as defined in Section 10.1. The physical process model is

$$\frac{dY}{dt} = H(Y, X) \qquad (10.2\text{-}2)$$

with initial and terminal conditions of

$$Y(t_0) = Y_0 \tag{10.2-3}$$

$$Y(t_f) = Y_f \tag{10.2-4}$$

The problem is to determine the control variable $X(t)$ that minimizes the objective function $\phi(Y, X)$ while the state variable $Y(t)$ satisfies the physical process model given by Equation 10.2-2. For purposes of illustration, consider the physical process model to be a linear function. Thus Equation 10.2-2 becomes

$$X = A_0 Y + A_1 \frac{dY}{dt} \tag{10.2-5}$$

where A_0 and A_1 are constant coefficients.

Substitution of Equation 10.2-5 into Equation 10.2-1 to eliminate X produces the following:

$$\phi(Y) = \int_{t_0}^{t_f} G\left(Y, \frac{dY}{dt}\right) dt \tag{10.2-6}$$

where G is a function of Y and dY/dt only.

Assuming that $\overline{Y}(t)$ is the function that minimizes $\phi(Y)$, the variational approach states that it is possible to add a function $\varepsilon Z(t)$ to $\overline{Y}(t)$, where ε is a constant and $Z(t)$ is equal to zero at $t = t_0$ and $t = t_f$. Thus

$$Y = \overline{Y} + \varepsilon Z \tag{10.2-7}$$

The integral ϕ will be a function of ε, and ϕ will be minimum when

$$\varepsilon = 0 \tag{10.2-8}$$

The objective function to be minimized becomes

$$\phi(\varepsilon) = \int_{t_0}^{t_f} G(\overline{Y} + \varepsilon Z, \dot{\overline{Y}} + \varepsilon \dot{Z}) \, dt \tag{10.2-9}$$

where $\dot{\overline{Y}} = d\overline{Y}/dt$ and $\dot{Z} = dZ/dt$.

Since $\phi(\varepsilon)$ is a minimum when $\varepsilon = 0$, it follows then that

$$\frac{d\phi(\varepsilon)}{d\varepsilon} = 0 \quad \text{at } \varepsilon = 0 \tag{10.2-10}$$

Next, examine the integrand G. The integrand G may be considered as a function of $\overline{Y}$ and $\dot{\overline{Y}}$, both having variations of ε. Taking the derivative of G with respect to ε yields

$$\frac{dG}{d\varepsilon} = \frac{\partial G}{\partial \overline{Y}} \frac{d\overline{Y}}{d\varepsilon} + \frac{\partial G}{\partial \dot{\overline{Y}}} \frac{d\dot{\overline{Y}}}{d\varepsilon} \tag{10.2-11}$$

From Equation 10.2-7

$$\frac{d\overline{Y}}{d\varepsilon} = -Z \tag{10.2-12}$$

and

$$\frac{d\dot{\overline{Y}}}{d\varepsilon} = -\dot{Z} \tag{10.2-13}$$

Substituting Equations 10.2-12 and 10.2-13 into Equation 10.2-11,

$$\frac{dG}{d\varepsilon} = -\left(\frac{\partial G}{\partial Y} Z + \frac{\partial G}{\partial \dot{Y}} \dot{Z}\right) \tag{10.2-14}$$

Bringing the differentiation under the integral and substituting Equation 10.2-14 into Equation 10.2-9,

$$\frac{d\phi}{d\varepsilon} = -\int_{t_0}^{t_f} \left(\frac{\partial G}{\partial \overline{Y}} Z + \frac{\partial G}{\partial \dot{Y}} \dot{Z}\right) dt \tag{10.2-15}$$

Integrating the second term of the integrand by parts, Equation 10.2-15 becomes

$$\frac{d\phi}{d\varepsilon} = -\int_{t_0}^{t_f} \left\{\frac{\partial G}{\partial Y} Z - \left[\frac{d}{dt}\left(\frac{\partial G}{\partial \dot{Y}}\right)\right] Z\right\} dt - \left(\frac{\partial G}{\partial \dot{Y}} Z\right)_{t_0}^{t_f} \tag{10.2-16}$$

By definition, $Z = 0$ at $t = t_0$ and $t = t_f$, which makes the last term of Equation 10.2-16 equal to zero. Thus

$$\frac{d\phi}{d\varepsilon} = -\int_{t_0}^{t_f} \left\{\frac{\partial G}{\partial \overline{Y}} Z - \left[\frac{d}{dt}\left(\frac{\partial G}{\partial \dot{Y}}\right)\right] Z\right\} dt \tag{10.2-17}$$

Since $d\phi/d\varepsilon = 0$, therefore

$$\frac{d\phi}{d\varepsilon} = -\int_{t_0}^{t_f} \left\{\frac{\partial G}{\partial \overline{Y}} - \left[\frac{d}{dt}\left(\frac{\partial G}{\partial \dot{Y}}\right)\right]\right\} Z \, dt = 0 \tag{10.2-18}$$

Since Z can be any arbitrary value except at $t = t_0$ and $t = t_f$, Equation 10.2-18 can be satisfied only if the terms within the braces are equal to zero. Thus

$$\frac{\partial G}{\partial \overline{Y}} - \frac{d}{dt}\left(\frac{\partial G}{\partial \dot{Y}}\right) = 0 \tag{10.2-19}$$

Equation 10.2-19, the Euler-Lagrange equation, corresponds to the *condition or optimum* for steady-state optimization. The Euler-Lagrange equation defines the necessary conditions for minimizing the objective function ϕ, subject to the constraints of the physical process model.

B. Example of Applying the Euler-Lagrange Equation

Let ϕ be the objective function to be minimized:

$$\phi(Y, X) = \int_0^T [Y^2(t) + X^2(t)]\, dt \tag{10.2-20}$$

and the physical process model is

$$\frac{dY}{dt} = C_0 X \tag{10.2-21}$$

with initial and terminal conditions

$$Y(0) = Q \tag{10.2-22}$$

$$Y(T) = 0 \tag{10.2-23}$$

The Euler-Lagrange equation is

$$\frac{\partial G}{\partial Y} - \frac{d}{dt}\left(\frac{\partial G}{\partial \dot{Y}}\right) = 0 \tag{10.2-24}$$

Substituting the value of G from the objective function into the Euler-Lagrange equation:

$$2Y - \frac{d}{dt}\left[2X\,\frac{dX}{d\dot{Y}}\right] = 0 \tag{10.2-25}$$

Since

$$X = \frac{1}{C_0}\frac{dY}{dt} \tag{10.2-26}$$

$$\frac{d}{dt}\left[X\,\frac{dX}{d\dot{Y}}\right] = \frac{1}{C_0{}^2}\left[\frac{d^2 Y}{dt^2}\right] \tag{10.2-27}$$

The Euler-Lagrange equation for the example is

$$Y - \frac{1}{C_0{}^2}\left(\frac{d^2 Y}{dt^2}\right) = 0 \tag{10.2-28}$$

The closed form solution to the differential equation in Equation 10.2-28 with boundary conditions $Y(0) = Q$ and $Y(t) = 0$ is

$$Y(t) = Q\{\cosh(C_0 t) - [\operatorname{ctnh}(C_0 T)][\sinh(C_0 t)]\} \tag{10.2-29}$$

and the corresponding optimal control variable X can be derived from the physical process model:

$$X(t) = Q\{\sinh(C_0 t) - [\operatorname{ctnh}(C_0 T)][\cosh(C_0 t)]\} \tag{10.2-30}$$

The solution may be verified by substituting Y into the Euler-Lagrange equation:

$$Q\{\cosh(C_0 t) - [\text{ctnh}(C_0 T)][\sinh(C_0 t)]\}$$

$$- Q\frac{C_0^2}{C_0^2}\{\cosh(C_0 t) - [\text{ctnh}(C_0 t)][\sinh(C_0 t)]\} = 0 \qquad (10.2\text{-}31)$$

Also

$$Y(0) = Q[1 - 0]$$

$$= Q \qquad (10.2\text{-}32)$$

and

$$Y(T) = Q\{\cosh(C_0 T) - [\text{ctnh}(C_0 T)][\sinh(C_0 T)]\}$$

$$= Q\left\{\cosh(C_0 T) - \left[\frac{\cosh(C_0 T)}{\sinh(C_0 T)}\right] \cdot [\sinh(C_0 T)]\right\} = 0 \qquad (10.2\text{-}33)$$

Equations 10.2-32 and 10.2-33 indicate that the optimal Y satisfy the boundary conditions.

C. Implementation of the Euler-Lagrange Equation for Optimal Control

The optimal X and Y given in Equations 10.2-29 and 10.2-30 are the optimal solutions to minimize the objective function given by Equation 10.2-20. The optimal control variable X can be directly implemented to minimize the objective function and to satisfy the terminal condition of $Y(T) = 0$. In practice, however, the physical process model can be assumed to carry an additional term U to represent the uncontrolled variable. Thus the actual physical model is

$$\frac{dY}{dt} = C_0 X + U \qquad (10.2\text{-}34)$$

The presence of the uncontrolled variable will cause the measured state variable Y^p at $t = t_p$ to deviate from the calculated optimal state variable $\hat{Y}^p$. A feedback technique to assess the system performance and to re-calculate the optimal X for the remaining path to be followed can be developed to implement the optimal control given by the example.

Figure 10.2-1 shows that the measured Y^p is compared against the calculated optimal $\hat{Y}^p$ at the end of every interval $t = t_p$. If the deviation is significant at the end of any interval, i.e., by comparing the difference of $\hat{Y}^p$ and Y^p with a known constant C_1,

$$|\hat{Y}^p - Y^p| \geq C_1 \qquad (10.2\text{-}35)$$

then a new optimal control variable X^p will be calculated by using

$$X^p(t) = Y^p C_0\{\sinh C_0(t - t_p) - [\text{ctnh} C_0(T - t_p)][\cosh C_0(t - t_p)]\}$$

$$\qquad (10.2\text{-}36)$$

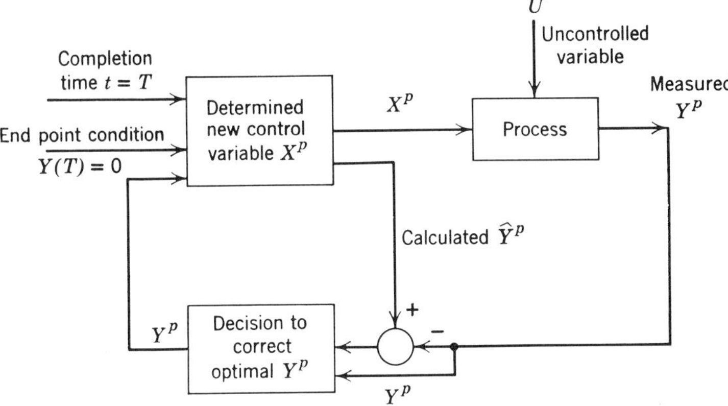

Fig. 10.2-1   Dynamic optimization with feedback.

The new control variable X^p calculated assumes that the system is starting at $t = t_p$ with initial condition equal to the measured state variable Y^p. The uncontrolled variable U can be ignored in the calculation of X^p because its effect is assumed to be included in the measured state variable. Figure 10.2-2 shows an exaggerated plot of the optimal state value $\widehat{Y}$. The solid line shows the initial calculated path for optimal $\widehat{Y}$. The longer dashed line shows the

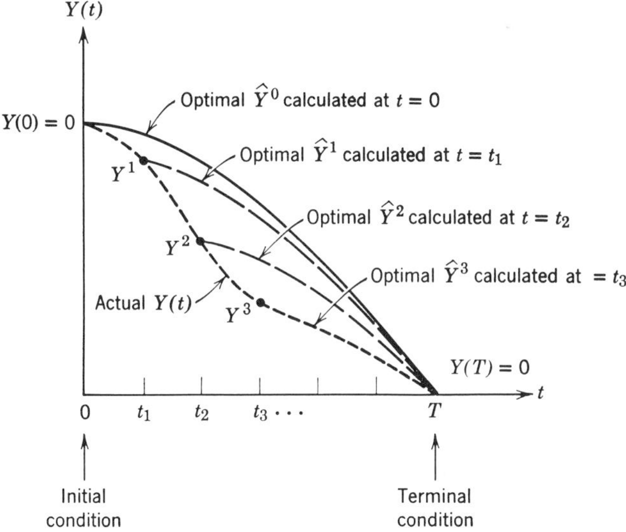

Fig. 10.2-2   Actual versus calculated optimal $Y(t)$ for dynamic optimization with feedback.

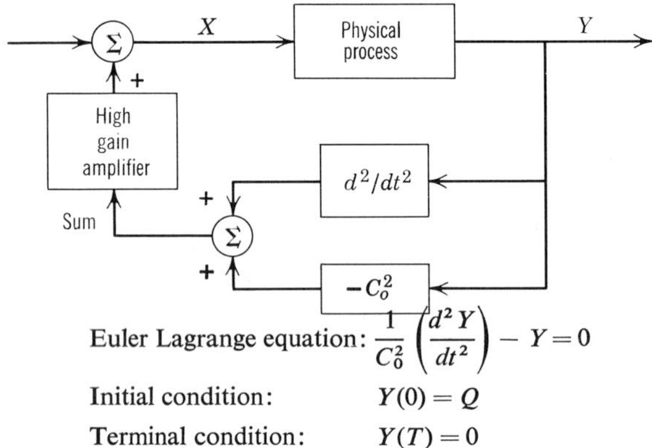

Euler Lagrange equation: $\dfrac{1}{C_0^2}\left(\dfrac{d^2 Y}{dt^2}\right) - Y = 0$

Initial condition: $Y(0) = Q$

Terminal condition: $Y(T) = 0$

Fig. 10.2-3 Direct implementation of the Euler-Lagrange equation.

succeeding recalculated paths for optimal $\hat{Y}$ by using feedback. The dashed line shows the actual path taken by Y.

The Euler-Lagrange equation can be solved directly and continuously to provide the solution for the optimal control variable $X(t)$. The Euler-Lagrange equation given by Equation 10.2-28 is the example used to illustrate one such method. Equation 10.2-28 can be formally solved by forcing the state and control variables of the physical process to satisfy the Euler-Langrange equation. Figure 10.2-3 shows a high gain amplifier taking the sum of $d^2 Y/dt^2$ and $-C_0^2 Y$ to create a change in the control variable X to force the sum to zero. Note, however, that there are *two problem areas in the direct solution method: the boundary conditions must somehow be satisfied and the stability problems cannot be ignored.*

As shown in Section 10.2B, the value of $X(t)$ depends on the following: the initial and terminal boundary conditions of Y and the time required to reach the terminal state. The initial condition of Y is contained in the process measurements. The terminal condition of Y is a given value and forms the *stopping criterion* ($Y_f = 0$ in the example). The time to reach the terminal condition cannot be explicitly included in the mechanized solution given in Figure 10.2-3. The difficulty of not being able to specify the time to reach the terminal condition can be partly circumvented by the approach shown in Figure 10.2-4. First, the ideal optimal control variable X^p is obtained by using Equation 10.2-36. Next, the effect of the uncontrolled variables and errors are compensated for by the feedback mechanized solution developed in Figure 10.2-3. Care must be taken to avoid instability. Finally, the stopping criterion terminates the calculations and sets X^p to the proper value for continued operation at the terminal condition.

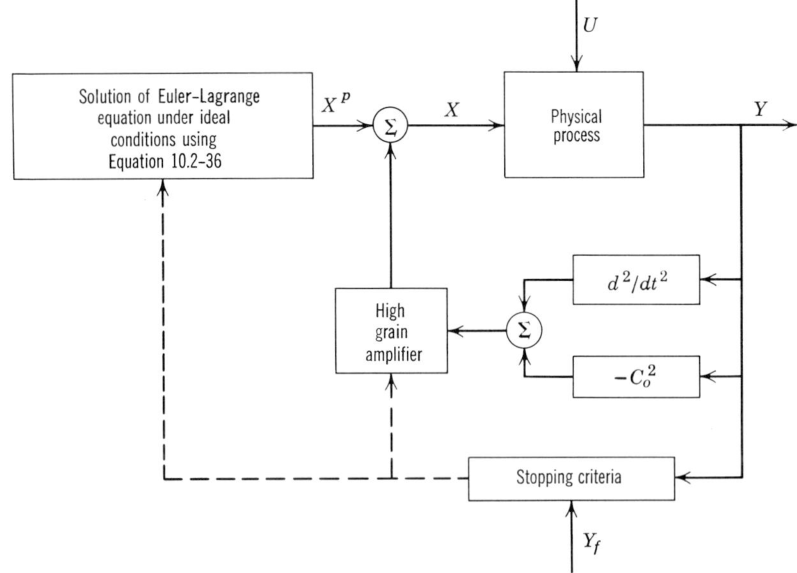

Fig. 10.2-4   Implementation of the Euler-Lagrange equation for optimal control.

10.3   DYNAMIC PROGRAMMING CONCEPTS

The theory of dynamic programming was developed by Bellman. It is a technique that is generally used for solving dynamic optimization problems in discrete-time form. The basic dynamic programming concept is based on the *recurrence relationship*, which allows each time-interval (or step) to be optimized independent of all the other time-intervals. This in effect reduces the computational problem by orders of magnitude, and transforms the problem from *simultaneously solving for the optimal control variables at all the time-intervals* to the problem of *sequentially solving for the optimal control variables at each time-interval*.

The recurrence relationship is based on the *principle of optimality*, which will be the first subject in this section. The decision tree approach to solving the multistage problem will then be introduced as an intuitive application of the principle of optimality.

A. The Principle of Optimality

The principle of optimality states that an optimal path (or policy) has the property that whatever the initial state and initial decisions are, the remaining decisions must constitute an optimal path (or policy) with regard to the state resulting from the first decision. By using the physical process model in discrete-time form with N-steps, the principle of optimality may be restated as: *The part of an N-step optimal policy leading from any intermediate state Y^p to the final state Y^N is an $(N - p)$ step optimal policy beginning from Y^p.*

The proof is by contradiction. Assume that the optimal path of the state variable is $(A \to B)$, as shown in Figure 10.3-1. Let C be a point on the path $(A \to B)$, then $(C \to B)$ is also the optimal path. Assume that an alternative optimal path $(C \to D \to B)$ exists, as shown in the dashed line. If $(C \to D \to B)$ is optimal, then the path $(C \to B)$ cannot be optimal. Therefore, there can only be one optimal path between $(C \to B)$, and that lies on the original path $(A \to B)$.

The principle of optimality can be applied to optimize dynamic objective functions effectively. Using the principle, it is possible to derive a recurrence relationship which will decompose a dynamic optimization problem from choosing a point in N dimensional state space into N choices of points in one dimensional state space. The principle of optimality is the foundation of the dynamic programming technique developed by Bellman.

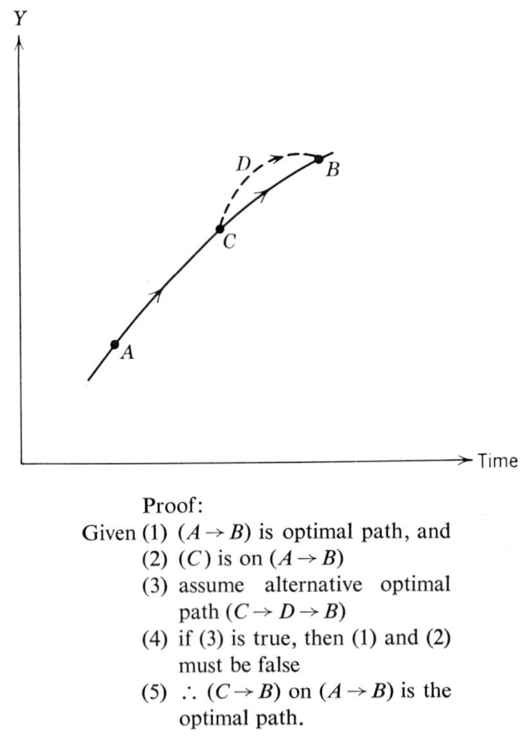

Proof:
Given (1) $(A \to B)$ is optimal path, and
     (2) (C) is on $(A \to B)$
     (3) assume alternative optimal path $(C \to D \to B)$
     (4) if (3) is true, then (1) and (2) must be false
     (5) $\therefore$ $(C \to B)$ on $(A \to B)$ is the optimal path.

Fig. 10.3-1   Proof of principle of optimality by contradiction.

B. The Decision Tree Approach

The decision tree approach to solving multistep dynamic optimization problems is intuitively based on the principle of optimality. The decision tree approach can provide insight into the recurrence relationship which will be derived in Section 10.4.

Any dynamic system with discrete-time and discrete-value state and control variables may be analyzed by the decision tree approach. Figure 10.3-2 shows a typical decision tree for a three-step dynamic optimization problem. Each *node* represents the state of the system at each step. Associated with each node are the state variables Y_j and the objective function ϕ. Each *path* represents a possible set of values for the control variables X_i at that step. Thus the nodes at the end of each path represent the state variables and the objective function which resulted from a specific set of adjustments of the control variables.

Using an example with one control variable and one state variable, let the resulting values of the objective function ϕ for the paths in Figure 10.3-2 be as follows:

$$Step\ 1$$

$$\phi^{A \to B} = 10$$

$$\phi^{A \to C} = 20$$

$$\phi^{A \to C} = 15$$

$$Step\ 2$$

$$\phi^{B \to E} = 1$$

$$\phi^{B \to F} = 6$$

$$\phi^{C \to F} = 5$$

$$\phi^{C \to G} = 4$$

$$\phi^{D \to G} = 3$$

$$\phi^{D \to H} = 2$$

$$Step\ 3$$

$$\phi^{E \to I} = 5$$

$$\phi^{F \to I} = 4$$

$$\phi^{F \to J} = 3$$

$$\phi^{G \to J} = 6$$

$$\phi^{G \to K} = 7$$

$$\phi^{H \to K} = 9$$

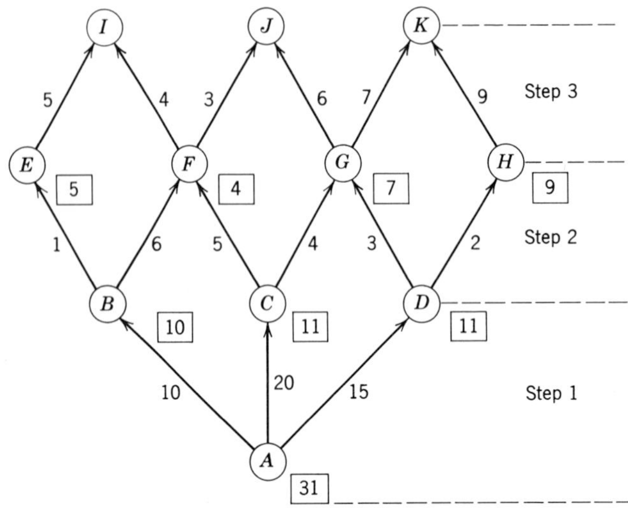

Notes. (1) □ indicates the optimal objective function at each node.
(2) The optimal path is $(A \rightarrow C)$, $(C \rightarrow G)$, and $(G \rightarrow K)$ for maximum objective function.

Fig. 10.3-2   The decision tree approach.

The decision tree approach will be used to solve the maximization problem backwards. The procedure is very simple and the example will be solved by visual inspection. Using the principle of optimality, each step must be optimal in order for the entire process to be optimal. For each step, the solution determines the maximum value of the objective function at each node, starting with Step 3. At Step 2 the maximum values of the input nodes of Step 3 may be used.

Starting from the input nodes of Step 3, i.e., nodes E, F, G, and H, the paths for Step 3 which will maximize the objective function ϕ are

$$E \rightarrow I$$
$$F \rightarrow I$$
$$G \rightarrow K$$
$$H \rightarrow K$$

The maximum objective function values at the input nodes to Step 3 are

$$\phi_{opt}^{E} = \phi^{E \rightarrow I} = 5 \tag{10.3-1}$$

$$\phi_{opt}^{F} = \phi^{F \rightarrow I} = 4 \tag{10.3-2}$$

$$\phi_{opt}^{G} = \phi^{G \rightarrow K} = 7 \tag{10.3-3}$$

$$\phi_{opt}^{H} = \phi^{H \rightarrow K} = 9 \tag{10.3-4}$$

For Step 2, the starting point can be the maximum values of the nodes E, F, G, and H, as listed in Equations 10.3-1 through 10.3-4. Starting at B, the possible paths are

$$\phi^B = \phi^{B \to E} + \phi^E_{\text{opt}} = 1 + 5 = 6 \qquad (10.3\text{-}5)$$

or

$$\phi^B = \phi^{B \to F} + \phi^F_{\text{opt}} = 6 + 4 = 10 \qquad (10.3\text{-}6)$$

By visual inspection, the maximum value of ϕ^B is

$$\phi^B_{\text{opt}} = \phi^{B \to F} + \phi^F_{\text{opt}} = 10 \qquad (10.3\text{-}7)$$

And the optimal path to B is

$$B \to F \to I$$

Similarly

$$\phi^C_{\text{opt}} = \phi^{C \to G} + \phi^G_{\text{opt}} = 11 \qquad (10.3\text{-}8)$$

And the optimal path to C is

$$C \to G \to K$$

Finally

$$\phi^D_{\text{opt}} = \phi^{D \to H} + \phi^H_{\text{opt}} = 11 \qquad (10.3\text{-}9)$$

And the optimal path to D is

$$D \to H \to K$$

Repeating the process for Step 1, the paths that will maximize the objective function at A are

$$\phi^A_{\text{opt}} = \phi^{A \to C} + \phi^C_{\text{opt}} = 20 + 11 = 31 \qquad (10.3\text{-}10)$$

The optimal paths for all the steps will be

$$A \to C$$
$$C \to G$$
$$G \to K$$

And the maximum objective function is

$$\phi^A_{\text{opt}} = \phi^{A \to C}_{\text{opt}} + \phi^{C \to G}_{\text{opt}} + \phi^{G \to K}_{\text{opt}} = 31 \qquad (10.3\text{-}11)$$

Equation 10.3-11 intuitively illustrates the principle of optimality and leads directly into the more rigorous recurrence relationship of Section 10.4.

The principle of optimality leads to a systematic and step-by-step method of solution for the optimum path. At each step in the calculation, only (1) the optimal cumulative value of the objective function at each node of the previous step, and (2) the value of the objective function contributed by the various paths of this step are needed. In contrast, the alternative is to first compute the objective function for all possible paths and then to choose the one having the optimal value. This procedure is illustrated below for the same example. Each path is defined by the nodes it traverses and the value of the objective function is also listed:

$$P_1: \quad A \to D \to H \to K, \quad \phi^1 = 15 + 2 + 9 = 26 \qquad (10.3\text{-}12)$$

$$P_2: \quad A \to D \to G \to K, \quad \phi^2 = 15 + 3 + 7 = 25 \qquad (10.3\text{-}13)$$

$$P_3: \quad A \to D \to G \to J, \quad \phi^3 = 15 + 3 + 6 = 24 \qquad (10.3\text{-}14)$$

$$P_4: \quad A \to C \to G \to K, \quad \phi^4 = 20 + 4 + 7 = 31 \qquad (10.3\text{-}15)$$

$$P_5: \quad A \to C \to G \to J, \quad \phi^5 = 20 + 4 + 6 = 30 \qquad (10.3\text{-}16)$$

$$P_6: \quad A \to C \to F \to J, \quad \phi^6 = 20 + 5 + 3 = 28 \qquad (10.3\text{-}17)$$

$$P_7: \quad A \to C \to F \to I, \quad \phi^7 = 20 + 5 + 4 = 29 \qquad (10.3\text{-}18)$$

$$P_8: \quad A \to B \to F \to J, \quad \phi^8 = 10 + 6 + 3 = 19 \qquad (10.3\text{-}19)$$

$$P_9: \quad A \to B \to F \to I, \quad \phi^9 = 10 + 6 + 4 = 20 \qquad (10.3\text{-}20)$$

$$P_{10}: A \to B \to E \to I, \quad \phi^{10} = 10 + 1 + 5 = 16 \qquad (10.3\text{-}21)$$

By visual inspection the optimal path is

$$A \to C \to G \to K$$

Although the effort involved in computing all the possible paths does not appear to be excessive in this simple example, the effort is usually prohibitive in any real-life problems.

10.4   DYNAMIC PROGRAMMING TECHNIQUES FOR DISCRETE-TIME FUNCTIONS

Section 10.3 introduced the basic dynamic programming concepts. These concepts reduced the dynamic optimization problem for discrete-time and discrete-value functions from the *simultaneous solution* for the optimal control variables for all time intervals to a *sequential solution* for the optimal control variables stepwise in time.

This section will apply these basic concepts to develop the computational algorithms for solving dynamic optimization problems for discrete-time and discrete-value functions. This section will first derive the *recurrence relationship* which was presented intuitively in Section 10.3. Then the general approach to the use of dynamic programming techniques will be given with a numerical example. The discussion will be limited to one control and one state variable for simplicity and clarity of presentation.

A. The Recurrence Relationship

The recurrence relationship is based on the principle of optimality. *The recurrence relationship relates the optimal objective function at any given step to the optimal objective function at the next step only.* Dynamic programming techniques for optimizing discrete-time functions use the recurrence relationship to optimize the objective function on a stepwise sequential basis. The computation may proceed forward or backward in time. The values of the optimal control variables over each time interval are expressed in terms of the state variables at the beginning of the interval. Let the following symbols be assigned:

$Y^0, Y^1, Y^2, \ldots, Y^P, \ldots, Y^N$ = values of the state variable at $(t_0), (t_0 + \Delta t)$, $(t_0 + 2\,\Delta t), \ldots, (t_0 + p\,\Delta t), \ldots, (t_0 + N\,\Delta t)$.

$X^0, X^1, X^2, \ldots, X^P, \ldots, X^N$ = values of the control variable at (t_0), $(t_0 + \Delta t), (t_0 + 2\,\Delta t), \ldots, (t_0 + p\,\Delta t), \ldots, (t_0 + N\,\Delta t)$.

ψ = physical process model function in discrete-time form.

$\phi_{\text{opt}}^{1 \to p}$ = optimal dynamic objective function for the 1 to p steps.

F_{opt}^{p} = steady-state objective function for the pth step, expressed in terms of X^{p-1} and Y^{p-1}.

G_{opt}^{p} = steady-state objective function for the pth step, expressed in terms of X^{p-1} only.

p = index for time intervals, $(p = 1)$ to $(p = N)$.

The dynamic physical process model in discrete-time form is

$$Y(t_0 + \Delta t) = \psi[(t_0 + \Delta t), Y(t_0), X(t_0), t_0] \tag{10.4-1}$$

and the objective function to be optimized is

$$\phi_{\text{opt}}^{1 \to N}[Y(t_0)] = \text{opt} \sum_{p=0}^{N-1} F[Y(t_0 + p\,\Delta t), X(t_0 + p\,\Delta t)] \tag{10.4-2}$$

The right-hand side of Equation 10.4-2 may be rewritten by taking the first term out of the summation sign. Thus:

$$\phi_{\text{opt}}^{1 \to N}[Y(t_0)] = F_{\text{opt}}^{1}[Y(t_0), X(t_0)] + \text{opt} \sum_{p=1}^{N-1} F[Y(t_0 + p\,\Delta t), X(t_0 + p\,\Delta t)]$$
$$\tag{10.4-3}$$

Observe that

$$\text{opt} \sum_{p=1}^{N-1} F[Y(t_0 + p\,\Delta t), X(t_0 + p\,\Delta t)] = \phi_{\text{opt}}^{2 \to N}[Y(t_0 + \Delta t)] \tag{10.4-4}$$

Therefore the *recurrence relationship* is

$$\phi_{\text{opt}}^{1\rightarrow N}[Y(t_0)] = F_{\text{opt}}^{1}[Y(t_0), X(t_0)] + \phi_{\text{opt}}^{2\rightarrow N}[Y(t_0 + \Delta t)] \qquad (10.4\text{-}5)$$

Using the notations X^0, X^1, ..., X^N and Y^0, Y^1, ..., Y^N as defined:

$$\phi_{\text{opt}}^{1\rightarrow N}[Y^0] = F_{\text{opt}}^{1}[Y^0, X^0] + \phi_{\text{opt}}^{2\rightarrow N}[Y^1] \qquad (10.4\text{-}6)$$

The steady-state objective function for the first step F_{opt}^{1} may be transformed into the G_{opt}^{1} form by using the physical process model to eliminate the state variable Y^0. Thus the recurrence relationship is

$$\phi_{\text{opt}}^{1\rightarrow N}[Y^0] = G_{\text{opt}}^{1}[X^0] + \phi_{\text{opt}}^{2\rightarrow N}[Y^1] \qquad (10.4\text{-}7)$$

$G_{\text{opt}}^{1}[X^0]$ may be solved for in terms of the control variable X^0 since it is only dependent on X^0 and not on any of the control variables at the other steps.

The *generalized forward recurrence relationship* at the pth interval is

$$\phi_{\text{opt}}^{1\rightarrow N}[Y^0] = G_{\text{opt}}^{1}[X^0] + G_{\text{opt}}^{2}[X^1] + \cdots + G_{\text{opt}}^{p-1}[X^{p-2}]$$

$$+ \phi_{\text{opt}}^{p\rightarrow N}[X^{p-1}] \qquad (10.4\text{-}8)$$

The *generalized backward recurrence relationship* at the pth interval may be similarly derived:

$$\phi_{\text{opt}}^{N\rightarrow 1}[Y^{N-1}] = G_{\text{opt}}^{N}[X^{N-1}] + G_{\text{opt}}^{N-1}[X^{N-2}] + \cdots + G_{\text{opt}}^{p+1}[X^{p}] + \phi_{\text{opt}}^{p\rightarrow 1}[X^{p\rightarrow 1}]$$

$$(10.4\text{-}9)$$

The forward recurrence relationship states that the optimal objective function for N steps is equal to the sum of the optimal objective function for $(p-1)$ steps plus the optimal objective function for the remaining $(N-p)$ steps. The backward recurrence relationship states that the optimal objective function for N steps is equal to the sum of the optimal objective function for the last $(N-p-1)$ steps plus the optimal objective function for the first p steps.

The optimization of a dynamic objective function of N steps is reduced by the recurrence relationship to a stepwise procedure of computing the optimal objective function for the first step, then for the second step, etc., ... At each step, the optimal objective function can be solved for only in terms of the control variable of that interval. Thus the optimal control variables can be selected independently, one at a time. The reduction in dimensionality is the main advantage of dynamic programming.

Time-dependent constraints on the state and control variables may be introduced. In other words, the constraints are introduced for specific time intervals or steps. This is a powerful tool, since it allows the designer freedom to adjust the constraint specifications and to see the effects on system performance. Note that the constraints must be specified in terms of the parameters and state and control variables for the specific steps involved. Otherwise the control variables for the different steps will become dependent and defeat the advantage offered by dynamic programming.

B. The General Approach for Numerical Solution

The general approach in numerically solving the optimal objective function is to apply the recurrence relationship of the optimal objective function either forward or backward in time. If the forward solution is used, the solution proceeds as follows:

$$\text{Step 1, Step 2, } \ldots, \text{ Step } N$$

If the backward solution is used, the solution proceeds as follows:

$$\text{Step } N, \text{ Step } (N-1), \ldots, \text{ Step 2, Step 1}$$

The following conditions must be satisfied at each step:

(1) Constraints for state and control variables, including terminal values.
(2) The physical process model for that step.
(3) The values of the state and control variables must be discrete and finite in number, so that an infinite combination of acceptable values do not have to be examined.

The selection of the optimal X^p, Y^p for that step will proceed as follows *if the backward solution is used*:

(1) Starting from Step N, construct a table of allowable X^p, Y^p for each Y^{p+1}. The set of X^p, Y^p, and Y^{p+1} must satisfy the physical process model.
(2) List the cumulative optimal values of the objective function $\sum_{p+1}^{N} F^p$ obtained before this step for all the values of Y^{p+1}.
(3) List the values of the objective function F^p for all allowable values of X^p, Y^p of this step.
(4) Add the objective function F^p of this step to the cumulative optimal objective function $\sum_{p+1}^{N} F^p$, obtained prior to this step to obtain the sum:

$$\sum_{p}^{N} F^p = F^p + \sum_{p+1}^{N} F^p \tag{10.4-10}$$

(5) Use search or table look-up techniques to select the optimal $\sum_{p}^{N} F^p$ for each allowable Y^p.
(6) Repeat Steps 1 through 5 until (X^0, Y^0) is reached.

Storage requirements for solving the dynamic optimization problem numerically can be prohibitive if all the intermediate values are saved. Let the additional symbols be assigned:

$Q =$ number of discrete values assigned to each variable.
$S =$ estimated number of storage locations.

The number of storage locations S required for each step for all the combinations of X^p, Y^p, and Y^{p+1} is estimated to be

$$S = 6Q^2 \qquad (10.4\text{-}11)$$

This is because six storage locations at each step are required for

$$X^p, \; Y^p, \; Y^{p+1}, \; F^p, \; \text{opt} \sum_{p+1}^{N} F, \; \left(F^p + \text{opt} \sum_{p+1}^{N} F^p \right) \qquad (10.4\text{-}12)$$

The total estimated number of storage locations S is

$$S = 6NQ^2 \qquad (10.4\text{-}13)$$

Thus if 100 steps are used ($N = 100$) and each variable is allowed 100 discrete values ($Q = 100$), the estimated number of storage locations required is

$$S = (6)(100)100^2 = (6)10^6 \qquad (10.4\text{-}14)$$

If only the optimal values of each interval are saved and all other values discarded, the storage requirement can be reduced considerably.

C. A Dynamic Programming Example

The recurrence relationship will be used to solve for the optimal objective function backward in time. The physical process model of the example in discrete-time form is

$$\frac{(Y^{p+1} - Y^p)}{\Delta t} = - X^p \qquad (10.4\text{-}15)$$

with initial and terminal conditions of

$$Y^0 = 5 \qquad (10.4\text{-}16)$$

$$Y^4 = 0 \qquad (10.4\text{-}17)$$

Let the control and state variables both assume six discrete values from 0 to 5. Also $\Delta t = 1$ will be used. Thus the physical process model becomes

$$Y^4 = Y^3 - X^3 \qquad (10.4\text{-}18)$$

$$Y^3 = Y^2 - X^2 \qquad (10.4\text{-}19)$$

$$Y^2 = Y^1 - X^1 \qquad (10.4\text{-}20)$$

$$Y^1 = Y^0 - X^0 \qquad (10.4\text{-}21)$$

A 6 by 4 grid may be formed for the state and control variables, as shown in Figure 10.4-1.

The objective function to be minimized is

$$\phi_{\min}^{4 \to 1}[Y^4] = \min \sum_{p=0}^{3} [(Y^p)^2 + (X^p)^2] \qquad (10.4\text{-}22)$$

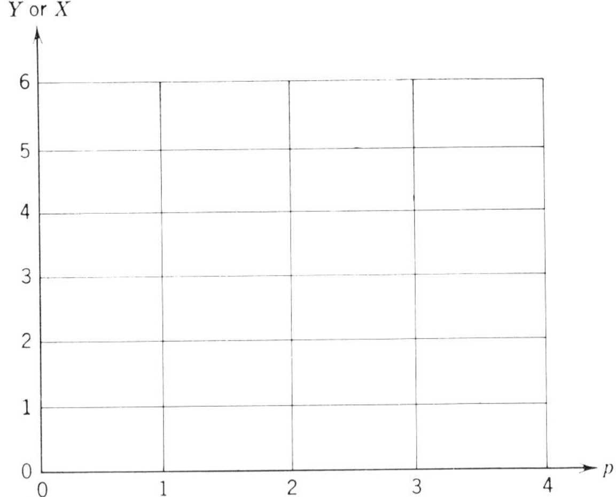

Fig. 10.4-1   Simplified grid for solving dynamic programming problems.

The parentheses in the equations are used to indicate that a quantity is raised to a power so as to distinguish it from the superscripts for the state and control variables.

The objective function to be minimized is a function of the terminal condition $Y^4 = 0$.

STEP 1. For $p = 4$, the recurrence relationship is

$$\phi_{min}^{4\rightarrow 1}[Y^4] = \phi_{min}^{3\rightarrow 1}[Y^3] + F_{min}^4[Y^3, X^3] \tag{10.4-23}$$

where

$$F_{min}^4[Y^3, X^3] = \min[(Y^3)^2 + (X^3)^2] \tag{10.4-24}$$

For $Y^4 = 0$, the allowable Y^3 and X^3 that satisfy the physical process model of Equation 10.4-18 are limited to $Y^3 = X^3$.

Table 10.4-1 shows a list of F_{min}^4 listed against the allowable set of Y^3 and X^3. Each will be the optimal value for the Y^3 it is listed against.

Table 10.4-1   The $p = 4$ Step

Y^4	X^3	Y^3	F^4	
0	0	0	0	←min for $Y^3 = 0$
0	1	1	2	←min for $Y^3 = 1$
0	2	2	8	←min for $Y^3 = 2$
0	3	3	18	←min for $Y^3 = 3$
0	4	4	32	←min for $Y^3 = 4$
0	5	5	50	←min for $Y^3 = 5$

STEP 2. For $p = 3$, the recurrence relationship for the optimal objective function is

$$\phi_{\min}^{4\to1}[Y^4] = \phi_{\min}^{2\to1}[Y^2] + F_{\min}^3[Y^2, X^2] + F_{\min}^4[Y^3, X^3] \qquad (10.4\text{-}25)$$

where $F_{\min}^4$ is listed in Table 10.4-1 for the various values of Y^3 and

$$F_{\min}^3[Y^2, X^2] = \min[(Y^2)^2 + (X^2)^2] \qquad (10.4\text{-}26)$$

At this step, the allowable values of X^2 and Y^2 are given by

$$Y^3 = Y^2 - X^2 \qquad (10.4\text{-}27)$$

The allowable X^2 and Y^2 and their corresponding F^3 and $F_{\min}^4$ are listed in Table 10.4-2.

Table 10.4-2   The $p = 3$ Step

Y^3	X^2	Y^2	F^3	$\min(F^4)$	$F^3 + \min(F^4)$	
0	0	0	0	0	0	←min for $Y^2 = 0$
0	1	1	2	0	2	←min for $Y^2 = 1$
0	2	2	8	0	8	
0	3	3	18	0	18	
0	4	4	32	0	32	
0	5	5	50	0	50	
1	0	1	1	2	3	
1	1	2	5	2	7	←min for $Y^2 = 2$
1	2	3	13	2	15	←min for $Y^2 = 3$
1	3	4	25	2	27	←min for $Y^2 = 4$
1	4	5	41	2	43	
2	0	2	4	8	12	
2	1	3	10	8	18	
2	2	4	20	8	28	
2	3	5	34	8	42	←min for $Y^2 = 5$
3	0	3	9	18	27	
3	1	4	17	18	35	
3	2	5	29	18	47	
4	0	4	16	32	48	
4	1	5	26	32	58	
5	0	5	25	50	75	

STEP 3. For $p = 2$, the recurrence relationship for the optimal objective function is

$$\phi_{\min}^{4\to1}[Y^4] = F_{\min}^1[Y^0, X^0] + F_{\min}^2[Y^1, X^1] + \min[F^3 + F^4] \qquad (10.4\text{-}28)$$

where $\min[F^3 + F^4]$ for each Y^2 are given in Table 10.4-2 and

$$F^2_{\min}[Y^1, X^1] = \min[(Y^1)^2 + (X^1)^2] \qquad (10.4\text{-}29)$$

At this step, the allowable values of Y^1 and X^1 are given by

$$Y^2 = Y^1 - X^1 \qquad (10.4\text{-}30)$$

The allowable Y^1 and X^1 and their corresponding F^2 and $\min[F^3 + F^4]$ are listed in Table 10.4-3.

Table 10.4-3   The $p = 2$ Step

Y^2	X^1	Y^1	F^2	$\min(F^3 + F^4)$	$F^2 + \min(F^3 + F^4)$	
0	0	0	0	0	0	←min for $Y^1 = 0$
0	1	1	2	0	2	←min for $Y^1 = 1$
0	2	2	8	0	8	
0	3	3	18	0	18	
0	4	4	32	0	32	
0	5	5	50	0	50	
1	0	1	1	2	3	
1	1	2	5	2	7	←min for $Y^1 = 2$
1	2	3	13	2	15	←min for $Y^1 = 3$
1	3	4	25	2	27	←min for $Y^1 = 4$
1	4	5	41	2	43	
2	0	2	4	7	11	
2	1	3	10	7	17	
2	2	4	20	7	27	
2	3	5	34	7	41	←min for $Y^1 = 5$
3	0	3	9	15	24	
3	1	4	17	15	32	
3	2	5	29	15	44	
4	0	4	16	27	43	
4	1	5	26	27	53	
5	0	5	25	42	67	

STEP 4. For $p = 1$, the recurrence relationship for the optimal objective function is

$$\phi^{4\to1}_{\min}[Y^4] = F^1_{\min}[Y^0, X^0] + \min[F^2 + F^3 + F^4] \qquad (10.4\text{-}31)$$

where $\min[F^2 + F^3 + F^4]$ for each Y^1 are given in Table 10.4-3 and

$$F^1_{\min}[Y^0, X^0] = \min[(Y^0)^2 + (X^0)^2] \qquad (10.4\text{-}32)$$

At this step the allowable values of Y^0 and X^0 are given by

$$Y^1 = Y^0 - X^0 \qquad (10.4\text{-}33)$$

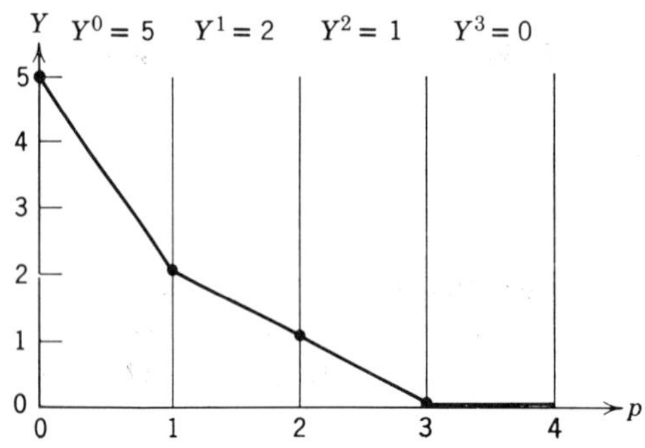

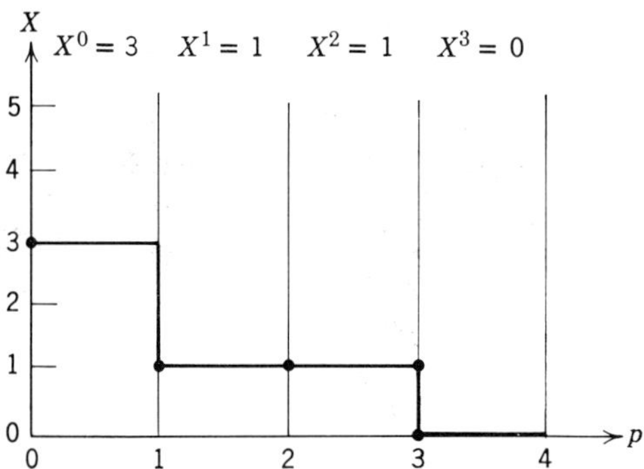

Fig. 10.4-2 Optimal X and Y solved by the dynamic programming approach.

subject to $Y^0 = 5$ as the initial boundary value.

The allowable Y^0 and X^0 and their corresponding F^1 and $\min[F^2 + F^3 + F^4]$ are listed in Table 10.4-4. Note that only $Y^0 = 5$ should be listed in Table 10.4-4.

Table 10.4-4 The $p = 1$ Step

Y^1	X^0	Y^0	F^1	$\min[F^2 + F^3 + F^4]$	$F^1 + \min[F^2 + F^3 + F^4]$	
0	5	5	50	0	50	
1	4	5	41	2	43	
2	3	5	34	7	41	←min for $Y^0 = 5$
3	2	5	29	15	44	
4	1	5	26	27	53	
5	0	5	25	41	66	

From Table 10.4-4 the minimum objective function is equal to

$$\phi_{\min}^{4\to1}[Y^4] = \min[F^1 + F^2 + F^3 + F^4] = 41 \qquad (10.4\text{-}34)$$

The set of state and control variables for $\phi_{\min}^{4\to1}$ can be obtained from Tables 10.4-1 through 10.4-4. Figure 10.4-2 shows the plot of optimal X^p and Y^p which minimizes the objective function $\phi^{4\to1}$.

10.5 IMPLEMENTATION OF OPTIMAL CONTROL BY DYNAMIC PROGRAMMING

This section will present the following approaches to the implementation of optimal control by using dynamic programming techniques:

Feedback optimal control using on-line dynamic programming computations.

Feedforward optimal control using on-line dynamic programming computations.

Feedforward optimal control using the results of off-line dynamic programming computations.

Feedback optimal control using the results of off-line dynamic programming computations.

The basic concepts of dynamic programming developed in Sections 10.3 and 10.4 will be the basis of the four approaches to optimal control implementation. The objective function to be minimized and the physical process model are the same as the example described in Section 10.4C.

A. Feedback Optimal Control Using On-Line Dynamic Programming Computations

Figure 10.5-1 shows the feedback optimal control approach using on-line dynamic programming computations. The computational procedure for solving the dynamic programming problem is developed off-line. The computational procedure is in turn based on the physical process model, the terminal condition Y^4, and the objective function. The terminal condition for the example is

$$Y^4 = \text{constant} \qquad (10.5\text{-}1)$$

which in turn defines the relationship of X^3 and Y^3 through the physical process model:

$$Y^{p+1} - Y^p = -X^p \qquad (10.5\text{-}2)$$

The computational procedure will allow the numerical computation for the value of X^p as a function of the terminal state variable Y^4, the step p,

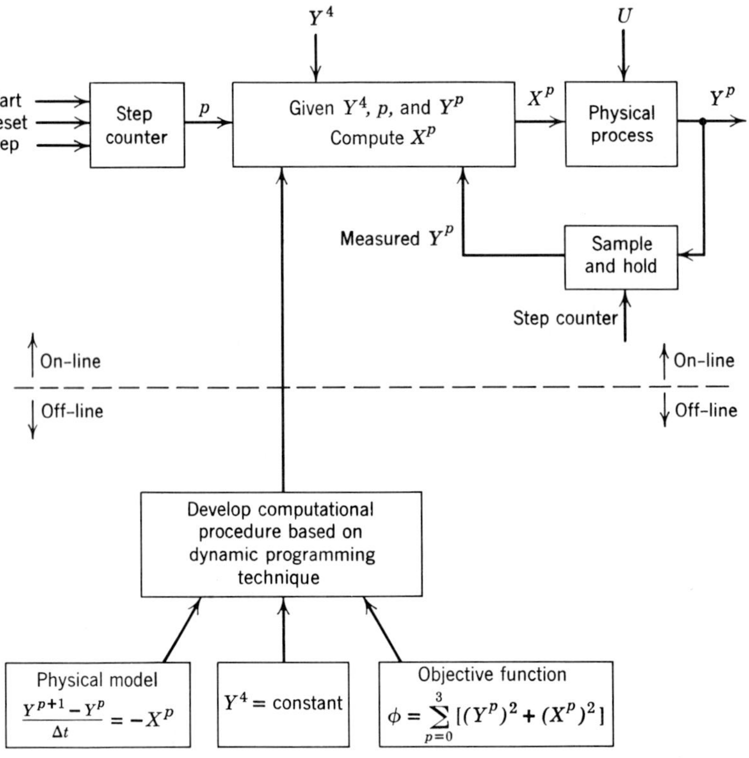

Fig. 10.5-1 Feedback optimal control using on-line dynamic programming computations.

and the state variable of that step Y^p. The procedure will be similar to that used to solve the example in Section 10.4C, except that the procedure used in this section can use any step p as the initial step. The procedure used in Section 10.4C uses Y^0 as the initial step.

The on-line portion of the optimal control approach given in Figure 10.5-1 is initiated by the step counter, which is always reset to $p = 0$ before the problem commences. When the start signal is received, the on-line computational procedure will first solve for the values of

$$X^3, X^2, X^1, \text{ and } X^0$$

as a function of Y^0 and Y^4. Then X^0 is applied to control the process.

When the next step signal is received, p will be stepped to $p = 1$, and the output of the sample and hold circuit* will be the measured Y^1. The measured state variable Y^1 will generally not agree with the predicted Y^1 because of the

* Without delays.

effect of the uncontrolled variables. A new computation will solve for the values of

$$X^3, X^2, \text{ and } X^1$$

as a function of Y^4 and the measured Y^1. The new X^1 will be applied to control the process.

The computational procedure may be repeated for steps $p = 2$ and $p = 3$. The summary of the various inputs, computed variables, and control actions are given in Table 10.5-1. The second superscript of the computed control

Table 10.5-1 Summary of Feedback Optimal Control Using On-Line Dynamic Programming Computations

Step	Input	Computations	Control Action
$p = 0$	Y^4, Y^0	$X^{30}, X^{20}, X^{10}, X^{00}$	Set $X^p = X^{00}$
$p = 1$	Y^4, Y^1	X^{31}, X^{21}, X^{11}	Set $X^p = X^{11}$
$p = 2$	Y^4, Y^2	X^{32}, X^{22}	Set $X^p = X^{22}$
$p = 3$	Y^4, Y^3	X^{33}	Set $X^p = X^{33}$

variables indicates the step at which the computations are made. The inputs are the given terminal state variable Y^4 and the measured inputs Y^1, Y^2, Y^3.

B. Feedforward Optimal Control Using On-Line Dynamic Programming Computations

In feedforward optimal control using on-line dynamic programming computations, the computations are only done once during the initial step $p = 0$. The state variables will not be measured and used for feedback purposes. Table 10.5-2 shows the summary of the inputs, computed variables, and

Table 10.5-2 Summary of Feedforward Optimal Control Using On-Line Dynamic Programming Computations

Step	Input	Computation	Control Action
$p = 0$	Y^4, Y^0	$X^{30}, X^{20}, X^{10}, X^{00}$	Set $X^p = X^{00}$
$p = 1$	$\cdots$	None	Set $X^p = X^{10}$
$p = 2$	$\cdots$	None	Set $X^p = X^{20}$
$p = 3$	$\cdots$	None	Set $X^p = X^{30}$

control actions of this approach. The feedforward approach using on-line dynamic programming computations requires much less computational effort than the feedback approach. The results of the feedforward approach would probably be less accurate than the feedback approach because the effects of uncontrolled variables and the inaccuracies of the physical process model are ignored.

C. Feedforward Optimal Control Using the Results of Off-Line Dynamic Programming Computations

In many cases on-line dynamic programming computations may require excessive computer storage and execution time. An alternative approach is to apply dynamic programming techniques in an off-line fashion to numerically solve the problem and to develop the control equations that fit the numerical solutions. These control equations are then used for on-line feedforward optimal control. The control equations are usually relatively simple to implement, and require modest computer storage and execution time.

The problem in Section 10.4C may be generalized to allow arbitrary values of Y^0 and Y^4 rather than $Y^0 = 5$ and $Y^4 = 0$. The control equations which are derived to fit the results of off-line dynamic programming computations are

$$X^0 = \frac{13}{21}(Y^0) - \frac{1}{21}(Y^4) \tag{10.5-3}$$

$$X^1 = \frac{5}{21}(Y^0) - \frac{2}{21}(Y^4) \tag{10.5-4}$$

$$X^2 = \frac{2}{21}(Y^0) - \frac{5}{21}(Y^4) \tag{10.5-5}$$

$$X^3 = \frac{1}{21}(Y^0) - \frac{13}{21}(Y^4) \tag{10.5-6}$$

Figure 10.5-2 shows the overall approach of feedforward optimal control using the results of the off-line dynamic programming computations. The off-line procedure developed the control equations given by Equations 10.5-3 through 10.5-6. The on-line procedure uses the control equations to solve for the optimal X^p as the step counter advances through the proper time interval. The value of Y at step $p = 0$ is measured to obtain Y^0.

The feedforward approach cannot take into account the inaccuracies of the physical process model and the effects of the uncontrolled variable. This

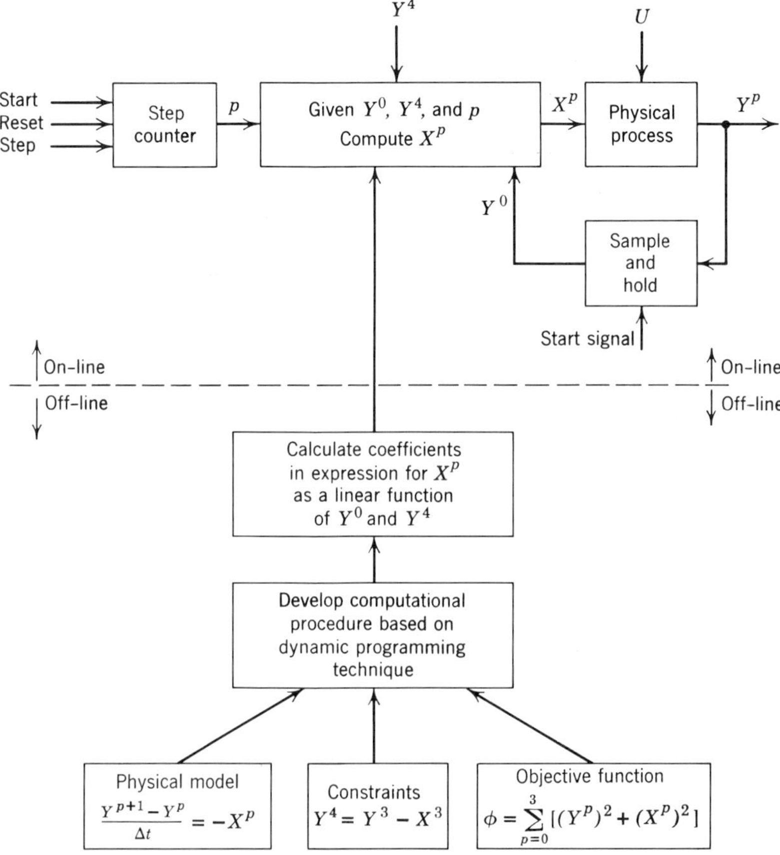

Fig. 10.5-2   Feedforward optimal control using results of off-line dynamic programming computations.

is the main disadvantage. For example, assume that the physical process model has the added term U for the uncontrolled variable:

$$Y^{p+1} = Y^p - X^p + U \qquad (10.5\text{-}7)$$

Let $Y^0 = 5$ and $U = 1$ for all the intervals. Table 10.5-3 shows the results of using the control equations given by Equations 10.5-3 through 10.5-6 to solve for the optimal X^p where $Y^p = 0$.

Table 10.5-3 may be compared with Table 10.5-4, which shows the results where the physical process model is identical with the actual plant. Note that the terminal boundary condition at $p = 4$ is no longer zero in Table 10.5-3, but is equal to 4. Also the effect of the uncontrolled variable was to increase the value of the objective function from $\phi = 40.47$ to $\phi = 62.56$.

Table 10.5-3 Feedforward Optimal Control Including the Effect of an Uncontrolled Variable

Step	Y^p	X^p	F
0	5	$\dfrac{65}{21}$	34.58
1	$\dfrac{61}{21}$	$\dfrac{25}{21}$	9.85
2	$\dfrac{57}{21}$	$\dfrac{10}{21}$	7.59
3	$\dfrac{68}{21}$	$\dfrac{5}{21}$	10.54
4	4		

Total $\phi = 62.56$

Table 10.5-4 Feedforward Optimal Control with No Uncontrolled Variable

Step	Y^p	X^p	F
0	5	$\dfrac{65}{21}$	34.58
1	$\dfrac{40}{21}$	$\dfrac{25}{21}$	5.04
2	$\dfrac{15}{21}$	$\dfrac{10}{21}$	0.74
3	$\dfrac{5}{21}$	$\dfrac{5}{21}$	0.11
4	0		

Total $\phi = 40.47$

D. Feedback Optimal Control Using Results of Off-Line Dynamic Programming Computations

Figure 10.5-3 shows that feedback may be used to improve the accuracy of optimal control to take into account the effects of the uncontrolled variables and the inaccuracies of the physical process model. The control equations

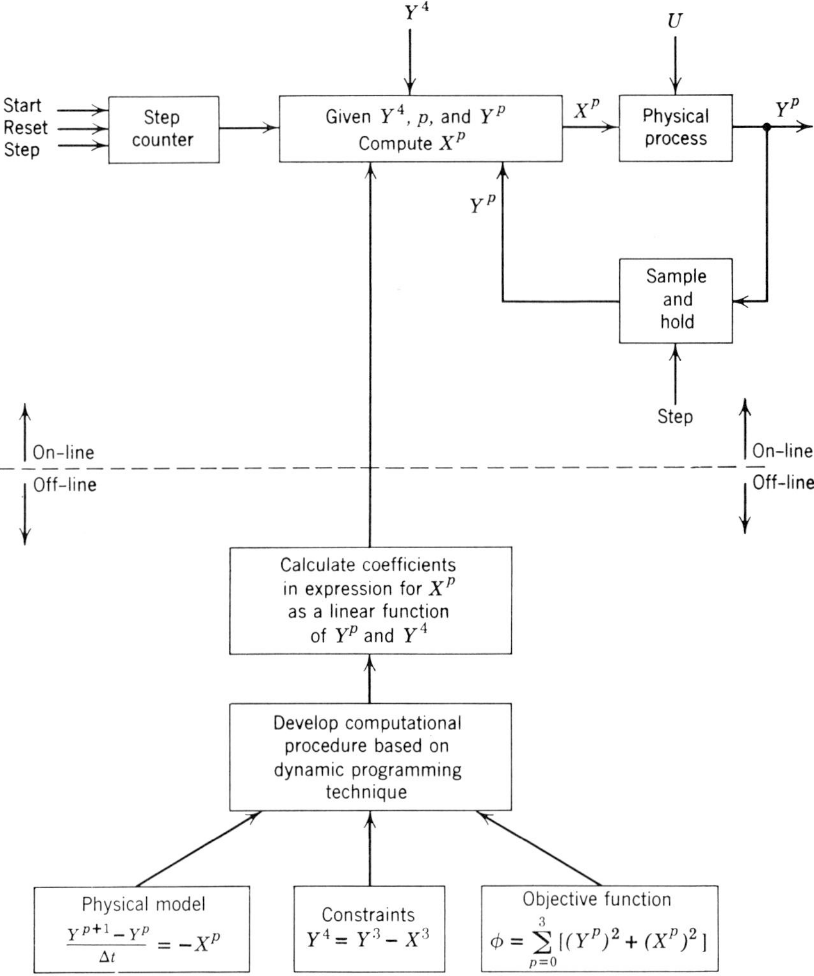

Fig. 10.5-3   Feedback optimal control using results of off-line dynamic programming computations.

developed from the off-line dynamic programming computations are as follows:

$$X^0 = \frac{13}{21}(Y^0) - \frac{1}{21}(Y^4) \tag{10.5-8}$$

$$X^1 = \frac{5}{8}(Y^1) - \frac{1}{8}(Y^4) \tag{10.5-9}$$

$$X^2 = \frac{2}{3}(Y^2) - \frac{1}{3}(Y^4) \tag{10.5-10}$$

$$X^3 = Y^3 - Y^4 \tag{10.5-11}$$

Note that these control equations express the optimal control variables as a function of the measured Y^p and the given Y^4. The measured Y^p at each step can take into account the inaccuracies of the physical process model and the effects of the uncontrolled variable. Table 10.5-5 shows the results

Table   10.5-5 Feedback Optimal Control Including the Effect of an Uncontrolled Variable

Step	Y^p	X^p	F
0	5	$\dfrac{65}{21}$	34.58
1	$\dfrac{61}{21}$	$\dfrac{305}{168}$	11.73
2	$\dfrac{351}{168}$	$\dfrac{117}{84}$	6.30
3	$\dfrac{71}{42}$	$\dfrac{71}{42}$	5.71
4	1		

Total $\phi = 58.32$

of feedback optimal control applied to the previous example. Note that the terminal boundary condition for Y^4 was reduced from $Y^4 = 4$ obtained by feedforward techniques to $Y^4 = 1$ by the use of feedback. Also, the optimal objective function with the uncontrolled variable was reduced from $\phi = 62.56$ for the feedforward case to $\phi = 58.32$ for the feedback case.

10.6 SUMMARY

The material presented in this chapter represents a minuscule portion of the available knowledge in the field of dynamic optimization. This chapter should be considered as an introductory survey of dynamic optimization techniques which are practical and reasonably easy to explain. It is hoped that this material will take away some of the mystery of the various techniques used for dynamic optimization. This in turn should enhance the willingness of engineers in industry to consider applying dynamic optimization techniques. Even if the engineer should conclude after the study that the payoff does not warrant applying dynamic optimization techniques to his problems, he has made the decision *consciously* and not *in ignorance*. This is particularly important as dynamic optimization problems are gradually moving from the realm of research and military applications to practical applications for industry.

This chapter also marks the end of the book, which has dealt in depth with the two important subjects of modeling and optimization of the automated process. It is hoped that some of the techniques of stating the problem—modeling—and controlling the process—optimization—will equip engineers with techniques to do creative investigations so that there will be less tendency to simply do what has been done before or to "automate the mess," i.e., automate existing operating techniques and procedures. Instead, as discussed in Chapter 1, each application should be analyzed in order to maximize the difference between the productivity and investment costs due to automation with process computers.

Bibliography

CHAPTERS 1 AND 2

Entries pertinent to more than one chapter are listed under the first chapter to which they apply.

Adams, G. E., W. M. Gaines, and L. A. Goshorn, "Computer Application to Process Control," *Proc. Twenty-Fourth Ann. Mining Symp.*, University of Minnesota, Minneapolis, 1963.

American Institute of Physics, *Temperature: Its Measurement and Control in Science and Industry*, Vol. 1, 1941; Vol. 2, 1955, Reinhold, New York.

Buckley, P. S., *Techniques of Process Control*, Wiley, New York, 1964.

Carroll, G. C., *Industrial Process Measuring Instruments*, McGraw-Hill, New York, 1962.

Ceaglske, N. H., *Automatic Process Control for Chemical Engineers*, Wiley, New York, 1956.

Cerni, R. H., and L. E. Foster, *Instrumentation for Engineering Measurement*, Wiley, New York, 1962.

Chestnut, H., *Systems Engineering Tools*, Wiley, New York, 1965.

Considine, D. M. (ed.), *Process Instruments and Controls Handbook*, McGraw-Hill, New York, 1957.

Coxon, W. F., *Flow Measurement and Control*, Macmillan, New York, 1959.

Coxon, W. F., *pH Measurement and Control*, Macmillan, New York, 1960.

Coxon, W. F., *Temperature Measurement and Control*, Macmillan, New York, 1960.

Ewing, G. W., *Instrumental Methods of Chemical Analysis*, McGraw-Hill, New York, 1954.

Gaines, W. M., "The Implications of Real Time, On-Line Industrial Applications on Digital Computer Design," *JACC* (1964).

Gaines, W. M., and G. E. Adams, "Use of Process Computers in the Cement Industry," *Mater. Process.* (September 1962).

Gaines, W. M., and P. P. Fischer, "Terminology for Functional Characteristics of Analog-to-Digital Converters," *Control Eng.* (February 1961).

Gaines, W. M., and T. Glass, "Applying Conventional Steam Power Plant Automation Technique to Nuclear Power Plants," *American Nuclear Society Symposium*, Pittsburgh, Pa., February 1962.

Gaines, W. M., L. Goshorn, and R. Livingston, "The Use of Analog Computers in the Dynamic Evaluation of On-Line Digital Control Computer Programs," *Third PICA Conference*, April 24–26, 1963.

Gaines, W. M., L. F. Kennedy, and A. M. Spielberg, "The Systems Approach to a Computer Control Generating Plant," *1960 National Power Conference*, Philadelphia, September 1960.

Gaines, W. M., and W. N. Patterson, "Accuracy in Process Computer Instrumentation Systems," *AIEE Winter General Meeting*, February 1961.

Gaines, W. M., and W. N. Patterson, "Tying the Digital Control Computer to an Industrial Process," *AIEE Petroleum Industry Conference*, New Orleans, La., September 20, 1961.

Gould, L. A., *Chemical Process Control*, Addison-Wesley, Reading, Mass., 1965.

Grabbe, E. M., S. Ramo, and D. E. Wooldridge (eds.), *Handbook of Automation, Computation and Control*, Vol. 1, *Control Fundamentals*, 1958; Vol. 2, *Computers and Data Processing*, 1959; Vol. 3, *Systems and Components*, 1961; Wiley, New York.

Gregory, R. H., and R. L. Van Horn, *Automatic Data-Processing Systems*, 2nd ed., Wadsworth Publishing Co., Belmont, Calif., 1963.

Hall, A. D., *A Methodology for Systems Engineering*, Van Nostrand, Princeton, N.J., 1962.

Harriott, P., *Process Control*, McGraw-Hill, New York, 1964.

Holzbock, W. G., *Automatic Control: Principles and Practice*, Reinhold, New York, 1958.

Lee, T. H., and T. J. Glass, "The Place of Process Computers," *Fall Instrument-Automation Conference and Exhibit, Instrument Society of America*, New York, September 26–30, 1960.

Leondes, C. T. (ed.), *Computer Control Systems Technology*, McGraw-Hill, New York, 1961.

McCracken, D. D., H. Weiss, and T. H. Lee, *Programming Business Computers*, Wiley, New York, 1959.

Savas, E. S., *Computer Control of Industrial Processes*, McGraw-Hill, New York, 1965.

Spink, L. K., *Principles and Practice of Flow Meter Engineering*, 8th ed., Foxboro Co., Foxboro, Mass., 1956.

Thomas, H. E., and C. A. Clarke, *Handbook of Electronic Instrumentation and Measurement Techniques*, Prentice-Hall, Englewood Cliffs, N.J., 1967.

Truxal, J. G., *Automatic Feedback Control System Synthesis*, McGraw-Hill, New York, 1955.

Truxal, J. G. (ed.), *Control Engineer's Handbook*, McGraw-Hill, New York, 1958.

Tucker, G. K., and G. M. Wills, *A Simplified Technique of Control Systems Engineering*, Minneapolis-Honeywell Regulator Co., Philadelphia, 1958.

Williams, T. J., "Studying the Economics of Process Computer Control," *ISA J.*, 8(1): 50–59 (1961).

Williams, T. J., *Systems Engineering for the Process Industries*, McGraw-Hill, New York, 1961.

Williams, T. J., and V. A. Lauher, *Automatic Control of Chemical and Petroleum Processes*, Gulf Publishing Co., Houston, Texas, 1961.

Wilson, I. G., and M. E. Wilson, *Information, Computers, and System Design*, Wiley, New York, 1965.

Young, A. J., *An Introduction to Process Control System Design*, Instruments Publishing Co., Pittsburgh, 1955.

CHAPTER 3

Bernard, J. W., and J. F. Cashen, "Direct Digital Control," *Instr. Control Systems*, **38**, 151–158 (September 1965).

Desmonde, W. H., *Real-Time Data Processing Systems: Introductory Concepts*, Prentice-Hall, Englewood Cliffs, N.J., 1964.

Gaines, W. M., and G. E. Adams, "Digital Computer Simulation of Mixing and Blending Operation in Cement Manufacture," *Computer Users' Symp.*, Utica, N.Y., May 4, 1961.

Gaines, W. M., and M. J. Tobias, "Signal Filtering in Digital Control Computer Systems," *IRE 7th Ann. Regional Conf.*, Phoenix, Ariz., April 1961.

Glass, T. J., "Current Trends in Process Computer Software," *Instrument Society of America, 22nd Ann. ISA Conf. Exhibit*, Chicago, Ill., September 11–14, 1967.

Hamming, R. W., *Numerical Methods for Scientists and Engineers*, McGraw-Hill, New York, 1962.

Hastings, C., *Approximations for Digital Computers*, Princeton University Press, Princeton, N. J., 1955.

Herriot, J. G., *Methods of Mathematical Analysis and Computation*, Wiley, New York, 1963.

Hildebrand, F. B., *Introduction to Numerical Analysis*, McGraw-Hill, New York, 1956.

1965 *IEEE Intern. Conv. Rec.* (Computers: Design Automation; Automatic Displays; Symposium on Direct Digital Control; Multiprocessor Systems; Techniques and Applications), *IEEE International Convention*, New York, March 22–26, 1965.

ISA Chemical and Petroleum Industries Division, "Guidelines and General Information on User Requirements Concerning Direct Digital Control" and "Questions and Answers on Direct Digital Computer Control," *Users' Workshop on Direct Digital Computer Control*, Princeton, N. J., April 3–4, 1963.

Jury, E. I., *Sampled-Data Control Systems*, Wiley, New York, 1958.

Jury, E. I., *Theory and Application of the Z-Transform Method*, Wiley, New York, 1964.

Kopal, Z., *Numerical Analysis*, 2nd ed., Wiley, New York, 1961.

Kuo, B., *Analysis and Synthesis of Sampled Data Control Systems*, Prentice-Hall, Englewood Cliffs, N. J., 1963.

Mandl, M., *Fundamentals of Electronic Computers (Digital and Analog)*, Prentice-Hall, Englewood Cliffs, N. J., 1967.

Martin, J. T., *Programming Real-Time Computer Systems*, Prentice-Hall, Englewood Cliffs, N. J., 1965.

McCracken, D. D., *A Guide to FORTRAN IV Programming*, Wiley, New York, 1965.

McCracken, D. D., and W. S. Dorn, *Numerical Methods and FORTRAN Programming: With Applications in Engineering and Science*, Wiley, New York, 1964.

Middlebrook, R. D., *Differential Amplifiers*, Wiley, New York, 1963.

Miller, W. E. (ed.), *Digital Computer Applications to Process Control*, Plenum Press, New York, 1965.

Monroe, A. J., *Digital Processes for Sampled Data Systems*, Wiley, New York, 1962.

Opie, S. R. B., "Direct Digital Control—A Total Systems Approach," *IEEE International Convention Record*, 1967.

Phister, M., Jr., *Logical Design of Digital Computers*, Wiley, New York, 1958.

Ragazzini, J. R., and G. F. Franklin, *Sampled-Data Control Systems*, McGraw-Hill, New York, 1958.

Scott, N. R., *Analog and Digital Computer Technology*, McGraw-Hill, New York, 1960.

Sprague, R. E., *Electronic Business Systems: Management Use of On-Line—Real-Time Computers*, Ronald Press, New York, 1962.

Tou, J., *Digital and Sampled-Data Control Systems*, McGraw-Hill, New York, 1959.

Williams, T. J., "What to Expect From Direct-Digital Control," *Chem. Eng.* (1964).

Ziegler, J. R., *Time-Sharing Data Processing Systems*, Prentice-Hall, Englewood Cliffs, N.J., 1967.

CHAPTER 4

Adams, G. E., "A Study of a Generalized Model for Digital Computer Control of Blending Processes," *AIEE Winter General Meeting*, CP 61–213, New York, 1961.

Adams, G. E., "Simulation Models for Digital Computer Control of a Plant," *Simulation Languages ISA Monograph, Joint Automatic Control Conf.*, Troy, N. Y., 1965.

Aris, R., *Introduction to the Analysis of Chemical Reactors*, Prentice-Hall, Englewood Cliffs, N. J., 1965.

Ayres, F., *Theory and Problems of Matrices*, Schaum, New York, 1962.

Bellman, R., *Introduction to Matrix Analysis*, McGraw-Hill, New York, 1960.

Bennett, C. A., and N. L., Franklin, *Statistical Analysis in Chemistry and the Chemical Industry*, Wiley, New York, 1954.

Bernstein, F., "Application of X-Ray Fluorescence to Process Control," 10th Ann. *Conf. Application X-Ray Anal.*, University of Denver, August 7–9, 1961.

Box, G. E. P., and P. W. Tidwell, "Transformation of the Independent Variables," *Technometrics*, **4**, 531–550 (1962).

Campbell, D. P., *Process Dynamics*, Wiley, New York, 1958.

Carlson, A., *Analog Simulation in Chemical Engineering*, Wiley, 1967.

Ceaglske, N. H., *Automatic Process Control for Chemical Engineers*, Wiley, New York, 1956.

Draper, N. R., and H. Smith, *Applied Regression Analysis*, Wiley, 1966.

Dwyer, P. S., *Linear Computations*, Wiley, New York, 1951.

Faddeeva, V. N., *Computational Methods of Linear Algebra*, Dover, New York, 1959.

Fisher, R. A., *Statistical Methods and Scientific Inference*, Oliver and Boyd, Edinburgh, New York, 1956.

Fisher, R. A., and F. Yates, *Statistical Tables for Biological, Agricultural, and Medical Research*, 6th ed., Hafner, New York, 1964.

Franks, R. G. E., *Mathematical Modeling in Chemical Engineering*, Wiley, 1967.

Freeman, H., *Discrete-Time Systems*, Wiley, New York, 1965.

Goldberg, S., *Introduction to Difference Equations*, Wiley, New York, 1958.

Graybill, F. A., *An introduction to Linear Statistical Models*, Vol. 1, McGraw-Hill, New York, 1961.

Gupta, S. C., *Transform and State Variable Methods in Linear Systems*, Wiley, New York, 1966.

Guy, A. M., "The Use of Supervisory and Direct Digital Control Computers in Automatically Controlling a Cement Plant," *IEEE Cement* Conf., 1966.

Hadley, G., *Linear Algebra*, Addison-Wesley, Reading, Mass., 1961.

Harriott, P., *Process Control*, McGraw-Hill, New York, 1964.

Hougen, O. A., and K. M. Watson, and R. A. Ragatz, *Chemical Process Principles*, Wiley, New York, 1964.

Johnston, J., *Econometric Methods*, McGraw-Hill, New York, 1963.

Kalman, R., "Mathematical Description of Linear Dynamical Systems", *J. Soc. Ind. Appl. Math., Ser. A: Control*, **1**, No. 2, 152–192 (1963).

Kalman, R. E., "On the General Theory of Control Systems," *Proc. 1st Intern. Congr. Intern. Federation Autom. Control*, **1**, 481–492, Moscow (1960).

Kalman, R. E., and R. W. Koepoke, "The Role of Digital Computers in the Dynamic Optimization of Chemical Reactions," *IBM Yorktown Heights Res. Rept. RC-77*, December 1, 1958.

Kalman, R. E., L. Lapidus, and E. Shapiro, "Computer Control of Chemical Processes." *Chem. Eng. Progr.*, **56**, 55–61 (1960).

Kalman, R. E., L. Lapidus, and E. Shapiro, "On the Optimal Control of Chemical and Petroleum Processes," *IBM Yorktown Heights Res. Rept. RC-76*, January, 1959.

Kalman, R. E., L. Lapidus, and E, Shapiro, *Proc. Joint Symp. Instrumentation Computation Process Devlpt. Plant Design,* Institute of Chemical Engineering, London, 1959, pp. B2–B13,.

Kammermeer, K., and J. O. Osburn, *Process Calculations*, Prentice-Hall, Englewood Cliffs, N.J., 1956.

Karplus, W. J., ed. ,*On-Line Computing*, McGraw-Hill, New York, 1966.

Lapidus, L., *Digital Computation for Chemical Engineers*, McGraw-Hill, New York, 1962.

Lapidus, L., "On the Dynamics of Chemical Reactors," *AIChE Reprint 1, Joint Autom. Contr. Conf.* M.I.T., Cambridge, Mass., September 1960.

Laranger, W. F., "On-Line Analysis by X-Ray Emission Technique," *12th Ann. Symp. Spectroscopy*, Chicago, Ill., May 16, 1961.

Levenspiel, O., *Chemical Reaction Engineering*, Wiley, New York, 1962.

Liebhafsky, H. A., H. G. Pfeiffer, E. H. Winslow, and P. D. Zemany, *X-Ray Absorption and Emission in Analytical Chemistry*, Wiley, New York, 1960.

MacFarlane, A. G. J., *Engineering Systems Analysis*, Addison-Wesley, Reading, Mass., 1964.

McAdams, W. H., *Heat Transmission*, 4th ed. McGraw-Hill, New York, 1954.

Olson, H. F., *Dynamical Analogies*, 2nd ed. Van Nostrand, Princeton, N.J., 1958.

Scheffe, H., *The Analysis of Variance*, Wiley, New York, 1959.

Shilling, G. D., *Process Dynamics and Control*, Holt, Rinehart and Winston, New York, 1963.

Smith, J. M., *Chemical Engineering Kinetics*, McGraw-Hill, New York, 1956.

Soroka, W. W., *Analog Methods in Computations and Simulation*, McGraw-Hill, New York, 1954.

Williams, E. J., *Regression Analysis*, Wiley, New York, 1959.

Williams, E. T., and R. C. Johnson, *Stoichiometry for Chemical Engineers*, McGraw-Hill, New York, 1958.

Young, A. J., *Plant and Process Dynamic Characteristics*, Academic, New York, 1957.

Zadeh, L. A., and C. A. Desoer, *Linear System Theory*, McGraw-Hill, New York, 1963.

CHAPTER 5

Allen, R. G. D., *Mathematical Economics*, Macmillan, London, 1956.

Bass, F. M., *Mathematical Models and Methods in Marketing*, Irwin, Homewood, Ill., 1961.

Baumol, W. J., *Economic Dynamics*, 2nd ed., Macmillan, New York, 1959.

Baumol, W. J., *Economic Theory and Operations Analysis*, Prentice-Hall, Englewood Cliffs, N.J., 1961.

Buzzell, R. D., *Mathematical Models and Marketing Management*, Harvard University Press, Cambridge, Mass., 1964.

Fabrycky, W. J., and P. E. Torgensen, *Operations Economy: Economic Applications of Operations Research*, Prentice-Hall, Englewood, Cliffs, N.J., 1966.

Forrester, J. W., *Industrial Dynamics*, M.I.T. Press, Cambridge, Mass., 1961.

Goetz, B. E., *Quantitative Methods: A Survey and Guide for Managers*, McGraw-Hill, New York, 1956.

Grant, E. L., and W. G. Ireson, *Principles of Engineering Economy*, 4th ed., Ronald, New York, 1960.

Happel, J., *Chemical Process Economics*, Wiley, New York, 1958.

Hein, L. W., *The Quantitative Approach to Managerial Decisions*, Prentice-Hall, Englewood Cliffs, N.J., 1967.

Horowitz, I., *An Introduction to Quantitative Business Analysis*, McGraw-Hill, New York, 1965.

Kemeny, J., G. A. Schleifer, Jr., J. L. Snell, and G. L. Thompson, *Finite Mathematics with Business Applications*, Prentice-Hall, Englewood Cliffs, N.J., 1962.

Kennedy, J. D., "Forecasting and Dynamic Business Modeling on the Electronic Analog Computer," *Midwestern Simulation Council Meeting*, South Bend, Ind., March 12, 1962.

Manne, A. S., *Economic Analysis for Business Decisions*, McGraw-Hill, New York, 1961.

Meier, R. C., and S. H. Archer, *An Introduction to Mathematics for Business Analysis*, McGraw-Hill, New York, 1960.

Springer, C. H., R. E. Herlihy, R. I. Beggs, and R. T. Mall, *Mathematics for Management Series: Basic Mathematics*, Vol. 1; *Advanced Methods and Models*, Vol. 2; *Statistical Inference*, Vol. 3; Irwin, Homewood, Ill., 1966.

Stern, M. E., *Mathematics of Management*, Prentice-Hall, Englewood Cliffs, N.J., 1963.

Stout, T. M., "Economics of Computers in Process Control," *Automation* (October–December, 1966).

Synder, L. R., *Essential Business Mathematics*, 4th ed., McGraw-Hill, New York, 1963.

Yamane, T., *Mathematics of Economists: An Elementary Survey*, Prentice-Hall, Englewood, Cliffs, N.J., 1955.

CHAPTERS 6, 7, AND 8

Ackoff, R. L., *Scientific Methods—Optimizing Applied Research Decisions*, Wiley, New York, 1962.

Adams, G. E., W. M. Gaines, J. H. Herz, J. R. Romig, "The Use of Process Control Computers in the Cement Industry," *AIEE Cement Industry Conference*, St. Louis, Mo., April 1962.

Beckenbach, E. F. (ed.), *Modern Mathematics for the Engineer*, McGraw-Hill, New York, 1956.

Bollinger, R. E., and D. E. Lamb, "The Design of a Combined Feedforward-Feedback Control System," *1963 Joint Automatic Control Conference*, University of Minnesota, Minneapolis, June 19–21, 1963.

Chang, S. S. L., *Synthesis of Optimum Control Systems*, McGraw-Hill, New York, 1961.

Charnes, A., and W. W. Cooper, *Management Models and Industrial Applications of Linear Programming*, Vols. 1 and 2, Wiley, New York, 1961.

Churchman, C. W., *Prediction and Optimal Decision*, Prentice-Hall, Englewood Cliffs, N.J., 1961.

Courant, R., and D. Hilbert, *Methods of Mathematical Physics*, Interscience, New York, 1953.

Dorfman, R., P. A. Samuelson, and R. M. Solow, *Linear Programming and Economic Analysis*, McGraw-Hill, New York, 1958.

Flagle, C. D., W. H. Huggins, and R. H. Roy (eds.), *Operations Research and Systems Engineering*, Johns Hopkins, Baltimore, Md., 1960.

Gass, S. I., *Linear Programming*, 2nd ed., McGraw-Hill, New York, 1964.

Kirchmayer, L. K., *Economic Operation of Power Systems*, Wiley, New York, 1958.

Lapidus, L., E. Shapiro, S. Shapiro, and R. E. Stillman, "Optimization of Process Performance," *J. Am. Inst. Chem. Engrs.*, 7, 288–294 (1961).

Lavi, A., and T. P. Vogl (eds.), *Recent Advances in Optimization Techniques*, Wiley, New York, 1966.

Leondes, C. T. (ed.), *Modern Control Systems Theory*, McGraw-Hill, New York, 1965.

Mishkin, E., and L. Braum, Jr. (eds.), *Adaptive Control Systems*, McGraw-Hill, New York, 1961.

Solodov, A. V., *Linear Automatic Control Systems with Varying Parameters*, American Elsevier, New York, 1966.

Symonds, G. H., *Linear Programming: The Solution of Refinery Problems*, Esso Standard Oil Company, New York, 1955.

Tou, J. T., *Optimum Design of Digital Control Systems*, Academic, New York, 1963.

CHAPTER 9

Beyer, W. G., H. J. Fiedler, and L. K. Kirchmayer, "Digital Computer Dispatching Systems," *Proc. PICA Conf.*, Phoenix, Arizona, April 1963.

Box, G. E. P., "The Effects of Errors in the Factor Levels and Experimental Design," *Bull. Intern. Statist. Inst.*, **38**, 339–355 (1961).

Box, G. E. P., "Evolutionary Operations, a Method for Increasing Industrial Productivity," *Appl. Stat.*, **6**, 3–23, (1957).

Box, G. E. P., and G. A. Coutie, "Application of Digital Computers in the Exploration of Functional Relationships," *Proc. Inst. Elec. Engrs.*, **103**, Part B, Suppl. No. 1, 100–107 (1956).

Box, G. E. P., and J. S. Hunter, "Condensed Calculations for Evolutionary Operation Programs," *Technometrics*, **1**, 77–95 (1959).

Box, G. E. P., and J. S. Hunter, "Multi-factor Experimental Design for Exploring Response Surfaces," *Ann. Math. Stat.* **28**, 195–241 (1957).

Box, G. E. P., and K. B. Wilson, "On the Experimental Attainment of Optimum Conditions," *J. Roy. Stat. Soc.* **13 (B)**, 1–45 (1951).

Box, G. E. P., and P. V. Youle, "The Exploration and Exploitation of Response Surfaces: An Example of the Link Between the Fitted Surface and the Basic Mechanism of the System," *Biometrics*, **11**, 287–323 (1955).

Box, M. J., "A New Method of Constrained Optimization and a Comparison with Other Methods," *Computer J.*, **8**, 42–52 (1965).

Brooks, S. H., "A Comparison of Maximum Seeking Methods," *Operations Res.*, **7**, 430–457 (1959).

Chew, V. (ed.), *Experimental Designs in Industry*, Wiley, New York, 1958.

Cochran, W. G., and G. Cox, *Experimental Designs*, 2nd ed. Wiley, New York, 1957.

Crocket, J. B., and R. Chernoff, "Gradient Methods of Maximization," *Pacific J. Math.*, **5**, 33–50 (1955).

Davies, O. L. (ed.), *The Design and Analysis of Industrial Experiments*, 2nd ed., Oliver & Boyd, Edinburgh, and Hafner., New York, 1956.

Fiedler, H. J., and L. K. Kirchmayer, "Digital Computer Control of System Operation," *Proceedings of the American Power Conference*, March 25–28, 1963

Fisher, R. A., *The Design of Experiments*, 6th ed., Oliver and Boyd, Edinburgh, New York, 1951.

Gaines, W. M., and H. Chestnut, "Automatic Optimizing of Poorly Defined Processes, Part I," *1962 Joint Autom. Contr. Conf.*, New York, 1962.

Gibson, J. E., K. S. Fu, and J. C. Hill, "Hill Climbing Using Piece-wise Cubic Approximation," *Control and Information Systems Laboratory*, Purdue University, Lafayette, Indiana; TR-EE64–7, 1964.

Goldfeld, S. M., R. E. Quandt, and H. F. Trotter, "Maximization by Quadratic Hill-Climbing," *Princeton University Econometric Research Program Memorandum No. 72*, January 19, 1965.

Kurshner, H. J., "A New Method of Locating the Maximum Point of an Arbitrary Multi-peak Curve in the Presence of Noise," *Lincoln Laboratory*, M.I.T., Lexington, Mass.

Kurshner, H. J., "Hill-climbing Method of the Optimization of Multi-parameter Noise Disturbed Systems," *Proc. Joint Autom. Contr. Conf.*, 1962

Read, D. R., "The Design of Chemical Experiments," *Biometrics*, **10**, 1–15 (1954)

Ringlee, R. J., "Economic Scheduling and Automatic Dispatching of a Hydro-Thermal Power System," *American Power Conf.*, **26**, 1116–1122, (1964).

Roberts, S. M., and H. I. Lyvers, "The Gradient Method in Process Control," *Ind. Eng. Chem.* (November 1961).

Rosenbloom, P. C., "The Method of Steepest Descent," *Numerical Analysis, Proc.* 6th *Sym. Appl. Math.*, McGraw-Hill, New York, 1956.

Schrage, R. W., "Optimizing a Catalytic Cracking Operation by the Method of Steepest Ascents and Linear Programming," *Operations Res.*, **6**, No. 4 (July–August 1958).

Wilde, D. J., *Optimum Seeking Methods*, Prentice-Hall, Englewood Cliffs, N.J., 1964.

CHAPTER 10

Aris, R., R. Bellman, and R. Kalaba, "Some Optimization Problems in Chemical Engineering," *Chem. Eng. Progr. Symp. Ser.*, No. 31, 95–102 (1960).

Balchen, J. G., *Dynamic Optimization of Continuous Processes*, Automatic Control Laboratory, Norwegian Institute of Technology, Trondheim, Norway, 1961.

Bellman, R., *Adaptive Control Processes: A Guided Tour*, Princeton University Press, Princeton, N.J., 1961.

Bellman, R., *Dynamic Programming*, Princeton University Press, Princeton, N.J., 1957.

Bellman, R. (ed.), *Mathematical Optimization Techniques*, University of California Press, Berkeley, 1963.

Bellman, R. E., and S. E. Dreyfus, *Applied Dynamic Programming*, Princeton University Press, Princeton, N.J., 1962.

Bellman, R., R. E. Glicksberg, and A. O. Gross, *Some Aspects of the Mathematical Theory of Control Process*, The Rand Corporation, Santa Monica, Calif., 1958.

Bernholtz, B., and L. J. Graham, "Hydro-Thermal Economic Scheduling: Part I and Part II," *AIEE Trans.*, **79**, Part III, 921–931 (1960).

Bliss, G. A., *Lectures on the Calculus of Variations*, University of Chicago Press, Chicago, 1946.

Boydston, R. E., "A Dynamic Solution to a Generalized Chemical Processing Model," *Joint Automatic Control Conference*, CP-60-979, M.I.T., Cambridge, Mass., Sept. 7–9, 1960.

Elsgolc, L. E., *Calculus of Variations*, Pergamon Press, London, 1961.

Fan, L. T., *The Continuous Maximum Principle*, Wiley, New York, 1966.

Fan, L. T., and C. S. Wang, *The Discrete Maximum Principle*, Wiley, New York, 1964.

Forsyth, A. R., *Calculus of Variations*, Dover Publications, New York, 1960.

Fox, C., *An Introduction to the Calculus of Variations*, Oxford University Press, New York, 1950.

Gibson, J. E. (ed.), "Proceedings of Dynamic Programming Workshop," *AIEE Conference*, Boulder, Colorado, June 1961.

Hestenes, M. R., *Calculus of Variations and Optimal Control Theory*, Wiley, New York, 1966.

Jeffreys, H., and B. S. Jeffreys, *Methods of Mathematical Physics*, Cambridge University Press, Cambridge, 1950.

Kipiniak, W., *Dynamic Optimization and Control: A Variational Approach*, The M.I.T. Press, Cambridge, Mass., and Wiley, New York, 1961.

Kipiniak, W., and L. A. Gould, "Dynamic Optimization and Control of a Stirred Tank Chemical Reactor," *Joint Automatic Control Conf.*, Cambridge, Mass., TP-60-982, September 1960.

Kirchmayer, L. K., *Economic Control of Interconnected Systems*, Wiley, New York, 1959.

Leitmann, G. (ed.), *Optimization Techniques with Applications to Aerospace Systems*, Academic Press, New York, 1962.

Merriam, C. W., III, *Optimization Theory and the Design of Feedback Control Systems*, McGraw-Hill, New York, 1964.

Nemhauser, G. L., *Introduction to Dynamic Programming*, Wiley, New York, 1966.

Neustandt, L. (ed.), *The Mathematical Theory of Optimal Processes*, Interscience (Wiley), New York, 1962.

Osterle, W. H., J. Gesier, A. DeSalvo, and K. M. Dale, "Computer Solution of Generating Units to be Operated: Part II," *AIEE Trans.*, **78**, Part III, 1959.

Pars, L. A., *An Introduction to the Calculus of Variations*, Wiley, New York, 1963.

Pontryagin, L. S., V. G. Boltyanskii, R. V. Gamkrelidze, and E. F. Mishchenko, *The Mathematical Theory of Optimal Processes*, Wiley, New York, 1962.

Reswick, J. B., and C. K. Taft, *Introduction to Dynamic Systems*, Prentice-Hall, Englewood Cliffs, N.J., 1967.

Roberts, S. M., *Dynamic Programming in Chemical Engineering Process Control*, Academic, New York, 1964.

Roberts, S. M., "Dynamic Programming Formulation of the Catalyst Replacement Problem," *Chem. Eng. Symp. Progr. Ser.*, **56**, No. 31, 1960.

Whittaker, E. T., and G. N. Watson, *A Course of Modern Analysis*, Cambridge University Press, Cambridge, 1950.

Problems

CHAPTER 1

1.1   What are the four stages of control developed prior to computer process control?

1.2   What are some of the objectives of process control?

1.3   What are the two different modes of operation for computer process control?

1.4   What are the four types of closed-loop computer process control?

1.5   Give the four definitions of productivity.

1.6   Which two out of the four definitions of productivity are more suitable for measuring the effectiveness of computer process control?

1.7   Table P1.1 shows the output and investment costs for the various proposed levels of automation.

(a) What is the optimal level?

(b) If the maximum investment allowable is 0.6 per unit, what is the optimal level?

Table P1.1   Output and Investment Cost of the Various Proposed Levels of Automation

Level of Automation	Output (per unit*)	Investment Cost (per unit*)
1	0.9	0.5
2	1.2	0.6
3	1.4	0.7
4	1.6	0.8
5	1.7	1.0
6	1.75	1.3

*1 unit is equal to $10,000,000.

1.8 Using Table P1.1, compute the incremental productivity at each level of automation. *Hint.* Incremental productivity is equal to the change in output divided by the change in investment cost.

1.9 What conclusions may be drawn by comparing the results of Problems 1.7 and 1.8?

1.10 What are the minimum prerequisites for automating an existing plant?

CHAPTER 2

2.1 Discuss the characteristics, strengths, and weaknesses of the various languages used for modeling.

2.2 Using block diagrams and English statements, prepare a brief functional model of a home kitchen.

2.3 A tank at level P is being filled with liquid at input flow rate F_1 and drained at the bottom with outflow F_2. The physical model can be described by the following English statements: (1) The rate of change of the level P is proportional to the inverse of the tank capacitance C times the difference between the input flow rate F_1 and the outflow F_2. (2) The outflow F_2 is equal to the tank level P divided by the fluid resistance R.

(a) Develop the mathematical model of the tank by using Statement 1.

(b) Using Statement 2 to eliminate the outflow F_2, develop the mathematical model of the tank. The tank level P should be expressed in terms of R, C, and F_1.

2.4 Develop the mathematical model of a mechanical system described by the following English statements (Newton's second law):

(1) The applied force F is equal to the mass M times the acceleration (second derivative of the displacement P),

(2) *plus* the viscous damping coefficient D times the velocity (derivative of the displacement P),

(3) *plus* the spring constant K times the displacement P.

2.5 A proposal to modernize an existing plant requires $10,000,000 fixed cost. This is expected to reduce the per unit cost of the product by $20,000.

(a) What is the break-even volume?

(b) If the product can be sold at an additional $500 per unit because of improved quality, what is the break-even volume?

2.6 Develop the economic model for the following problem:

(1) A diet that utilizes 12 foods and 5 nutrients is being planned. Each food has a price and contains certain percentages of each nutrient.

(2) Find the daily mix that will minimize the cost of the diet, subject to satisfying the daily minimum requirement for each nutrient.

(3) The food quantities cannot be negative.

2.7 Prepare a procedural model of "How to go to work in the morning" in flowchart form. *Hint.* See p. 35, D. D. McCracken, H. Weiss, and T. H. Lee, *Programming Business Computers*, Wiley, New York, 1959.

2.8   Prepare a procedural model in flow-chart form for a thermostat that will control home heating.

2.9   Prepare a procedural model in flow-chart form for a three-mode analog sub-loop controller.

2.10   What are the characteristics of information?

2.11   Discuss the information characteristics of the various portions of a computer process control system.

2.12   Given a service station gasoline pump, identify the control variables, parameters, state variables, uncontrolled variables, and economic variables.

CHAPTER 3

3.1   Discuss the horizontal and vertical functional models of the computer process control system.

3.2   In the vertical functional model of the computer are there firm lines of demarcation between hardware and software? Why (or why not)?

3.3   What are the resources required by a program or subroutine?

3.4   Determine the ending location of a program of 2039 words with a starting location of 4041 for the following conditions:

(a) The program is contiguously located in memory.

(b) The program is being interrupted by a series of programs that requires 451 words of storage.

3.5   A real-time operating program has an average overhead factor of 40%; i.e., 40% of the computer time is devoted to nonproductive work in interrupts, housekeeping, and so on. Determine the completion time of a program that requires a net execution time of 450 msec and begins execution at 08:00 for the following conditions:

(a) No interrupt

(b) Total interrupt of 500 msec

(c) If the average overhead factor is 30%, what are the results of (a) and (b)?

3.6   (a) List the input stimuli of the real-time operating system.

(b) List the output stimuli of the real-time operating system.

(c) What are the subfunctions of the real-time operating system?

3.7   (a) Describe the functional model of the computer process control input system.

(b) Describe the functional model of the computer process control output system.

3.8   What are the main functions of a typical computer process control system?

3.9   Name the functions of a computer process control system that are necessary for direct digital control.

3.10   Draw the functional block diagram of direct digital control.

3.11   Write the algorithm for the output signal by using direct digital control with the following types of control:

(a) Proportional

(b) Integral

(c) Derivative

CHAPTER 4

4.1 What are the two basic laws of systems analysis?

4.2 Express the two basic laws of systems analysis for the following types of system and include the physical units used:

(a) Pneumatic
(b) Hydraulic
(c) Thermal (conduction)
(d) Mechanical (translation and rotation)
(e) Electrical

4.3 Write the equations for a blending and homogenizing process in which three state variables A, B, and C are the percent compositions of certain chemical compounds. The blending involves four raw materials, each having percent compositions of the chemical compounds corresponding to the state variables. Assume a continuous blending process with a constant flow rate R and a homogenizer with a holding volume U; the prices of the raw materials are P_1, P_2, P_3, and P_4. Derive the following:

(a) The steady-state equations relating the ratio of the relative flow rates of the individual materials to the output mixture
(b) The cost of the output mixture as a function of the flow rates of the individual materials

4.4 For the system in Problem 4.3 assume the following values:

$A = 10\%$
$B = 20\%$
$C = 15\%$
$T = 3$ hours, homogenizer holding time (defined as the time to change completely the contents at the operating flow rate)
$M_a = 10\%$, percent of compound A in the input stream from the blender
$M_b = 20\%$, percent of compound B in the input stream from the blender
$M_c = 15\%$, percent of compound C in the input stream from the blender
Assume that one of the input materials in the blender has a change in composition and that the material composition from the blender is changed by having M_a change from 10% to 12%.
Assuming that steady-state conditions existed at the time the material-composition change occurred, plot the output of the homogenizer for each hour for 10 hours after the change.

4.5 Derive the linear regression model for two variables to *go through the mean*.

4.6 Derive the linear regression model for one state variable and two control variables to *go through the mean*.

4.7 Perform the Fisher F-test for the problem given in Figure 4.3-3 for 95% confidence level. Use Table 4.4-2 for the F-distribution.

4.8 Develop a linear regression model for the electrical element with the measurements of voltage and current given in Table P4.1.

Table P4.1   Experimental Data for an Electrical
Circuit Element

Point No.	Voltage (volts)	Current (milliamperes)
1	39.0	5.0
2	45.0	10.0
3	60.0	15.0
4	71.5	20.0
5	85.0	25.0
6	99.0	30.0
7	113.0	35.0
8	125.0	40.0

4.9   Do the following for Problem 4.8:

(a) Translate the regression model to *go through the mean.*
(b) Compute the standard estimate of error.
(c) Compute R^2.
(d) Test the hypothesis that the resistance $A_1 = 0$, with an α-risk of 0.05, using Table 4.4-2 for the F-distribution.
(e) Test the hypothesis that the resistance $A_1 = 0$, with an α-risk of 0.01, using Table 4.4-3 for the F-distribution.

4.10   Table P4.2 shows the effect of temperature on the yield of a chemical process. (The data are shown in dimensionless form.)

(a) Determine the coefficients of a linear regression model.
(b) Construct an analysis of variance table.
(c) Test the hypothesis that the regression coefficient $A_1 = 0$, with an α-risk of 0.05, using Table 4.4-2 for the F-distribution.

Table P4.2   Experimental Data on Yield
Versus Temperature for a Chemical Process

Point No.	Yield	Temperature
1	1.0	4.90
2	5.3	5.20
3	3.8	6.90
4	6.8	8.50
5	10.5	8.95
6	8.2	9.95
7	9.1	10.95
8	13.3	12.10
9	13.8	12.80
10	17.5	14.90

4.11 (a) Develop a second-order polynomial regression model for the data shown
      in Table P4.3.
  (b) Test the hypothesis that the regression coefficients are equal to zero, with
      an α-risk of 0.05, using Table 4.4-2 for the F-distribution.

Table P4.3   Experimental Data

Point No. 1	X	Y
1	100	22.15
2	101	24.60
3	102	26.45
4	103	26.70
5	104	28.10
6	105	28.85
7	106	30.05
8	107	30.35
9	108	31.38
10	101	22.10
11	102	26.15
12	104	28.45
13	106	30.25
14	108	31.60
15	110	33.15
16	100	22.90
17	103	27.25
18	106	29.90
19	109	31.95

4.12 An experiment is to be conducted to determine the adiabatic curve of a gas by
measuring the pressure P_n and volume V_n. Using the techniques of regression
analysis, derive the equations that will make the best estimates of γ and c to
minimize the sum of the square of the error between the regression model and
the data; γ and c are regression coefficients. *Hint.* The normal relationship for
an adiabatic process is $PV^\gamma = c$.

4.13 (a) Develop a second-order polynomial regression model for the data shown
      in Table P4.4.
  (b) Set up the regression matrix by using $Y = 0$ and $X = 0$ as the origin.
  (c) How can the origin be changed to make the matrix easier to manipulate,
      i.e., to make several of the terms equal to zero?

Table P4.4   Experimental Data

Point No.	X	Y
1	0	-3.0
2	1	0.2
3	2	0.8
4	3	4.5
5	4	5.6
6	5	6.2
7	6	6.7

4.14 A process is estimated to have a first-order dynamic response, but the coefficients of the differential equation describing the relationship between control and state variables are not known. Determine the best estimate of the coefficients of the differential equation by using the experimental data in Table P4.5.

Table P4.5   Experimental Data Taken at Equally Spaced Intervals of Time

Control Variable	State Variable
0.46	0.48
0.36	0.47
0.25	0.40
0.28	0.30
0.21	0.28
0.32	0.23
0.36	0.28
0.38	0.35
0.31	0.38
0.20	0.36

4.15 Develop the equations for a second-order process in terms of the measurement of X_i and Y_i, in which the index i represents samples at equally spaced time intervals Δt. Assume that the basic form of the continuous differential equation is

$$\frac{d^2Y}{dt^2} + B_1 \frac{dY}{dt} + B_0 Y = AX$$

4.16 Develop the dynamic physical process model with one state variable Y and one control variable X by using the experimental data given in Table P4.6.

Table P4.6 Experimental Data

t	X	Y
t_i	1.0	-0.100
$t_i + \Delta$	1.0	0.100
$t_i + 2\Delta$	1.0	0.240
$t_i + 3\Delta$	1.0	0.471
$t_i + 4\Delta$	1.0	0.444
$t_i + 5\Delta$	1.0	0.310
$t_i + 6\Delta$	1.0	0.269
$t_i + 7\Delta$	0.8	0.722
$t_i + 8\Delta$	0.8	0.350
$t_i + 9\Delta$	0.8	0.425
$t_i + 10\Delta$	0.7	0.648
$t_i + 11\Delta$	0.7	0.758
$t_i + 12\Delta$	0.7	0.467
$t_i + 13\Delta$	0.8	0.735
$t_i + 14\Delta$	0.8	0.802
$t_i + 15\Delta$	0.8	0.617
$t_i + 16\Delta$	0.9	0.680
$t_i + 17\Delta$	0.9	0.652

4.17 Develop the dynamic physical process model with one state variable Y and one control variable X by using the experimental data given in Table P4.7.

Table P4.7 Experimental Data

t	X	Y
t_i	1.0	0.150
$t_i + \Delta$	1.0	-0.050
$t_i + 2\Delta$	1.0	0.050
$t_i + 3\Delta$	1.0	0.000
$t_i + 4\Delta$	1.0	0.190
$t_i + 5\Delta$	1.0	0.271
$t_i + 6\Delta$	1.0	0.244
$t_i + 7\Delta$	0.8	0.360
$t_i + 8\Delta$	0.8	0.619
$t_i + 9\Delta$	0.8	0.372
$t_i + 10\Delta$	0.7	0.350
$t_i + 11\Delta$	0.7	0.625
$t_i + 12\Delta$	0.7	0.448
$t_i + 13\Delta$	0.8	0.808
$t_i + 14\Delta$	0.8	0.717
$t_i + 15\Delta$	0.8	0.735
$t_i + 16\Delta$	0.9	0.852
$t_i + 17\Delta$	0.9	0.467

CHAPTER 5

5.1   A proposed plant has a $30,000,000 fixed cost. The unit price of the product is $85 and the unit variable cost is $30. What is the break-even volume?

5.2   A typical make-or-buy decision requires a choice of two alternatives:

Alternative 1: To make the product would require a $10,000,000 fixed cost and the product can be manufactured at $4000 per unit.

Alternative 2: The product may be purchased with no fixed cost but at a price of $6500 per unit.

(a) If the sale price of the product is $12,000 per unit, what is the break-even volume of the two alternatives?

(b) What is the break-even volume if the sale price is $13,500 per unit? *Hint.* The unit price does not affect the break-even volume.

5.3   Compute the net income, return-on-sales, return-on-investment, and discounted cash flow at 6% for the venture listed in Table P5.1. Use a 50% rate for federal-income-tax.

Table P5.1   Financial Results of Proposed Investment Opportunity

Years	Revenue	Expense	Net Investment
1	$ 0	$2.0	$3.0
2	0	3.0	5.0
3	5.0	6.0	7.0
4	10.0	5.0	4.0
5	10.0	4.0	3.0
6	7.0	3.0	2.0
7	0	0	1.0

5.4   A plant is initially capitalized at $25,000,000. Compute the annual depreciation and residual value for a 12-year depreciation period by using the following formulas:

(a) Straight line
(b) Sum of the years' digits
(c) Double declining balance

5.5   A business had an investment at the beginning of the year of $12,600,000. During the year the following changes took place:

Net income = $850,000
Δ Depreciation = $1,000,000
Δ Accounts receivable = $500,000
Δ Inventory = $250,000
Δ Accounts payable = $135,000

(a) What is the cash flow for the year?
(b) What is the investment at the end of the year?

5.6 A computer system rents for $10,000 a month and has a sale price of $450,000. The service cost per month (if purchased) is $1000.

  (a) What is the break-even period in number of months between rental and purchase?

  (b) If a 6% discount rate is used, what is the break-even period?

5.7 Determine the uniform rate of return within $\pm 1/2\%$ for the cash flow schedules shown in Table P5.2. *Hint.* Use trial and error.

Table P5.2 Cash Flow Schedules

Years	A	B	C
1	$ −1000	$ −1000	$ −1000
2	0	0	200
3	0	0	200
4	0	0	200
5	1300	600	200
6	0	700	200
7	0	0	200

5.8 Develop the linear programming model for a company that manufactures four products from the same raw material. The raw material, labor requirements, and prices for each of the products are given in Table P5-3. The production schedule has a daily limit of 100 units of raw material and 125 units of labor.

Table P5.3 Material, Labor Requirements, and Price for Four Products

Product	Material	Labor	Price
A	5	3	2
B	1	2	1
C	15	6	11
D	4	4	4

5.9 Plot the following linear elastic demand curves:

  (a) $Y = 100 - 5X$

  (b) $Y = 200 - 10X$

  (c) $Y = 1000 - 20X$

5.10 What are the elasticities for the demand curves in Problem 5.9? *Hint.* Elasticity is the change in demand for one unit change in price.

5.11   Given the demand curve $Y = 120 - (0.0001)X$, solve for the optimal price, resulting volume, and profit-before-taxes if the fixed cost is $5,000,000 and the unit variable cost is $500,000.

5.12   Use the nonlinear elastic demand curve shown in Figure 5.7-3 to compute the optimal price and profit-before-taxes for a product with fixed cost of $6000 and unit variable cost of $6/unit.

CHAPTER 6

6.1   Discuss the place of the physical process model in the optimization procedure.

6.2   Is it desirable to optimize more than one objective function?

6.3   Name two methods of combining several objective functions into a single objective function.

6.4   Do linear objective functions have an optimum?

6.5   Do linear objective functions with constraints have an optimum?

6.6   Can physical process model equations be considered constraints?

6.7   Suggest two methods of imposing soft constraints on the objective function.

6.8   What is the effect of constraints on the quality of process optimization?

6.9   Does feedforward optimal control always require a physical process model?

6.10   Is a physical process model absolutely necessary for the optimization procedure?

6.11   Discuss three approaches to optimal control implementation.

6.12   Chose a process with which you are familiar and develop the following in quantitative terms:

(a) The physical process model
(b) The objective function
(c) The constraints on state and control variables

6.13   Select an article from the available literature and discuss the following:

(a) The physical process model
(b) The objective function
(c) The constraints on state and control variables

Prepare a term paper or report summarizing and interpreting the article.

CHAPTER 7

7.1   Given the following physical process model Y and objective function F,

$$Y = 50 + 40X$$

$$F = (\overline{Y} - Y)^2 + 2(\overline{X} - X)^2$$

(a) Find the optimal feedforward control equation and draw the corresponding block diagram.
(b) Find the value of the objective function F for $\overline{X} = 2$ and $\overline{Y} = 130$.
(c) Repeat for $\overline{X} = 2$ and $\overline{Y} = 100$.

7.2 (a) For the example in Problem 7.1 plot the value of the objective function F as a function of $\bar{X}$ for $\bar{Y} = 130$.

(b) Plot the actual value of Y obtained as a function of $\bar{X}$.

7.3 Repeat Problems 7.1 and 7.2 but change the value of the coefficients of the objective function so that

$$F = (\bar{Y} - Y)^2 + 60(\bar{X} - X)^2$$

7.4 Prepare a summary of the results of Problems 7.1, 7.2, and 7.3 and include any conclusions that can be drawn.

7.5 Given the following physical process model in matrix notation,

$$\begin{pmatrix} Y_1 \\ Y_2 \end{pmatrix} = \begin{pmatrix} 50 & 40 \\ 10 & 20 \end{pmatrix} \cdot \begin{pmatrix} X_1 \\ X_2 \end{pmatrix}$$

The objective function is

$$F = \psi_1(\bar{Y}_1 - Y_1)^2 + \psi_2(\bar{Y}_2 - Y_2)^2 + \theta_1(\bar{X}_1 - X_1)^2 + \theta_2(\bar{X}_2 - X_2)^2$$

and the coefficients of the objective function are

$$\psi_1 = 1, \qquad \psi_2 = 8, \qquad \theta_1 = 100, \qquad \theta_2 = 300$$

(a) Find the optimal control equations to satisfy

$$\bar{Y}_1 = 46, \qquad \bar{Y}_2 = 14, \qquad \bar{X}_1 = 0.7, \qquad \bar{X}_2 = 0.3$$

(b) Draw the corresponding block diagram.

(c) Find the resultant values of Y_1, Y_2, X_1, and X_2.

7.6 Given the same physical process model as in Problem 7.5, but with the following conditions,

$$\psi_1 = 1, \qquad \psi_2 = 8, \qquad \theta_1 = 0, \qquad \theta_2 = 0$$

$$\bar{Y}_1 = 46, \qquad \bar{Y}_2 = 14$$

(a) find the optimal control equations,

(b) draw the corresponding block diagram,

(c) find the resultant values of Y_1, Y_2, X_1, and X_2.

7.7 Given the same physical process model as in Problem 7.5 but with the following conditions,

$$\psi_1 = 1, \qquad \psi_2 = 0, \qquad \theta_1 = 0, \qquad \theta_2 = 0$$

$$\bar{Y}_1 = 46, \qquad X_2 = X_{20} \text{ (a constant)}$$

(a) Find the optimal control equations.

(b) Draw the corresponding block diagram.

(c) Find the resultant values of Y_1, Y_2, and X_1.

7.8 Modify the physical process model of Problem 7.5 by including an additional equation that defines an additional state variable Y_3:

$$Y_3 = 25X_1 + 10X_2$$

(a) For the objective function

$$F = (\bar{Y}_1 - Y_1)^2 + 8(\bar{Y}_2 - Y_2)^2 + 4(\bar{Y}_3 - Y_3)^2$$

find the optimal control equations and draw the corresponding block diagram.

(b) For $\bar{Y}_1 = 46$, $\bar{Y}_2 = 14$, and $\bar{Y}_3 = 20$ find the resultant values of Y_1, Y_2, Y_3, X_1, and X_2.

(c) For $\bar{Y}_1 = 46$ and $\bar{Y}_2 = 14$ find the value of $\bar{Y}_3$ that will result in perfect control; i.e., $Y_1 = \bar{Y}_1$, $Y_2 = \bar{Y}_2$, and $Y_3 = \bar{Y}_3$.

7.9 Solve Problems 7.1 through 7.8 by using the Lagrange multiplier technique.

7.10 Given the following objective function to be minimized,

$$F = \psi_1(Y_0 - Y)^2 + \dot{\theta}_1 X^2$$

where Y is the state variable and X is the control variable. The physical process model is

$$Y = AX$$

(a) Derive an expression for X that will minimize the objective function.
(b) Draw the corresponding block diagram.

7.11 Given the physical process model,

$$Y_1 = X_1$$

$$Y_2 = X_1 + X_2^2$$

$$Y_3 + Y_3^2 = X_1 + X_3^2$$

The cost of not operating at the desired point $\bar{Y}_1$, $\bar{Y}_2$, and $\bar{Y}_3$ is

$$F = (\bar{Y}_1 - Y_1)^2 + (\bar{Y}_3 - Y_3)^2$$

(a) Determine the simultaneous nonlinear equations for the conditions of optimum.
(b) Develop the optimal control equations by solving for the optimum values of X_1, X_2, and X_3 in terms of $\bar{Y}_1$, $\bar{Y}_2$, and $\bar{Y}_3$.

7.12 Find the values of X_1 and X_2 that will minimize

$$F = 2X_1 + X_2 + 63$$

subject to the constraints

$$X_1 \geq 0$$

$$X_2 \geq 0$$

$$10 - X_1 - X_2 \geq 0$$

$$6 - X_1 \geq 0$$

$$7 - X_2 \geq 0$$

$$X_1 + X_2 - 5 \geq 0$$

7.13 In a cement plant two ores are blended to meet magnesium and calcium composition objectives. The quantities of the two ores blended are HC for *high calcium*, and HM for *high magnesium*. The calcium and magnesium content of these two ores are (in fractions)

	HM	HC
Fraction calcium	0.40	0.46
Fraction magnesium	0.04	0.01

(a) Write the equations relating HC, HM and output composition M and C.
(b) Determine the required quantities of the ores HC and HM if the desired blend is

$$M_0 = 0.03, \qquad C_0 = 0.42$$

Consider the following:

(1) What criterion would you use for balancing errors?
(2) What relative weighting would you give errors in magnesium? Calcium? Why?

7.14 A mixing operation is used to blend three raw materials, M_1, M_2, and M_3, to yield a final product of a desired consistency. The flow rates, in pounds per hour, of the raw materials M_1, M_2, and M_3, are V_1, V_2, and V_3, respectively. The flow rates are controlled individually by valves that can be independently adjusted. There is no storage in mixing and the flow rate of the final product is

$$V_q = V_1 + V_2 + V_3$$

The flow rate of the final product must remain constant and the production rate has been set at 10,000 pounds per hour. The raw materials are composed of two compounds, X_1 and X_2, and a residual that has not been defined. The composition of the raw materials on the average is as follows:

$$M_1: 10\% \text{ of } X_1, \text{ with } 90\% \text{ residual}$$
$$M_2: 20\% \text{ of } X_2, \text{ with } 80\% \text{ residual}$$
$$M_3: 60\% \text{ of } X_1, \text{ with } 40\% \text{ residual}$$

It is desired that the final product be composed of 20% X_1 and 10% X_2. Calculate the value of V_1, V_2, and V_3 to meet the requirements of the final product composition and production rate.

CHAPTER 8

8.1 Find the optimal control equations and draw the corresponding block diagram for Problem 7.5 by using the incremental physical process model instead of the total physical process model.

8.2 Repeat Problem 8.1 for the conditions corresponding to Problem 7.6.

8.3 Repeat Problem 8.1 for the conditions corresponding to Problem 7.7.

8.4 Repeat Problem 8.1 for the conditions corresponding to Problem 7.8.

8.5 Verify that

$$[\mathbf{\Delta Y}]^T \; \mathbf{\Phi}[\mathbf{\Delta Y}] = \psi_{11}\,(\Delta Y_1)^2 + \psi_{22}\,(\Delta Y_2)^2 + \psi_{33}\,(\Delta Y_3)^2$$

where

$$\mathbf{\Phi} = \begin{pmatrix} \psi_{11} & 0 & 0 \\ 0 & \psi_{22} & 0 \\ 0 & 0 & \psi_{23} \end{pmatrix}$$

and

$$\mathbf{\Delta Y} = \begin{pmatrix} \Delta Y_1 \\ \Delta Y_2 \\ \Delta Y_3 \end{pmatrix}$$

8.6 (a) Develop the simultaneous equations that will give the best solution (in the least squares sense) for the following underdetermined system for a feed-forward control:

$$\mathbf{Y} = \mathbf{AX}$$

where

$$\mathbf{A} = \begin{pmatrix} A_{11} & A_{12} \\ A_{21} & A_{22} \\ A_{31} & A_{32} \end{pmatrix}$$

The objective function to be minimized is

$$G = \psi_{11}(\overline{Y}_1 - Y_1)^2 + \psi_{22}(\overline{Y}_2 - Y_2)^2 + \psi_{33}(\overline{Y}_3 - Y_3)^2$$

(b) Revise the simultaneous equations so that their solution can be implemented by *feedback control*; i.e., solve the system equations in terms of ΔX_1, ΔX_2.

8.7 (a) Develop the optimal control equation for *feedback control* for a simplified paper process. Assume that the weighting of errors is equal. Give numerical values for all constants in the system. The state variables for a paper-mill dryer section are

$$Q_1 = \text{weight of base, in weight per unit of dried paper}$$
$$Q_2 = \text{``points Mullin,'' a measure of burst strength}$$
$$Q_3 = \text{percent in finished paper}$$

The control variables are

$$M_1 = \text{weight per unit of feed stock to dryer}$$
$$M_2 = \text{horsepower to dryer drive}$$
$$M_3 = \text{steam pressure, a measure of steam flow to the dryers}$$

A paper-mill dryer can be represented by the following physical process model:

$$\mathbf{Q} = \mathbf{BM}$$

where

$$\mathbf{B} = \begin{pmatrix} 1.02 & 0.0084 & -0.05 \\ -0.0168 & 0.326 & 0.6335 \\ -0.52 & 0.160 & -0.960 \end{pmatrix}$$

The desired values of the state variables are

$$Q_1 = 0.04, \qquad Q_2 = 0.12, \qquad Q_3 = 0.2$$

(b) Develop an iterative *feedback* control for the above if M_1 is fixed at -0.03. Explain your choice of a criterion for selecting the values of ΔM_1 and ΔM_2.

8.8 Given the following *actual* physical process model,

$$Y_1 = 50 + 10X_1 + 25X_2$$

$$Y_2 = 5 + 4X_1 + 10X_2$$

and the following *approximate* physical process model

$$Y_1 = A_1 + B_1 X_1 + C_1 X_2$$

$$Y_2 = A_2 + B_2 X_2 + C_2 X_2$$

If feedforward optimal control techniques are applied to the process by using the approximate model, estimate the accuracy of the coefficients A_1, B_1, C_1, A_2, B_2, and C_2 necessary to control Y_1 and Y_2 within $\pm 5\%$ of the desired values $\overline{Y}_1 = 63$ and $\overline{Y}_2 = 10.2$. Assume that the objective function is

$$F = (\overline{Y}_1 - Y_1)^2 + 10(\overline{Y}_2 - Y_2)^2$$

8.9 Find the optimal control equations for the conditions of Problem 8.8 and draw the corresponding block diagram by using feedforward control techniques that include provisions for adapting to changes in the coefficients of the model.

8.10 Do Problem 8.8 by using feedback control techniques.

8.11 Do Problem 8.9 by using feedback control techniques.

8.12 Using the model of Problem 8.8, work out the procedure for adapting to the following constraints on the control variables:

$$0 \leq X_1 \leq 1$$

$$0 \leq X_2 \leq 1$$

$$X_1 + X_2 = 1$$

8.13 Using the results of Problem 8.11, work out the procedure for adapting to the following constraints on the control variables:

$$0 \leq X_1 \leq 1$$

$$0 \leq X_2 \leq 1$$

8.14 Search the literature for a discussion of an example that uses combined feedforward and feedback control techniques to adapt to an uncontrolled variable. Compare the performance of such a system with that of another system that uses feedback alone.

CHAPTER 9

9.1   Use the data given in Table P9.1 to determine the path of steepest ascent.

Table P9.1   Experimental Data

X_1	X_2	F
2.1	2.1	0.3574
1.9	2.1	0.3683
1.9	1.9	0.3709
2.1	1.9	0.3606

9.2   Using the data given in Table P9.1,
   (a) fit the function

$$Y = B_0 + B_1 \, \Delta X_1 + B_2 \, \Delta X_2$$

   by using a least squares criterion for best fit.
   (b) What is the squared error of the best fit?

9.3   With the data given in Table P9.1,
   (a) fit the function

$$Y = B_0 + B_1 \, \Delta X_1 + B_2 \, \Delta X_2 + B_{12} \, \Delta X_1 \, \Delta X_2$$

   by using a least squares criterion for best fit.
   (b) What is the squared error of the fit?

9.4   Using the coefficients determined from Problems 9.1 through 9.3:
   (a) Compare the path of steepest ascent calculated in Problem 9.1 with that obtained by use of the equation in Problem 9.2.
   (b) Repeat for the equation in Problem 9.3.
   (c) Explain the differences.

9.5   A response surface is described by the following polynomial:

$$F = 0.5 + 0.13 X_1 - 0.308 X_1{}^2 + 0.13 X_1{}^3 - 0.166 X_1{}^4 + 0.015 X_2$$
$$- \, 0.0392 X_2{}^2 + 0.0175 X_2{}^3 - 0.003 X_2{}^4 + 0.04 X_1 X_2 - 0.035 X_1{}^2 X_2{}^2$$

Starting at an operating point at $X_1 = 2.0$ and $X_2 = 2.0$, calculate the path of steepest ascent.

9.6   If the optimal gradient method were being used as the control strategy in Problem 9.5, *estimate* where the second gradient calculation would have to be made if the point of reversal could be exactly determined.

9.7   (a) Calculate several (greater than 3) points of the response surface along the direction vector for Problem 9.5. Select the points so that at least one falls beyond the maximum value along the direction vector. Fit a quadratic curve to the data. Use regression analysis techniques.

Hint. Define a new coordinate system in which one axis is parallel to the direction vector and the other is perpendicular to the direction vector. Then

$$F = A_0 + A_1 \zeta_1 + A_{11} \zeta_1{}^2$$

where ζ_1 is a function of X_1 and X_2.

(b) Calculate the peak value along the response vector.

(b) How does this compare with the value estimated in Problem 9.6?

9.8   Table P9.2 lists the values of the response surface measured at several different values of the control variable. The control variables are selected to form a 2^2 factorial design.

(a) Calculate the gradients from the first, second, third, and fourth sets of data.

(b) Calculate the gradient that is the average of the data taken.

(c) Use statistical techniques to determine whether the data represents a significant gradient at risk of $\alpha = 0.5$, using Table 4.4-2 for the F-distribution. Make a confidence check by assuming that the interaction effect is negligible.

Table P9.2 2^2 Factorial Design of Experiments

Treatment	X_1	X_2	F^1	F^2	F^3	F^4
1	0.425	1.53	0.5074	0.4992	0.5080	0.5077
2	0.225	1.53	0.5132	0.5100	0.5051	0.4936
3	0.225	1.33	0.5108	0.5024	0.5095	0.5085
4	0.425	1.33	0.5016	0.5137	0.5149	0.5036

Notes. (1) The central point of design is $X_1 = 0.325$ and $X_2 = 1.43$.

(2) F^1, F^2, F^3, and F^4 are the response surface measurements on different replications of the design.

9.9   Consider the process with an objective function described by the steady-state response surface

$$F = \frac{10 X_1{}^{0.1}}{X_3{}^{0.8}} + 3 X_1 X_2$$

Assume that the process is presently operating at

$$X_1 = X_2 = X_3 = 1$$

Use gradient techniques to determine the direction of steepest ascent.

9.10   Determine the new operating point in Problem 9.9 if the step size is

$$S = [\Delta X_1{}^2 + \Delta X_2{}^2 + \Delta X_3{}^2]^{\frac{1}{2}} = 0.1$$

9.11 The data listed in Table P9.3 resulted from a series of factorial designs in a "hill-climbing" experiment that used a simple path of steepest ascent strategy. On the basis of these data:

(a) Would you say the step size was proper? Why?

(b) Would a modified gradient strategy work better? Which one? Why? List the steps in the strategy that you have selected and the equations to use.

(c) Would a quadratic fit or a similar multipoint curve fitting strategy be useful at this stage of the experiment? Which one? Why?

(d) Would larger spacing on the factorial design result in more rapid convergence on the optimum?

Table P9.3   Data From Hill-Climbing Experiment

Centroid				
$\bar{X}_1$	$\bar{X}_2$	X_1	X_2	F
0.190	0.275	0.290	0.375	0.5201
		0.090	0.375	0.5116
		0.090	0.175	0.5114
		0.290	0.175	0.5183
0.380	0.295			
0.185	0.350			
0.400	0.350	0.500	0.450	0.5122
		0.300	0.450	0.5202
		0.300	0.250	0.5192
		0.500	0.250	0.5096
0.1850	0.395	0.2850	0.495	0.5202
		0.0850	0.495	0.5103
		0.0850	0.295	0.5113
		0.2850	0.295	0.5197
0.410	0.390			

9.12 A static optimization of a process is to be made with gradient techniques. It is suspected that the initial operating point will be somewhat remote from the optimum, the response surface will be asymmetrical, and the response measurement contaminated with noise near the optimum.

(a) Describe briefly what strategies you would provide for this case and why.

(b) Draw a *functional* flow chart of the procedure that would be followed in such an optimization strategy.

9.13 Given the data in Table P9.4, fit the following form of equation to the data

$$Y = B_0 + B_1 \Delta X_1 + B_2 \Delta X_2$$

for

(a) one replication,

(b) two replications.

   (c) Are the coefficients in (a) or (b) significant at the 90% confidence level in an F-ratio test?

Table P9.4   Experimental Data

		First Replica	Second Replica
X_1	X_2	F	F
1	1	0.5074	0.4992
1	-1	0.5132	0.5100
-1	-1	0.5108	0.5024
-1	1	0.5016	0.5137

9.14   Consider the first-order interaction term and fit the equation

$$Y = B_0 + B_1\,\Delta X_1 + B_2\,\Delta X_2 + B_{12}\,\Delta X_1\,\Delta X_2$$

to the data in Table P9.4 for two replications.

9.15   (a) Are any of these coefficients significant at the 95% confidence level? Use Table 4.4-2 for the F-distribution.

     (b) At the 99% level?

CHAPTER 10

10.1   Use the method of dynamic programming to find the best straight-line approximation to the curve $Y = X^2$ over the range $0 \le X \le 2$.

   (a) Find the best approximation by using a single straight-line segment.

   (b) Repeat for two straight-line segments.

   (c) Repeat for three straight-line segments.

     Hint. For a single segment let the equation of the straight line be $Y_1 = MX + B$ and the total error of the approximation be

$$\text{Error} = \int_0^2 (Y - Y_1)^2\,dX$$

10.2   Given a physical process model,

$$Y_1^{p+1} = 0.8Y_1^p + 0.2(50 - 10X^q)$$

$$Y_2^{p+1} = 0.8Y_2^p + 0.2(1 + 4X^q)$$

where q denotes the interval between consecutive instants of time p and $p + 1$. Find the values of X^q that minimize the expression

$$F = \sum_{q=0}^{4}[(\bar{X} - X^q)^2 + \psi_1(\bar{Y}_1 - Y_1^q)^2 + \psi_2(\bar{Y}_2 - Y_2^q)^2]$$

Thus five values of X, X^0, X^1, X^2, X^3, and X^4, corresponding to the values over the intervals $(q=0)$ to $(q=1)$, $(q=1)$ to $(q=2)$, $(q=2)$ to $(q=3)$, $(q=3)$ to $(q=4)$, and $(q=4)$ to $(q=5)$, are to be found. The value of X is constant over an interval, whereas the values of Y_1 and Y_2 vary over the interval. Solve this problem in two ways by using dynamic programming.

(a) Find the numerical solution for the X's that depend only on the conditions at $q=0$ and $q=5$.

(b) Find the solutions for each X^q that depend on the conditions of the beginning of the corresponding interval. Assume that

$$\bar{X} = 0.5, \qquad \bar{Y}_1 = 45, \qquad \bar{Y}_2 = 3, \qquad \psi_1 = 1, \qquad \psi_2 = 20$$

10.3 (a) Determine a differential equation to describe the trajectory that minimizes the integral $\phi(X)$:

$$\phi(X) = \int_0^T [\tfrac{1}{2}X^2 + \exp(Y)] \, dt$$

with the dynamic constraint

$$\frac{dY}{dt} = X + AY$$

and

$$Y(0) = C$$

$$\frac{dY}{dt}\bigg|_{Y=T} = 0$$

(b) How would you solve this differential equation?

10.4 A power system consists of three generating units, A, B, and C. Their maximum capacity is

$$R_a = 1, \qquad R_b = 1, \qquad R_c = 2 \qquad \text{(in per unit power)}$$

The cost of operating the units is

$$F_a = 1, \qquad F_b = 0.83, \qquad F_c = 1.25 \qquad \text{(in per unit cost/per unit power)}$$

The cost of start-up is

$$S_a = 2, \qquad S_b = 2, \qquad S_c = 2.85 \qquad \text{(in per unit cost)}$$

At time zero only unit A is operating. The scheduling of units on-line must be done one hour in advance, for that much lead time is required.

(a) Draw the decision tree for the next 4 hours.

(b) Assume that the system load is predicted to vary as follows:

Time	System Load
0	0.9
1	1.2
2	2.1
3	1.8

If 10% excess capacity is required for safety considerations, determine the optimal scheduling to minimize cost.

10.5   The forecasted monthly production requirements of a plant are shown in Table P10.1. The cost of changing the level of production is the square of the difference of the monthly productions.

(a) Determine the monthly production to minimize the cost of changing the level of production and the cost of excess production. Assume that excess production does not carry over to the next month. The penalty for excess production is the square of the excess production.

(b) Assume that the excess production can be carried into subsequent months and the penalty for excess production is 0.1 times the square of the excess. What is the optimal monthly production?

Table P10.1   Forecasted Production Requirements

Month	Production Requirements
0*	1000
1	1300
2	1100
3	900
4	1300

* Assume that the production level for the preceding month is 800 units.

10.6   Do Problem 10.5 by using a monthly production requirement, starting at 800 units and increasing by 100 units per month.

Index